Metabolic Structure and Regulation

A Neoclassical Approach

Metabolic Structure and Regulation

A Neoclassical Approach

Raymond S. Ochs
Department of Pharmaceutical Sciences
St. John's University, Queens, NY

CRC Press
Taylor & Francis Group
Boca Raton London New York

CRC Press is an imprint of the
Taylor & Francis Group, an **informa** business

CRC Press
Taylor & Francis Group
6000 Broken Sound Parkway NW, Suite 300
Boca Raton, FL 33487-2742

First issued in paperback 2020

ISBN-13: 978-1-4822-3608-8 (hbk)
ISBN-13: 978-0-367-50300-0 (pbk)

Library of Congress Cataloging-in-Publication Data

Names: Ochs, Raymond S., author.
Title: Metabolic structure and regulation : a neoclassical approach / Raymond S. Ochs.
Description: Boca Raton : CRC Press, 2018.
Identifiers: LCCN 2017036757 | ISBN 9781482236088 (hardback)
Subjects: LCSH: Metabolism.
Classification: LCC QP171 .O25 2018 | DDC 612.3/9--dc23
LC record available at https://lccn.loc.gov/2017036757

Visit the Taylor & Francis Web site at
http://www.taylorandfrancis.com

and the CRC Press Web site at
http://www.crcpress.com

To my wife Jessica for all her support, as always.

Contents

Introduction

There is new interest in metabolism [1]. Even the hard core molecular biologist James Watson recently said "I never thought, until about two months ago, I'd ever have to learn the Krebs cycle" [2].

In 1985, Richard Hanson predicted a return of investigators towards metabolism: "Students in the modern era regard metabolic pathways as a solved problem" [3]. He went on to predict a *metabolism redux*. Unfortunately, his prescience was off by about thirty years. Much of our information today comes from textbooks.

Yet metabolism is not a strong part of current biochemistry textbooks; the information is largely structural. Many of the ideas are handed down from earlier textbooks, and do not constitute a body of understanding that can be applied to the analysis of pathways.

The first and oldest pathway, glycolysis, has currency because of the Warburg effect: cancer cells generally have very high rates of glycolysis. Since most are aware that this pathway is relatively inefficient as an energy generator and yet cancer cells are surely survival specialists, it appears to make little sense. The Warburg effect dates from the early twentieth century; it is still under debate!

A paucity of systematic understanding of metabolism has led to a piecemeal approach, and some popular ideas that are not well founded. For example, many consider nicotinamide adenine dinucleotide (NAD) as a regulatory molecule, and adenosine triphosphate (ATP) as an energy barometer.

In this work, I present a metabolic viewpoint, mostly putting forth ideas that were at one time well established but at present not well known. The work is more holistic than reductionist, with the goal of establishing a set of principles as free as possible from contradiction. I name this construct *neoclassical* as it presents a largely orthodox view of metabolic ideas, which are valuable enough to be revived in the climate of a renewed interest and important advances in metabolic thinking as more investigators are becoming interested in the area.

One very popular idea that also has historical roots is the compartmentation[i] issue. This is advanced in different ways, but a popular thought is that pathway intermediates or ions can become concentrated within specific regions of a water space. This was recently hailed as a "hot area" in a cover story of metabolism and disease in *Cell* [4]. The discussants suggested that glycolysis enzymes were all bound together, just like the respiratory complexes. Overall, many view metabolism as a pile of old literature that seems to have nothing systematic apart from the connection diagrams that define pathways.

This notion is, in my view, unacceptable: it can explain anything. When concentrations don't change, it could be argued that they are in a local aqueous subspace that defies measurement. When they do change, it is not taken as a contradiction of compartmentation. No one investigating changes in mRNA levels would take data showing no change and argue that it must have changed but couldn't be measured because of localization. If the diffusion-defying hypothesis were really to be accepted, any hypothesis could be adopted. It would obviate an understanding of metabolic patterns, any use for equilibrium constants, or kinetic constants. We would also have to discard the global metabolic measurements approach—metabolomics—which makes the assumption that in fact metabolites do equilibrate with the cell water. I will show that belief in compartmentation is nonetheless pervasive; eschewing the concept is essential if we are to attempt a consistent hypothesis.

In the following, I offer what Mandelbrot described in his monograph [5] as a personal essay rather than a cataloging of events. The first pathway is glycolysis, familiar to most, but the ideas presented may well be new, as the fundamental notions of a pathway, cofactors in general and ATP in particular are presented in a metabolic context. Next, a treatment of mitochondria is provided that, again, is fairly familiar territory, but also provides some distinct principles, one of which is importantly the distinction between isolated mitochondrial behavior and mitochondrial function in the context of an intact cell.

I next consider pathway connectivity, revisit compartmentation, and show how metabolism is more easily interpreted if we accept uniform distribution throughout the water space. I have also included some issues of stereochemistry as they apply to metabolism, and how to view transporters across membranes.

In the following chapter on enzymes, the presentation is orthodox from the standpoint of a kineticist, but unfamiliar if the reader's viewpoint is shaped by general textbook treatments. In particular, kinetic constants and enzyme inhibition analysis are presented in a way that assists interpretation of metabolic events.

Next, a focus on cell specialization puts pathways in context. Thus, a liver cell and a muscle cell have virtually identical pathways for energy formation; minor alterations in regulators or connected reactions account for their distinction. Following this are considerations of a few signaling systems. The distinctive treatment here is to consider signal systems as pathways as well, which leads us to unique viewpoints, in particular for that of calcium ions.

A separate chapter considers computers and metabolism. The discussion includes metabolomics as well as some earlier ideas that provide unique insight into metabolic problems. The combination of computing and various fields of scientific endeavor has invariably provided new outlooks, and metabolism is no exception.

The final chapter is a consideration of medical issues that are directly related to metabolism. This is not to say that genetic approaches are not crucial; James Watson overstated the case when he opined that genetics was not useful for solving key disease puzzles. What I hope to demonstrate in this work is that "learning the Krebs cycle" is not the solution. Rather, we need an introduction to a few basic principles of metabolism. A first step is to consider the metabolic approach as a systems view, which is taken in Chapter 1.

REFERENCES

1. McKnight, S.L. (2011) Back to the future: Molecular biology meets metabolism. *Cold Spring Harb Symp Quant Biol* 76, 403–411.
2. Apple, A. (2016) An old idea, revived: Starve cancer to death. in *The New York Times*, New York.
3. Ochs, R.S., Hanson, R.W., and Hall, J. (1985) *Metabolic Regulation*, Vol 1, Elsevier, Amsterdam.

4. Szewczak, L. (2016) Targeting disease through metabolism, *Cell* 165: 1561–2.
5. Mandelbrot, B.B. (1982) *The Fractal Geometry of Nature*, W.H. Freeman, San Francisco.

ENDNOTE

i. This term is an unfortunate conglomerate, occasionally shortened to *compartment*, or lengthened to *compartmentalization*. Its meaning is also somewhat fluid, as it refers both to physical compartments in cells delimited by membranes, a definition which is universally accepted, and conceptual compartments in which different pools of metabolites exist in the same water space. As described in this treatise, the second type is at odds with known chemical description, and typically invoked to explain specific data anomalies.

Author

Raymond S. Ochs is a biochemist with a career-long specialty in metabolism spanning 30 years. Previously, he wrote the textbook *Biochemistry*, contributed the metabolism chapters to another text, *Principles of Biochemistry*, and co-edited a collection of articles published as *Metabolic Regulation*. His research interests concern major pathways of liver and muscle, including glycolysis, gluconeogenesis, ureogenesis, fatty acid metabolism, glycogen metabolism, and control by cAMP, Ca^{2+}, diacylglycerol, and AMPK. He is currently professor of pharmacy at St. John's University in New York, teaching biochemistry, physiology, and medicinal chemistry.

1 Metabolic Perspective

1.1 SYSTEM BEHAVIOR

A system is a set of rules that is internally consistent. For example, thermodynamics is a systematic approach to reaction changes, which has broad application and just a few rules. Some of these have been modified and incorporated into metabolic thinking, which is discussed in this book. Physiologists broadly categorize interacting tissues and organs into functioning wholes, such as the "cardiovascular system" or the "digestive system." The benefit is that the essential interactions of the components can be both understood and predicted at a higher level than those of a single cardiomyocyte or hepatocyte. Within such systems, there are rules of interaction that are sometimes mathematical and other times descriptive. Thus, for the digestive system, the physiologist can state that the liver produces bile, which travels through the bile duct, may be stored in the gall bladder, and is released to the duodenum where it combines with lipids to produce a mixed micelle for later hydrolysis. The details of these processes may still be active, unsettled issues. System boundaries may be indistinct; we could consider the cardiovascular–respiratory grouping as a system as these interact. Still the benefit is that within the rules for functions and connections, consistent results can be expected.

Metabolism does have such underpinnings. For the most part they have not really been in dispute for many years, but now that metabolism itself has become of current interest, it would be useful to know some of those rules that have been developed.

An early systems approach to biology was proposed by Von Bertalanffy [1], who attempted to expand the reach of thermodynamics beyond the equilibrium realm. A recent application of systems to metabolism has been the development of *metabolomics*, a descendent of previous cataloging of large data sets of DNA (genomics) and protein (proteomics). However, this means of hoping to discover principles by collecting large amounts of data has its own complications (see Chapter 9), and does not take into account a large number of principles that have been uncovered by metabolic biochemists that have been found to be valuable guides.

Weinberg [2] provides an understanding of systems analysis that is commensurate with Von Bertalanffy, but more practical and of direct relevance to the study presented here. In this view, *large-number* problems are analyzed by a statistical approach, and *small-number* problems are handled by enumeration. Thus, metabolomics would fit with the large-number idea, and the biological classifications such as the enzyme commission numbers or the hierarchy of biological species with the small-number problems. A system is neither of these, but rather a *middle-level* problem: the observations are too numerous to keep track of by enumeration, and yet intractable to statistical analysis. For these, developed principles are needed as guideposts to understand the subject. Several of these have been developed over the last decades, are not particularly controversial, but are not widely known.

For many years, solving the sequence of the human genome provided an over-arching goal, with the promise of solving almost any problem in biology. Most biological scientists became molecular biologists. While other fields such as protein chemistry and analytical chemistry remained somewhat populated, metabolism as a discipline was left with few adherents.

The renaissance of interest in metabolism and its regulation means that an increasing number of scientists are coming to terms with this subject. A systematic treatment of metabolism provides the ability to compare disparate sets of data, and a framework not only for interpreting increasingly large datasets, but for designing the experiments to collect them in the first place.

Consistency is a key requirement in formulating our ideas about a system. While many ideas, taken in isolation, may appear to explain data, if they are not consistent with other notions, they cannot be considered as part of the system.

As an example, consider the hypothesis of *compartmentation*. This may seem to be a well-established idea in metabolism as it is so commonly invoked to explain what might otherwise be anomalous trends in data. In fact, it is so commonly cited that it might be considered unassailable. The idea here is that there are subspaces within a defined cellular region, so that measured values are not relevant. If this were a system principle, then the entire notion of metabolomics would need to be discarded, as those measurements would not reflect the "local" concentrations of metabolites. So too would any measurement of metabolic concentrations in cells; the idea of a compartmentation that somehow defies solution diffusion renders values of measured intermediates irrelevant. It also contradicts the long-established observation that in many cases, the Km values for enzymes measured *in vitro* matches the measured concentration of their substrates in intact cells [3]. Cases can be cited where the notion of compartmentation was proposed for metabolites, like oxaloacetate and fructose-1,6-bisP, which were subsequently shown to be erroneous, and these, too, needed to be removed from the list of oddities that appeared to support the compartmentation idea. Even the theoretical basis for diffusion difficulties has been brought into serious doubt.

There is a sense in which compartments do exist which can be shown to have consistent behavior. This extends beyond the major cellular spaces, such as specific membranes, and the water spaces that are defined by those membranes (e.g., cytosol and mitochondrial matrix). This alternative compartmentation arises not because of a problem with diffusion, but rather due to strong binding, as in the case of ADP to cellular proteins.

In the remainder of this opening chapter, I will explore some of the principles that meet the qualifications of system ideas for understanding metabolism. Their adoption will help form a metabolism perspective.

1.2 HIERARCHIES OF ANALYSIS

The two views of biological systems, holistic and reductionist, are extremes of a hierarchy of analysis. As whimsically yet poignantly discussed by Hofstadter [4], each has significant virtues: the holistic view has global reach; the reductionist is more specific and mechanistic. Neither can claim superiority in understanding a system.

In order to analyze a system, a variety of viewpoints must be adopted. For a biological system, the more holistic viewpoints of cells and organelles are important in understanding the physiological significance of observations. Reductionist notions of binding constants, energetics, and kinetics are important in understanding the basis of control systems. The systems approach to biology itself was originally a reaction to an overly reductionist climate [1]. Moving between the study of individual and collections of enzymes is the most traveled part of the hierarchy in the present work. Here, the notion of collections of enzymes and new properties that emerge from different viewpoints are essential to understanding metabolic systems.

Control in biological systems can be considered at different levels of a hierarchy. For example, in the realm of physiology, there is a clear notion of negative feedback. For example, estrogen travels from its production site to the hypothalamus to control release of other hormones into the systemic blood, which ultimately control estrogen levels themselves. Within many cells, citrate is produced by the mitochondria, crosses to the cytosol, and decreases phosphofructokinase activity. This can be taken as an explanation of the Pasteur Effect; alternatively, the Pasteur Effect can be viewed as a higher level of the hierarchy, with citrate inhibition as one underlying mechanism. Indeed, citrate is also known as an activator of fatty acid biosynthesis, or at least of acetyl CoA carboxylase, which is a rate-limiting step in the biosynthesis of fatty acids. This presents some complications for a consistent view of the meaning of citrate export. Even the notion of a "rate-limiting step" itself has been modified in recent years, providing an added subtlety to regulatory systems both at the level of pathways as well as at the level of enzyme mechanisms. Thus, the movement between hierarchies can reveal some complications in both underlying assumptions and the generality of ideas.

1.3 STEADY STATE

The steady state is one of two great models of reaction flow, applicable to both metabolic pathways and enzymatic reactions within those pathways. It is important to distinguish the *steady state* from an *equilibrium*, the more widely understood notion. Often the two are used interchangeably, to the detriment of understanding metabolic systems.

This is confounded by the use of the unusual term, "dynamic equilibrium," which is sometimes loosely taken as a synonym for both steady state and equilibrium. Add to this the physiological notion of *homeostasis*, which is also casually interpreted as either steady state or equilibrium.

Each notion describes some sort of balance, but we must carefully distinguish them. We can consider equilibrium in terms of physical models: a ball suspended in a two-dimensional valley, or a dual-pan balance at rest each represent a reaction in which forward and reverse rates are the same. Another one might be two people who exchange gifts, with no further transactions. Even with more than two, so long as the group is closed, we can translate these notions to reactions: this is an equilibrium model. Equilibrium is a good model for acid–base dissociations, or for binding of a hormone to a receptor. It is also a good description of the specific binding interaction between enzyme subforms and inhibitors.

Steady state requires a different sort of thinking. Here, there is also an exchange, but there is, in addition, a direction. The flow of material is not balanced. Suppose we use the last example, of people making exchanges. If two people exchange gifts unevenly, and this continues indefinitely, then, at some point, one person will have all the gifts. This is not a model of equilibrium, but it isn't a model of steady state either because nothing is steady. This is an endpoint situation. If we have three people, the first could hand a gift to the second, who then hands it to the third, who sets it down. Suppose the first has a stack of gifts, and hands them off at a constant rate to the second, who is able to pass it to the third and accept the next at a constant rate, and the third accepts the gift and puts it down at a constant rate. At this point, we have a steady state. It is possible that the folksy nature of the example belies the significance of coming to terms with the steady state. In order to be clear, the rate of person one to person two exchange equals the rate of person two to person three exchange. That is really the key. What is constant (steady) is the number of gifts held by person two. This represents the intermediate, standing in for one or more intermediary concentrations that are unchanged with time over the span in which the steady state is in force.

There is one more way to consider the distinction between equilibrium and steady state and that requires us to take a slightly more rigorous thermodynamic view. If we introduce the thermodynamic definition of a *state* as the object of our inquiry, and the corresponding definitions of *surroundings* as all other parts of the universe, and the *boundary* as the border between system and surroundings, we can consider a system in which a reaction appears to be static, with no change in its components. This could be an equilibrium or a steady state: the distinction is entirely due to the surroundings. In an equilibrium, in the absence of a change in the surroundings, the system will remain unchanged. By contrast, the steady-state system *requires* a constant change in the surroundings to maintain system constancy. For our pathway example, this is a continual decrease in substrate and a continual accumulation of product.

One of the reasons for insisting on a clear separation between equilibrium and steady state is apparent in our understanding of reversible inhibitors of enzymatic reactions. Extending our consideration of the Km above, the common textbook interpretation of the actions of inhibitors invokes effects on Vmax and Km. Such a treatment implies that this imparts some understanding about how inhibitors act. This is a fundamental error, well known to enzyme kineticists (see Cornish-Bowden [5]); it is too often passed over as one of those minor points that might have interpretations that could go either way. However, consider this statement made in a monograph on metabolic regulation [6], for which the theme is to introduce control strength (Chapter 9). In a discussion of the inhibition of inositol phosphatase by Li+, the author attempts to interpret what uncompetitive inhibition means in terms of how such an inhibitor affects Km (which is, after all, a textbook interpretation). It is described as remarkable, and unusual, as the inhibitor causes a decrease in Km. Since this is taken as a way of improving the binding of substrate to enzyme, then it means that the presence of an inhibitor must lead to activation of the enzyme! This error is the result of a confusion between equilibrium and steady state, and is the reason that it is not possible to use Km to interpret enzyme inhibition. In the analysis, Km is treated

as an equilibrium constant (it is not), and then the interpretation of a simple inhibition mode becomes impossible to understand. We can go further with this: the use of
Km as an equilibrium constant to interpret enzyme inhibition is not correct for any
inhibition mode. However, the apparent answer seems correct in the case of competitive inhibition. The fact that it falls apart in the other cases lead virtually all of those
who follow this line with a complete lack of understanding of uncompetitive or noncompetitive inhibition modes. In fact, the competitive mode only seems comfortable
because of the common idea of what *competitive* means. Further discussion of this
point is found in Chapter 4; the confusion is dissipated when steady-state rather than
equilibrium models are used.

1.4   PATHWAYS: INTERCONNECTED REACTIONS

It is important to distinguish between a *reaction view* and a *pathway view*. The *reaction view* is by far the more common in a textbook consideration of biochemistry,
and is the vantage point of much of biochemical analysis. However, distinct properties appear when we adopt a *pathway view* of interconnected reactions that exist
in intact cells. To show how these are distinct, consider the pathway of glycolysis,
defined as a route from glucose to lactate. If we strictly assume that the pathway is a
sum of its parts, then the NAD/NADH cofactor pair cancels out, and ATP accumulates. However, much of the kinetic behavior of the individual reactions is different in
cells from those measured in isolation: *in vitro* kinetic analysis requires that product
formation be zero, but in a steady state it is decidedly nonzero. Some reactions in
pathways are near to equilibrium, some are greatly displaced from equilibrium; in
a kinetic analysis, all of them are irreversible by design. Another property that cannot be extended from a reaction view to a pathway view is regulation by allosteric
effectors. These may or may not have significance in a pathway. For example, it is not
always the case that an effector that has activity *in vitro* actually changes concentration appropriately in cells. A distinct set of tools is thus required for pathway analysis.

The distinction between *near-equilibrium* and *metabolically irreversible* is a useful one that applies only to reactions that are part of pathways. Another generalization can be made, which is a new one advanced here: the distinction between
pathway-reversible and *pathway-irreversible*. Consider the reaction c → d that
occurs as part of a pathway sequence:

$$\ldots \to \underline{c \to d} \to \ldots$$

in which c and d are pathway intermediates. In order to be considered for the new
category of *pathway reversible*, two properties of this reaction are necessary:

1. The enzyme must be metabolically irreversible.

     The notion of *near equilibrium* as opposed to being *metabolically irre-
     versible* is considered further in virtually all parts of this book. The concept
     has a long history, and is widely used by Krebs and colleagues. We take the
     overall reaction involved, which includes any cofactors of the reaction and

compare the value of the measured components to the equilibrium constant for the reaction. If these have similar values, roughly within an order of magnitude, then they are *near equilibrium* in the cell. Further characteristics of these reactions are discussed in Chapter 2; the key property is that such reactions can be used for opposing metabolic pathways, such as glycolysis and gluconeogenesis in the liver. *Metabolically irreversible* reactions show a mass action ratio that is two (and usually more) orders of magnitude less than the reaction equilibrium constant.

2. There must be mobile cofactor(s) for the reaction.

Intermediary metabolites that are not connected in the pathway sequence, but are free to participate in other reactions within the cell are known as *mobile cofactors*. Only a few of these are known, such as the pair NAD/NADH or ATP/ADP. A further example is inorganic phosphate (Pi), which is released from one side or the other of the reaction.

If the two conditions are satisfied, it is possible that this step may be traversed in the opposing direction by a separate reaction in which the substrate is d and the product c:

$$\ldots \rightarrow \underline{d \rightarrow c} \rightarrow \ldots$$

If the reaction d → c has cofactors, they must be distinct from those of the reaction c → d. Thus, these are two distinct reactions, catalyzed by distinct enzymes, which share pathway intermediates. This reversibility of c with d I call *pathway reversible*; both reactions are metabolically irreversible, but the pathway can be reversed. In glycolysis, the enzyme pair phosphofructokinase/fructose bisphosphatase in liver provides an example (Chapter 2).

As an example of a *pathway-irreversible* step, reactions involving proteolysis cannot satisfy the requirements above. For example, a regulatory system that leads to the production of certain inflammatory cytokine involves the activation of a transcription factor called NF-kB. This factor is maintained in its inactive form by binding an inhibitor protein, I-kB. When the latter is phosphorylated, it becomes a substrate for addition of several units of ubiquitin, which targets I-kB for proteolysis through the proteosome. The released NF-kB is active and can enter the nucleus and acts as a transcription activator.

Two immediate advantages accrue from the adoption of the categorization of *pathway-reversible* and *pathway-irreversible*. First, it provides a link between enzymes of distinct pathways. For example, cells contain numerous protein kinases and far fewer protein phosphatases; however, the latter still display substrate preferences. By collecting the protein kinase and protein phosphatase for a particular step, we have an instance of a pathway-reversible process that is characteristic for this event, and distinct from any pathway-irreversible event. As an example, the kinase LKB1 has at about a dozen known protein targets, including AMPK. At least one protein phosphatase, PP2A, is known to dephosphorylate AMPK-P, restoring it to an inactive form. There is some ambiguity about whether other phosphatases, such as PP2C, might instead act on AMPK-P. With c assigned to the dephosphorylated AMPK and d the phosphorylated AMPK, we can state that this conversion of c → d is pathway-reversible, with LKB1 catalyzing the forward conversion and PP2A or PP2C the reverse. Another use for the concept arises from steps that are *pathway-irreversible*. The above example of NF-kB regulation clearly requires ongoing protein synthesis, so it definitely fits into this category. What is intriguing is that because the this protein can be rapidly accumulated once proteolysis is halted, if we were to rely only upon the time frame of regulation we may not be able to distinguish this from short-term regulation. Adopting the concept of this being a *pathway-irreversible* event makes this clear. Further points concerning the distinction between rapid metabolic events and those involving appearance of new protein are discussed in Section 5.5.

1.5 THERMODYNAMICS AND KINETICS: A METABOLIC PERSPECTIVE

Thermodynamics and kinetics—two guiding principles of chemistry—are somewhat modified when applying them to biological systems. Much of what has been described above is actually an application of one or the other of these ideas in order to establish the principles. For example, the notion of *near equilibrium* is clearly of thermodynamic origin. Yet, the concerns of an organic chemist are distinct from those of a metabolic biochemist, so that there are distinct uses for thermodynamics. Indeed, much of the art of thermodynamics is taking a widely agreed-upon principle and narrowing down the circumstances into something that can be employed for specific conditions.

Four metabolic uses of thermodynamics are

1. Energy coupling in reactions
2. Energy production by mitochondria
3. Redox reactions and their relationship to energy
4. Energy production and utilization by metabolic pathways

Energy coupling in reactions is analogous to half-reactions in redox reactions: a reaction is theoretically dissected into energy-favored and energy-forbidden parts, with the sum producing a favorable direction (see Ingraham and Pardee [7]), usually involving ATP as the molecule serving as the driving force. Consider the reaction catalyzed by hexokinase:

$$\text{Glucose} + \text{ATP} \rightarrow \text{glucose-6-P} + \text{ADP}$$

Taking the *pathway view*, the components of the half-reaction glucose → glucose-6-P are pathway intermediates. Those of the half-reaction ATP → ADP are mobile cofactors. This means that the reaction is driven by separate reactions in the cell that continuously supply ATP and remove ADP. Another way of stating this is that the cofactors of this reaction step are balanced when it is part of a pathway (perhaps glycolysis). As a more elaborate example, consider the activation of fatty acids, which can take place in three separate metabolic spaces in cells: the peroxisome (very long-chain fatty acids), the cytosol (short chain fatty acids), or the mitochondria (long-chain fatty acids) (see Chapter 6). In each case, the overall reaction

$$\text{fatty acid} + \text{CoA} + \text{ATP} \rightarrow \text{fatty acyl-CoA} + \text{AMP} + \text{PPi}$$

cannot be said to be the entire process in the cell. It must also be coupled to cellular pyrophosphatase:

$$\text{PPi} \rightarrow 2\text{Pi}$$

We can identify the pathway intermediate half-reaction for this process as

$$\text{fatty acid} \rightarrow \text{fatty acyl-CoA}$$

with the remaining reactions as the other half-reaction:

$$\text{CoA} + \text{ATP} \rightarrow \text{AMP} + \text{PPi}$$
$$\text{PPi} \rightarrow 2\text{Pi}$$

The similarity of this example to the first is evident: the pathway intermediates are energetically driven by separate reactions that are mobile cofactors, produced elsewhere, which must be regenerated for the process to continue in a pathway. However it is also clear that the notion of "half-reaction" must be stretched in order to identify the driving force reactions that accomplish the activation of fatty acids, and similar reactions which are involved in metabolic energy coupling.

Kinetics is the other pillar for analysis of metabolic regulation. The two parameters of the Michaelis–Menten equation—Vmax and Km—are invariably the most common ones used to analyze reactions both *in vitro* and within cells. These were introduced at the beginning of this chapter; a more detailed treatment is presented in Chapter 5. A further important use of the Vmax is to compare the value for an enzymatic reaction to the flux rate of a pathway, a common theme of Newsholme and Start [8]. This is often a dramatic means of distinguishing reactions that are likely to be rate-limiting from those that are not. For example, in liver, phosphofructokinase has a Vmax close to the rate of glycolytic flux, yet lactate dehydrogenase (a near-equilibrium enzyme) has a Vmax two orders of magnitude greater. While the values can indicate the total amount of enzyme protein available, the technique can be overdone; some enzymes display intermediate values and the decision becomes less clear. It should also be pointed out that Vmax and Km are derived strictly from an idealized Michaelis–Menten equation; most regulatory enzymes instead display sigmoidal kinetics. As pointed out by Atkinson [9] assuming a Vmax control is a defect in the analysis of control theory since most enzymes are not acutely regulated by changes in Vmax. In fact, the common shift in the infection point of sigmoidal enzymes (meaning a change in $S_{0.5}$ rather than Km) is a key regulatory feature of control point enzymes. While there are in fact mathematical descriptions of sigmoidal response, the tug-of-war between two opposing views (Koshland–Nemethy–Filmer versus Monod–Wyman–Changeux), raging since the 1960s has yet to be resolved; the issue is not further explored here (see Newsholme and Start [8] for a still-current description). What has been revealed are detailed molecular structures of enzymes, and identification of important metabolic control molecules that act allosterically; these will be considered for how they regulate metabolic pathways.

The other widespread consideration of enzyme kinetics, also touched upon above, is that of enzyme inhibition (and activation). The underlying process is explored in Chapter 4, with applications throughout the work. It is important to mention that the kinetic analysis is also simplistic: only a Michaelis–Menten-type derivation is applied, and yet the ideas extend to all enzymes, for some of which it is undoubtedly not correct. Since we are without a general equation for the more complicated case, we are left with this simplistic approach. A separate analysis is the control strength approach (Chapter 9); this has a distinct difficulty of assuming Vmax control as has been just pointed out. While neither is ideal, they offer an important insight into problems of regulation that are central to understanding metabolism. Another notion related to control strength is simply a titration with an inhibitor selective for what is believed to be the key control point in a pathway, first proposed by Rognstad [10]. This approach is detailed in Chapter 5.

1.6 SHORT- VERSUS LONG-TERM CONTROL

Control of metabolic pathways may be broadly divided between short-term and long-term regulatory systems. The divide is clear: a long-term event requires the formation of new protein. There are many reviews available for long-term control, which is achieved both at the transcriptional and translational level. While some of these events overlap with regulatory systems that are important in short-term regulation, this book will consider the long-term events as locking in a new steady state that can be analyzed under conditions that the protein levels do not change.

It is important to hold the distinction between these events in mind when considering incubations of cells that can progress over many hours or days as is possible in cell cultures. Thus, comparing results of incubations of an hour or two with those that take place over 24 h or longer are often impossible as genetic expression is likely in the latter situation.

There is also a hierarchy of metabolic alterations. For example, changes in cytosolic Ca^{2+} are elicited in most cells on a time frame of seconds or less; for excitable cells, the time frame is milliseconds. Metabolic intermediates of pathways adjust to new steady states upon a perturbation within minutes; this is a similar time scale for protein phosphorylation. Consider the significance of observing the phosphorylation state of a protein over a 1-h time frame as opposed to several days, or what it would mean to measure Ca^{2+} in cells after even a few minutes. The reverse is equally problematic: any regulatory system that requires a protein synthetic event or involves a protein degradative event cannot be considered in the short term, as the ability to reverse the regulation is not present. This invokes a general rule of regulatory systems: each involves a trigger to initiate the event, and a system to turn it off. Thus, we have the pairing of adenylate cyclase with phosphodiesterase (cyclic AMP control);

sarcoplasmic reticulum calcium release with sarcoplasmic reticulum CaATPase pump (contractile Ca^{2+} signaling); phospholipase activation to form diacylglycerol with either phosphorylation or deacylation of diacylglycerol (protein kinase C activation); and G-protein acquisition of GTP with GTPase activity (G-protein coupled signaling). It is important to recognize that both activation and inactivation are critical for the operation of all signaling systems. In particular, the rapidity that the signaling event can achieve is dependent upon the degradation of the signal.

It may seem simplistic to focus our attention on the short-term control of an event when the long-term controls have been fixed. This may be particularly unsuited to those schooled in thinking primarily about events in the long term: genetic alterations, changes in protein production rates, or alterations in their breakdown. It may seem more nuanced—and accurate—to perceive of a smooth alteration of states as protein levels change continuously and consider metabolic changes that occur as a consequence. Considering that most biologists are trained in some aspect of molecular biology, it is likely to be a majority view. However, if our goal is to come to terms with metabolism itself, then we must strictly separate the long- and short-term situations for one reason: the steady-state model does not apply when we are moving between different long-term situations. The notion that a more elaborate, less simplistic analysis *should* be available means that we are not recognizing the division between long- and short-term control. Once the division is recognized, the next question we must contemplate is, how well do we really understand control systems that are operating in the short term?

The ideas outlined in this first chapter are presented in more detail and with examples through this book. It is in fact the examples themselves that provide the inspiration for the principles in the first place. To begin with, we consider the first pathway that was elucidated in cells, the one that exists in virtually all of them, and for which we have unquestionably the greatest fund of information: glycolysis.

REFERENCES

1. Von Bertalanffy, L. (1975) *General Systems Theory*, George Braziller, Inc., New York.
2. Weinberg, G. M. (1985) *An Introduction to General Systems Thinking*, Wiley-Interscience.
3. Cleland, W. W. (1970) Steady state kinetics. In *Enzymes* (Boyer, P. D. ed.), Academic Press, New York.
4. Hofstadter, D. R. (1979) *Gödel, Escher, Bach: An Eternal Golden Braid*, Basic Books, New York.
5. Cornish-Bowden, A. (1995) *Fundamentals of Enzyme Kinetics*, Portland Press, London.
6. Fell, D. (1997) *Understanding the Control of Metabolism*, Portland Press, London.
7. Ingraham, L. L., and Pardee, A. B. (1967) Free energy and entropy in metabolism. In *Metabolic Pathways* (Greenberg, D. M. ed.), Academic Press, New York.
8. Newsholme, E. A., and Start, C. (1973) *Regulation in Metabolism*, Wiley, London.
9. Atkinson, D. E. (1990) *Control of Metabolic Processes*, Plenum Press, New York.
10. Rognstad, R. (1979) Rate-limiting steps in metabolic pathways, *J. Biol. Chem.*, 254: 1875–8.

2 Glycolysis
The Reference Pathway

Metabolic systems start with glycolysis; it is the first pathway discovered, the first learned, and the most thoroughly analyzed. It is self-contained: glycolysis can provide energy for a cell on its own, and connect with other needed routes. In the case of red blood cells (here, *red cells*), only a few other pathways are required, such as a pentose phosphate shunt to produce NADPH reductive power ultimately for antioxidant protection, and a side route to produce 2,3-bisphosphoglycerate to regulate oxygen affinity towards hemoglobin. In most cells, however, glycolysis contributes a minor amount of cellular energy, dominated by oxidative phosphorylation. Glycolysis serves in the critical role of a connector to other pathways. The beginning stages are entry points for glucose and other sugars, as well as glycogen (Figure 2.1). The pathway end product is lactate, but some pyruvate is converted to other products, such as alanine. Because of near-equilibrium at lactate dehydrogenase, some lactate formation is inevitable. Pyruvate is the connection point to oxidative metabolism, so it may be considered a branch point between these outcomes.

Considering glycolysis as a route connecting multiple pathways, the question arises: What exactly *is* the pathway of glycolysis? This is not as simple a question as it might seem. We might take the pathway to be

$$\text{glucose} \rightarrow \rightarrow \text{lactate} \tag{2.1}$$

but the sequence may not begin with glucose. The endpoint seems reasonable, but then again, what if we end it instead with pyruvate? Is it still glycolysis? In yeast, the terminating metabolite is ethanol; is there some reason other than semantics that this is called *fermentation*? It is appropriate that we begin our investigation of metabolism with the definition of a pathway.

2.1 DEFINITION OF A PATHWAY

A pathway is a sequence of enzyme-catalyzed reaction steps that has as its primary quality that the product of one reaction is the substrate of the next. Once we treat this as a connected sequence rather than an individual reaction, we must change our viewpoint, as new features emerge. This we will call the *pathway view*. A *pathway-substrate* is a metabolite that initiates a pathway. The term was introduced by Newsholme and Start [1] and Newsholme and Leech [2] to represent the substrate of the flux-generating step of a pathway. Considering Figure 2.1, it is evident that several different pathway-substrates are possible. To define a particular route, we would

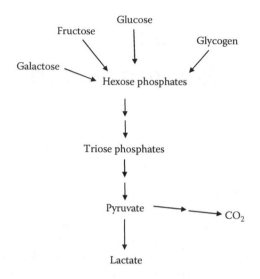

FIGURE 2.1 Connections to glycolysis.

need to specify this substrate. For example, we could say we are considering "glycolysis from glycogen," as opposed to "glycolysis from glucose."

It is evident, however, that if we accept "glycolysis from glucose" as a pathway, then we would also have to specify the origin of the glucose. We might consider that the pathway starts with extracellular glucose. This means that the glycolytic pathway would have to include the glucose transporter, which then becomes the flux-generating step.

The end product of a pathway is also less clearly defined than we might expect. Most commonly, we consider that glycolysis ends with lactate. This view is encouraged by naming the yeast pathway, which ends with ethanol *fermentation*. What if we end the route at pyruvate, and remove that product to the mitochondria for complete oxidation to CO_2, or else to other fates depending on the cell and the metabolic condition? What of the fact that branch points of the route depicted in outline in Figure 2.1 may remove intermediates?

One simple answer to all of these concerns is to say that it is the height of pedantry; after all, we know a pathway when we see it and all of the above can be called *glycolysis*. This is the accepted practice in two particular arenas: in textbooks and in electronic databases. In the former, metabolic pathways are actually consensus reactions (canonical sequences) that reflect broadly similar routes in most living cells. For the electronic databases, pathways are defined by convenience. There must be names to assign stored reaction sequences. The set of reactions commonly classified under the label of glycolysis serve as links to other information. The purpose of the pathway in a database is to allow the software algorithm to associate an enzyme (itself perhaps identified from its DNA sequence) with a function. By set-theory logic, any enzyme in the "glycolytic set" must have as its function the glycolytic pathway. However, this need not be the case. Considering the possibilities of distinct beginnings, endings, and branch points, this automatic assignment may be erroneous. The connections themselves are an important consideration, since every glycolytic intermediate connects with another reaction that is *not* glycolysis.

In order to provide a working definition of a pathway, we will take the following statements as necessary ingredients:

1. A sequence of reactions must exist in which the product of one reaction is the substrate of the next.
2. The first substrate in this sequence is flux-generating; a change in its concentration will change the flux, the rate through the entire pathway.
3. When a pathway is in a defined *steady state*—the assumed condition for pathways under metabolic consideration—then all intermediates in the pathway are at a constant concentration so long as this steady-state remains. In addition, the rates of each reaction within a linear sequence are the same.
4. In the steady state, the product concentration can change; in fact, this is an indication that it *is* a product.
5. Aside from pathway intermediates, there is a separate category of intermediates called **mobile cofactors**. These molecules—which include ATP and NAD—belong to multiple pathways. Another way to view these metabolites is that they connect pathways in parallel.
6. A pathway must balance all of its mobile cofactors. For glycolysis, that is widely recognized for the NAD/NADH pair, but it is equally true of the others: ATP, ADP, and Pi. Since there is a limited pool of these mobile cofactors, the net production of ATP, for example, must be exactly balanced by other pathways that consume ATP, or glycolysis cannot proceed.

With these considerations, we can now consider the reactions common to most glycolytic pathways, or canonical glycolysis.

2.2 CANONICAL GLYCOLYSIS

The sequence illustrated in Figure 2.2 shows the steps common to the route of glucose to lactate in many cell types. The value of considering a generalized pathway is both the appreciation of the logic of glycolysis as well as the broad commonality of this pathway in many different cell types.

The pathway intermediates run down the center of the page, apart from the singular example of the aldolase cleavage, in which the 6-carbon fructose-1,6-P_2 is converted to the two three-carbon sugar phosphates, DHAP is converted to GAP, catalyzed by triose isomerase (the enzyme is indicated as an asterisk to the side). GAP and DHAP are the first of the triose portion of the pathway, which has twice the flux of the hexose portion. Each conversion step is a small chemical transformation, but after the catalysis by the 11 enzymes shown, a profound difference is evident between the starting glucose and ending lactate.

The enzymes are indicated in ovals to the left of the individual reactions, and, as far as is known, all exist in the cell cytosol. The center arrows have two distinct qualities symbolized by a double-headed arrow (near-equilibrium) or a single

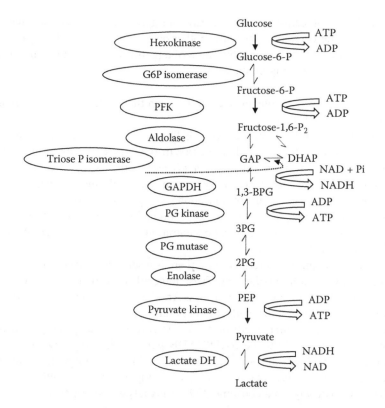

FIGURE 2.2   Glycolysis. GAP, glyceraldehyde-3-P; DHAP, dihydroxyacetone-P; 1,3-BPG, 1,3-bisphosphoglycerate; 3PG, 3-phosphoglycerate; 2-PG, 2-phosphoglycerate; PEP, phosphoenolpyruvate; GAPDH, glyceraldehyde phosphate DH; DH, dehydrogenase; PG, phosphoglycerate; P, phosphate.

directional arrow (metabolic irreversibility). These are discussed briefly in this chapter and in more detail in Chapter 3. To the right of the reaction sequence are the *mobile cofactors* of glycolysis, which are ATP, ADP, P_i, NAD, and NADH. This topic is also considered in the present chapter and in a subsequent one (Chapter 4). We will first consider the nature of the overall pathway.

2.3   LOGIC OF GLYCOLYSIS

The glucose provided to the pathway arises from outside the cell, and enters through one of a number of glucose transporters. These are of two types: a non-energy-linked exchange abbreviated GLUT [3] (appended with a number indicating the isoform), and an energy-linked exchange with sodium ions, abbreviated SGLUT [4]. Most cells have just the GLUT type.

Both types of transporters are present in two epithelial cells: those in the kidney proximal tubule, and enterocytes. In these, glucose entering the cell is largely destined for removal rather than metabolism through glycolysis. The entrance is driven by the SGLUT in the apical membrane, and exit is through a GLUT of the basolateral membrane. This achieves the return of kidney-filtered glucose in the first case, and glucose uptake from the gut in the second.

There are numerous studies of GLUT, and their distinguishing features [4,5]. Many focus on known genetic and protein differences, some their unique cell biological behavior, and some their tissue differences. The present discussion is limited to specific metabolic consequences of different glucose transporters. Beyond the transport step all of the glycolytic intermediates up to the pyruvate kinase step are phosphorylated. This has the effect of trapping those metabolites in the cytosol. The reason for this is not because they are more water-soluble than glucose, but because they have no transporter to move them across a membrane to another space.

In the first phosphorylation reaction, catalyzed by hexokinase, the kinetic feature of product inhibition has long been noted, suggesting a means of regulating the early portion of the pathway [6,7]. This inhibition, however, is not present in the isozymes present in liver, kidney, intestine, and endocrine cells of the pancreas. In those cells, the isozyme form is known as glucokinase (hexokinase IV), and has a very high (10 mM) K_m^i for glucose [8], and is not product inhibited by G6P. Thus, while this enzyme is metabolically irreversible in all cells, apart from its product inhibition in some isoforms, its control is not well established. Some have proposed that the enzyme itself might redistribute in the cell, as it can be found to adhere to mitochondria in fractionation studies [6,7]. The idea here is that ATP concentration is limiting far from the mitochondria so that hexokinase must come close to the source of production. There are several reasons to be skeptical of this idea; however, a more modern explanation for the association of hexokinase with membranes is related to apoptosis, as discussed in Chapters 3 and 4.

Following phosphorylation of glucose on the 6-position, an isomerase operates to convert the carbonyl in the 1-position to an alcohol. The isomerase reaction requires ring opening (Figure 2.3), which has been suggested to contribute to metabolic regulation, largely on the basis of kinetic analysis [9]. However, the isomerase catalyzes a reaction close to its equilibrium point [10,11]. In terms of chemical logic, the isomerization to F6P produces a free hydroxyl group at the 1-carbon, which is phosphorylated in the ensuing step by phosphofructokinase (PFK). With the end carbons phosphorylated, the molecule is split into two trioses at the aldolase step, marking the end of the hexose stage and beginning of the 3-carbon stage of glycolysis. The two trioses are consolidated into GAP (glyceraldehyde-3-P) through triose phosphate isomerase. From this point forward, the flux is double that of the hexose stage. Conversion of GAP

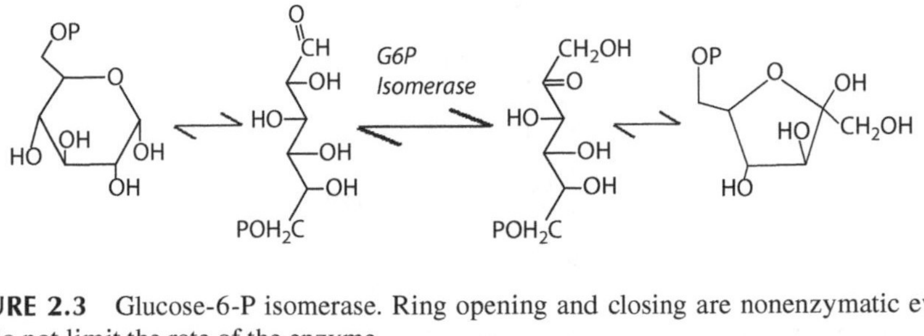

FIGURE 2.3 Glucose-6-P isomerase. Ring opening and closing are nonenzymatic events and do not limit the rate of the enzyme.

to 1,3-BPG (1,3-bisphosphoglycerate) is the sole oxidative step in glycolysis, and also accomplishes a phosphorylation using P_i (inorganic phosphate). In a sense, this single reaction, combining oxidation with phosphorylation, is reminiscent of the phrase usually reserved for mitochondrial energy production. What is commonly considered *oxidative phosphorylation* is the mitochondrial pathway for ATP formation, detailed in the next chapter. In glycolysis, ATP formation occurs in the phosphoglycerokinase step in which 1,3-BPG is converted to 3-PG. Overall, the reactions of GAPDH and PGK can be viewed as a pair of coupled reactions that achieve what is known as a *substrate level phosphorylation*. In the early days of study of mitochondrial energy formation, it was widely believed that the mitochondria used a similar mechanism to make ATP. As we will appreciate, it is entirely distinct.

The next two reactions, a mutase and an enolase, lead to the 3-PGA (3-phosphoglycerate), which is next converted to the very similar 2-PGA (2-phosphoglycerate) and then to PEP (phosphoenolpyruvate). These reactions bring about the rearrangement of a phosphoryl bond to produce PEP, the eponymous high-energy component (Figure 2.4). The chemical modifications involve merely a movement of the phosphate group and a dehydration, yet the result is an electron arrangement that produces a high-energy bond in PEP that can engage in phosphoryl transfer to ATP in the following reaction. This ATP-forming step of glycolysis is pyruvate kinase. Unlike PGK, pyruvate kinase is a regulated step of the pathway.

The final step is the only other redox reaction in the sequence, LDH (lactate DH). For glycolysis to perform as the sole energy-generating enzyme in the cell, as in the red cell example, the NADH produced at GAPDH must be matched by its utilization at LDH.

Not all of the enzymes of this or any pathway should be considered on an equal basis when it comes to regulation, or their connections to other pathways. As indicated above, just three of the steps are potentially regulatory: HK, PFK, and PK. In the next section, I detail the major division of metabolic reactions between two types: near-equilibrium and the metabolically irreversible.

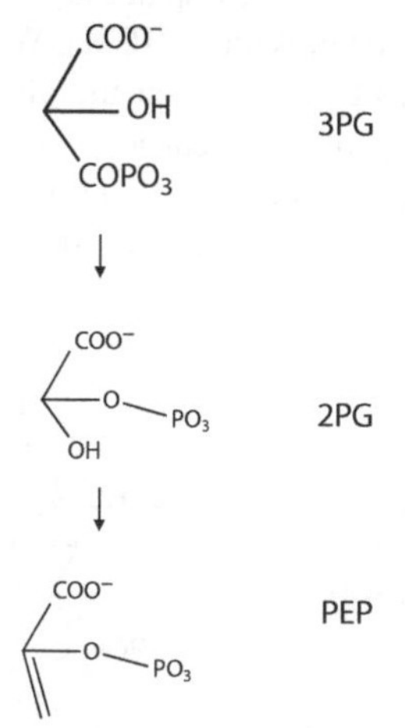

FIGURE 2.4   Triose rearrangements forming PEP (phosphoenolpyruvate).

2.4 NEAR-EQUILIBRIUM VERSUS METABOLIC IRREVERSIBLE REACTIONS

A powerful concept that cuts the Gordian knot of many metabolic problems is the division of enzymes into two camps: either *near-equilibrium* or *metabolically irreversible* [1,10]. The first category of enzymes catalyze their reactions close to the equilibrium constant, that is, the mass-action ratio is within one or two orders of magnitude of the equilibrium constant. The second category—which contains far fewer enzymatic reactions—have mass-action ratios so much smaller than the reaction equilibrium constant that the reaction is not reversed under cellular conditions.

A unique property of near-equilibrium (NEQ) reactions is that they can be used for the forward reaction for one pathway, and operate in reverse for a separate pathway within the cell. This switch, of course, requires a significant change in cellular conditions. As an example, G6P isomerase is a glycolytic enzyme catalyzing F6P formation, but in fasting conditions in liver it is a gluconeogenic enzyme catalyzing G6P formation. Enzymes that catalyze reactions close to equilibrium are invariably found in large quantities. This is due to the fact that the fundamental action of catalysts is to bring a reaction towards equilibrium, and enzymatic reactions are proportional to the concentration of enzyme.

A second property of NEQ reactions is that they are not subject to hormonal or allosteric regulation in cells. Rather, they are exquisitely sensitive to changes in substrate and product concentration. Thus, NEQ reactions can be reversed in direction. The reason these enzymes are insensitive to cellular regulators is simply that they are already near equilibrium, and it is the fundamental action of enzyme catalysis to bring reactions close to equilibrium. It follows that it is of no importance to investigate molecules that activate or inhibit these steps. The fact that they have this behavior in *in vitro* enzyme assays is strictly a consequence of how enzymes are assayed (see Chapter 5). As a practical matter, it is not rewarding to consider this class of enzymes as potential drug targets either. The same feature that makes them achieve near-equilibrium—the high concentration of enzyme—also means that only a relatively high concentration of a drug could achieve blockade of these steps. This makes off-target actions of these inhibitors more likely.

Metabolically irreversible (mIRR) reactions, such as the PFK (phosphofructokinase) step, are displaced from equilibrium by two or more orders of magnitude. This enzyme catalyzes the reaction

$$\text{fructose-6-P} + \text{ATP} \rightarrow \text{frucose-1,6-P}_2 + \text{ADP} \qquad (2.2)$$

but not the reverse reaction in cells. It is among the metabolically irreversible steps that we find potential regulatory sites for pathways.

Much of the analysis that led to this finding was performed by H.A. Krebs and colleagues and first formulated clearly in Newsholme's [1] *Regulation in Metabolism* in 1973. It could be restated that the mass-action ratio *tau* is close to K_{eq}, at the temperature and ionic strength of cells [12].

A direct way of viewing the equilibrium position of a reaction in a metabolic sequence is to write the expression for the free energy of a reaction

$$\Delta G = \Delta G^0 - RT \ln \tau \qquad (2.3)$$

where τ is the mass-action ratio. Combining this with the equation for standard free energy

$$\Delta G^0 = -RT \ln K_{eq} \qquad (2.4)$$

we arrive at the free energy equation comparing τ to K_{eq}:

$$\Delta G = -RT \ln K_{eq} + RT \ln \tau \qquad (2.5)$$

Most readers will not find the equations particularly surprising as they are part of every standard treatment of elementary chemistry. The form of Equation 2.5 may be a bit unusual, but it demonstrates the clear relationship between the equilibrium constant and the mass-action ratio, and does so in terms that are useful for metabolic analysis. The NEQ and mIRR enzymes correspond to two conditions that can be summarized as

$$\Delta G \cong 0 \text{ when } K_{eq} \cong \tau \qquad (2.6)$$

$$\Delta G \ll 0 \text{ when } K_{eq} \gg \tau \qquad (2.7)$$

The inequality and the approximation are required because every reaction in the pathway must satisfy the condition

$$\Delta G < 0 \qquad (2.8)$$

The distinction between Equations 2.6 and 2.7 provides us with the definition for the two great classes of enzymes: mIRR (*metabolically irreversible*) and NEQ (*near-equilibrium*). Let us consider examples relevant to glycolysis.

2.5 MEASURED INTERMEDIATES IN GLYCOLYSIS

The intermediates of glycolysis have been measured extensively in three tissues: red cells, liver, and skeletal muscle. The red cell data [13,14] demonstrated a close correspondence between the mass–action ratios and equilibrium constants for all of the glycolytic enzymes apart from hexokinase, phosphofructokinase, and pyruvate kinase. Even in these early studies, the authors were careful to point out that any one of these three mIRR enzymes cannot simply be said to be the rate-limiting step, but that they participate in rate control. This later evolved into more nuanced ideas about control strength (see Chapter 9). In a similar way, the same results for glycolytic reactions were obtained for liver, heart, and skeletal muscle [10].

Both technical and conceptual issues must be considered in the analysis of metabolic data taken to establish the mass–action ratio. For example, obtaining exact equilibrium constants that pertain under *in vivo* conditions can be a difficulty [15]. Another significant problem is an error in determination due to the significant binding of a metabolite to proteins; this is further considered in a separate section below (total versus free concentrations).

An alternative view in response to anomalous measurements is that they indicate that *all* intermediates of various pathways are *channeled* through pathways, passing between enzymes, but not exchanging with cell water (e.g., refs. [16,17]). In some cases where the data has been re-evaluated, the challenge has been met (e.g., refs. [18–20]). Yet this has not diminished interest in this mechanism, which is a form of compartmentation (see Chapter 4). The idea of compartmented metabolites is nonsystematic, in conflict not just with the physical meaning of diffusion, but with other established ideas of metabolism.

2.6   ENERGY COUPLING IN GLYCOLYSIS

A common view of reaction energetics is that one substrate of higher energy drives another of lower energy so that the overall free energy change is negative. This is correct, but it is also easy to slip into the erroneous idea that standard free energies for partial reactions are appropriate surrogates to explain the overall reaction *in vivo*. In the last section, I have shown how the actual free energy and the equilibrium situation can be profitably used to better understand metabolism by the simple yet powerful concept of NEQ and mIRR enzyme classes. How does the standard free energy contribute to our understanding of metabolism?

Standard free energy is useful in the deconstruction of a reaction into parts in which we imagine an overall negative ΔG^0 can occur even if one part is positive. Thus, for the hexokinase step, the overall ΔG^0 is negative (about -17 kJ/mol). However, the partial reaction involving just the pathway intermediates must have a positive ΔG^0 (that is, the formation of glucose-6-P from glucose [21]). In the *formal* sense in which we consider energy coupling, it is the loss of the phosphoryl moiety from ATP that makes the overall ΔG^0 negative.

It is important to use the different free energy terms precisely, as it is common to draw incorrect conclusions about energy coupling. For example, standard conditions never occur under physiological circumstances. In the **standard** state, all concentrations are at 1 M, for example. In the **equilibrium** condition, by definition the free energy change is zero. Under actual cellular condition, we require that all reactions have a negative free energy change. What is not possible is for any reaction that is part of a pathway to have a free energy that is positive or zero.

An exposition of energy coupling was clearly laid out in the 1960s [22]. The formal addition of standard free energies divides a reaction into half-reactions to illustrate how it is possible to use the energy of a portion of a reaction to enable bond formation that would otherwise not be thermodynamically favorable. However, these half-reactions don't actually exist. It is strictly a thermodynamic argument for how the overall process could function. Since thermodynamics is path-independent and the mechanism for a reaction requires kinetics instead, the actual steps require a different and specific route. It is also required that the overall *actual* free energy change be negative. The concept of energy coupling applies not just to an enzymatic mechanism, but to a metabolic pathway. A comparison of standard and actual free energies can be illustrated for glycolysis.

2.7   A PATHWAY-FREE ENERGY PLOT FOR GLYCOLYSIS

The same laboratory that provided measurements of red cell glycolysis intermediates [14] analyzed the metabolite concentrations to provide a free energy plot of the pathway. This was modified to compare both ΔG and ΔG^0 values for each reaction of glycolysis as a function of the pathway intermediates. This energy diagram is presented as Figure 2.5.

The plot was constructed by assigning the value of 0 to G^0 for glucose, and arbitrarily moving the G value downwards to represent the lower free energy of a cellular glucose. The absolute G values are arbitrary; it is the difference between these values that is important, and these were derived from measured intermediates in red cells. Next, the values of ΔG and ΔG^0 were plotted as the slopes connecting the intermediates of the glycolytic pathway. The intermediates are presented as equidistant points comprising the abscissa.

The changes in free energies can be visually compared from the slopes of the lines connecting the intermediates. A steep slope represents a large change in free energy for that reaction. In the upper curve, values of ΔG^0 are presented. These are

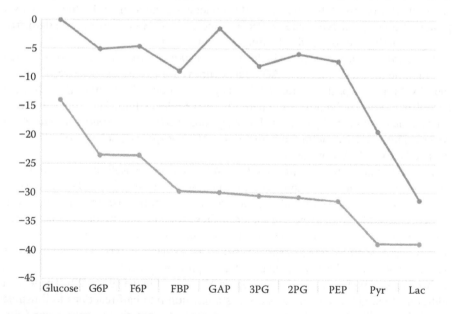

FIGURE 2.5   Free energy pathway diagrams for glycolysis. The ordinate for the upper curve is G^0, or standard free energy, and for the lower curve, G, or actual free energy. Both are in units of kcal/mol. Abbreviations: G6P, glucose-6P; F6P, fructose-6P; FBP, fructose-1,6-P_2; GAP, glyceraldehyde-3P; 3PG, 3-phosphoglycerate; 2PG, 2-phosphoglycerate; PEP, phosphoenolpyruvate; pyr, pyruvate; lac, lactate.

in clear contrast with the lower curve, which displays values of ΔG. The graph shows that there is some correspondence between the metabolically irreversible steps of glycolysis—hexokinase, phosphofructokinase, and pyruvate kinase—evident as the steep slopes in the bottom curve with corresponding large negative slopes in the top, the standard free energy curve. However, further examination of the graph shows that this is more coincidental than insightful. For example, the reactions of aldolase and phosphoglucomutase show a positive slope in the standard free energy profile. This means that under standard conditions, these reactions are impossible, and as also pointed out by Hamori [23], glycolysis cannot proceed. It is a misconception of energy coupling to imagine that one enzymatic reaction that is unfavorable can be "pulled forward" by a subsequent one that has a very large and negative free energy change. The reaction steps are not imaginary ones that can be manipulated to obtain possible overall free energies, but actual reaction steps in the pathway. For this reason, *every* step in a steady state pathway must have a negative free energy change. This is clearly the case in the lower plot of Figure 2.5. Most of the reactions have a very low free energy change, indicated by the near-zero slope of the connecting line. These represent the near-equilibrium reactions. As for the overall free energy changes, each of these may be obtained by the differences between the free energy of the first and last metabolite. In the case of the standard states, this value is -131 kJ/mol; for the actual free energy change of the pathway (lower curve), the value is -104 kJ/mol. Thus, overall both conditions are favorable (and similar), but the pathway is not possible under standard conditions.

Suppose we inhibit the flux of glycolysis by blocking the PFK (phosphofructo-kinase) reaction. One way this could be done is by introducing fatty acids, which lead to an increase in the cytosolic concentration of citrate an allosteric inhibitor of PFK[ii]. Alternatively, PFK activity may be altered by its cellular regulator, fructose-2,6-bisphosphate [24,25]. A lowering of PFK lowers the rate of glycolysis, so that a new steady state would be formed. Once again, every step in this new steady state must have a constant rate. There are two rates for each portion of the pathway, with the triose portion having twice the rate of the hexose portion. The near-equilibrium steps—which includes most of the reactions—change by mass-action effects. Consider, for example, the G6P isomerase reaction. With a lowered PFK activity, F6P will increase in concentration, which will tend to push the G6P isomerase in the backward direction from glycolysis. Assuming that glycolysis is still operating (a reasonable one apart from a few conditions in a few cell types like liver), the net direct of G6P isomerase is the same, but its rate is lowered due to the adjustment of its substrate G6P and product F6P, controlled indirectly by events affecting PFK.

Of course, in order to understand how the metabolically irreversible enzymes are regulated we must look to external control mechanisms, as mIRR reactions are *not* directly controlled just by their substrate and product levels.[iii] For example, the rate of pyruvate kinase is under the control of the allosteric regulator fructose-1,6-P_2, which is the product of PFK. This control is known as *feed-forward activation*. Still, it is not the only control exerted over the PK reaction. In the case of liver PK, the enzyme is subject to phosphorylation with subsequent inactivation. In the muscle, while there is no phosphorylation, fructose bisphosphate feed forward activation still regulates pyruvate kinase [26].

An important quality of near-equilibrium enzymes is that they exist in large quantities, as it takes a good deal of total enzyme in order to achieve the near-equilibrium state. Simply by inspecting the tables of enzyme quantity for a given cell type for each of the glycolytic enzymes this point is evident (as summarized in ref. [1]. As a corollary, the induction of cellular protein, while it occurs for all enzymes, is far more critical for the metabolically irreversible ones, as these are the ones where it makes a large difference when a small increase or decrease is affected.

2.8   MOBILE COFACTORS

There is a metabolic distinction between the intermediates arrayed (with a single exception of DHAP) down the center of Figure 2.2 and those off to the right-hand side attached by curved arrows. This distinction is apparent in a *pathway view* as opposed to a *reaction view*; these were defined in the first chapter, and now investigated in the context of a complete pathway, glycolysis.

Consider the reaction catalyzed by hexokinase. The reaction view is

$$\text{Glucose} + \text{ATP} \rightleftarrows \text{G6P} + \text{ADP} \tag{2.9}$$

and all components on the left of the equilibrium sign are substrates and all on the right are products. In this view, there is no distinction between glucose and ATP; both are substrates for the forward reaction. However, if this reaction is considered from the perspective of the *pathway view*, then glucose is no longer equivalent to ATP. In this case, ATP is a mobile cofactor. Similarly, in the pathway view, ADP is no longer a product, but rather another mobile cofactor. These are of course the energy transfer molecules, but they are also broadly communicating molecules that connect hundreds of metabolic reactions within pathways.

Aside from these there are a few other mobile cofactors in glycolysis: NAD, NADH, and Pi. All share the quality that, within a pathway, they communicate with separate, multiple points of metabolism.

2.9   ATP AND ADP OF GLYCOLYSIS

A major role for glycolysis is the production of ATP from ADP. In the absence of mitochondrial pathways, this is the only route for the provision of energy for a cell. For all cells, the use of glycolytic intermediates as intersection points (pathway intermediates) of other pathways is an equally important function.

Glycolysis is often divided into two parts: the six-carbon sequence and the three-carbon sequence. It happens that this also preserves a distinction that exists between the direction of the ATP and ADP cofactors. For the six-carbon sequence, ATP is the cofactor on the substrate side, which means this sequence utilizes ATP energy. For the three-carbon sequence, ATP appears as the cofactor on the product side, which means that this sequence produces ATP energy. The net increase is due to the fact that there is twice the flux through the three-carbon side, so that ATP formation exceeds ATP utilization and we have a total energy formation of 2 ATP per glucose converted to lactate.

Since there are four reactions that involve reactions with the mobile cofactors ATP and ADP in glycolysis, we say that they *share* a common pool of these molecules. The number of cells that use ATP as a cofactor is large, and most of these reactions are in the cytosol. The KEGG database lists over 350 such reactions [28]. We say that all of these reactions, and all of the pathways in which they are involved are in this way connected, but not in the same sense as the sequential run of glycolysis. These hundreds of reactions are connected in *parallel*, just as the four under consideration in this pathway are.

The belief by many investigators that ATP concentrations fluctuate within cells stems at least in part from the fact that some targets of current investigation are known to be sensitive to ATP *in vitro*. One prime candidate is the ATP-sensitive potassium channel (K-ATP channel) prominent in insulin secreting [29] and glucagon-secreting [30,31] cells of the endocrine pancreas. There is also an unfortunate harkening back to the notion of the "energy charge" first introduced in the early 1960s by Atkinson [32], and since reborn in various forms. At the time, the regulatory kinases of glycolysis, hexokinase [33], phosphofructokinase [34], and pyruvate kinase [35] were known to inhibited *in vitro* by ATP. Thus, the idea that ATP and other adenine nucleotides were both cellular energy intermediates as well as cellular regulators seemed reasonable; perhaps some ratio of nucleotides provide an index of regulation.

2.10   HISTORICAL VIEW OF ATP

It may be useful to step back and consider earlier studies of ATP; after all, accurate measurement of this nucleotide has been available since the 1920s, and it is present in all cells at high, millimolar concentration. By 1950, the notion that ATP concentration was a constant in cells was considered a settled issue. In Hill's "challenge to biochemists" he points to the fact that, even during active muscle contraction, preparations show a constancy of ATP levels [36]. The challenge was to explain how ATP could actually be the intermediate energy provider for muscle contraction! The answer to this paradox is that muscle contains another high-energy phosphate that can rapidly replete ATP when its level begins to dip: phosphocreatine (P-Cr). In the resting state the creatine phosphokinase reactions catalyzes the formation of this compound from creatine (Cr) through the forward reaction of the near-equilibrium step:

$$Cr + ATP \rightleftharpoons Cr \sim P + ADP \qquad (2.10)$$

The Cr ~ P builds up during the resting phase, and reaches a level that is much greater than the total concentration of ATP. During periods of active contraction, the reverse of equation 10 occurs, so that ATP remains constant. This notion that ATP is constant in cells is sometimes rediscovered and published as if it is a surprise that cells can consume ATP but not change its level (e.g., ref. [37]).

A related issue is a summary of cellular energy reserves that treats ATP and creatine phosphate as if they could be categorized as substrates, such as the glycogen

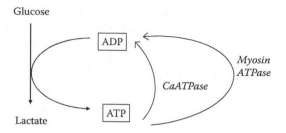

FIGURE 2.6   Energy balance in fast muscle. Glycolysis provides the ATP as the sole source. Essentially just two reactions utilize it: CaATPase, the enzyme that is required for recycling Ca^{2+} back into the ER, and the myosin ATPase, producing the contraction itself.

and fat depots [38,39]. This is a misconception that prevents clear understanding of energy metabolism. To say that we can survive 3 seconds on ATP and 8 seconds on creatine phosphate, longer on glycogen, and longer still on fat reserves puts these on an equal footing when they are not. Exercise physiology treatments even suggest that activities can be associated with each of these; thus the shot put can be performed with just ATP; a sprint with just creatine phosphate [40,41]. Our cells do not contain some sort of automatic shifting device that can smoothly transition between ATP and sugar. Rather, it is the carbohydrate and lipid that are the actual substrates; ATP and creatine phosphate are intermediate molecules. The finer distinction between them is that ATP does not fluctuate in concentration, whereas creatine phosphate does, in order to maintain a relatively constant ATP. What is correct is that we have sufficient ATP in cells to last about 3 seconds, which would drop precipitously if there were not reactions rapid enough to ensure that it does not appreciably decrease in concentration.

It is clear that ATP is the intermediate in contraction. To focus on a specific event, consider the overall energetics of contraction for an anaerobic muscle cell (that is, fast, white, or glycolytic fiber). The entire ATP production is accomplished by glycolysis, and the utilization can be accounted for by just two enzymatic reactions: myosin ATPase (which drives the contraction) and the endoplasmic reticular CaATPase (which drives reuptake of Ca^{2+} to the endoplasmic reticular lumen to terminate the contractile signal). This system is illustrated in Figure 2.6. It should be pointed out that, in the resting state, this does not account for the more complex situation of energy utilization. However, since the overall rate of glycolysis (matching the rate of ATP utilization) is extremely low at rest, it is not a significant detraction. Muscle has the greatest range of energy utilization between its active and less-active states.

2.11   ADENINE NUCLEOTIDES

The three principal adenine nucleotides are ATP, ADP, and AMP. In the glycolytic pathway itself, there is no role for AMP. However, all participate in the balance of adenine nucleotides.

In some reactions in which the phosphoryl bonds of ATP are utilized, the overall splitting is at the *beta* rather than the *alpha* phosphoryl of the molecule. This event can be formally summarized by considering the half-reaction

$$ATP \rightarrow AMP + PPi \tag{2.11}$$

As just stated, AMP is not involved in glycolysis in most organisms, but in some bacteria, PPi can replace ATP as the high energy donor [42], for example

$$\text{Fructose-6-P} + PPi \rightarrow \text{fructose-1,6-P}_2 + Pi^{\,iv} \tag{2.12}$$

In some metabolic reactions, such as fatty acyl CoA formation (Chapter 6) AMP and PPi are formed. In order to maintain steady state, these also have to be regenerated, as AMP and PPi are also mobile cofactors. These require ancillary reactions. The principal reaction by which AMP is converted to the other adenine nucleotides is catalyzed by adenylate kinase:

$$ATP + AMP \rightleftarrows 2ADP \tag{2.13}$$

a reaction known to be near-equilibrium in cells [45]. Thus, this reaction can also produce AMP when a building of ADP results, a feature that plays a prominent role in regulation that we will investigate in Chapter 7.

The reaction involving breakdown of PPi is the metabolically irreversible pyrophosphatase:

$$PPi \rightarrow 2Pi \tag{2.14}$$

Thus, the PPi will resupply the Pi pool, and AMP can be returned to ADP. We already have a means to reconvert ADP to ATP—the glycolytic pathway. We next consider a second major pair of mobile cofactors: NAD and NADH.

2.12 REDOX COFACTORS OF GLYCOLYSIS: NAD AND NADH

The redox cofactor pair of nicotinamide nucleotides operates in a similar way to the adenine nucleotide pair of glycolysis. Like the pair ATP/ADP, the pair NAD/NADH are also mobile cofactors and can be considered in the *reaction view* to be substrates and products, but in the *pathway view* to be cofactors that connect separate reactions in parallel. Also like the ATP/ADP pair, there are numerous dehydrogenases (over 200 listed in ref. [46]) that draw on the same pair.

The formation of NADH is due to just one enzyme in glycolysis: glyceraldehyde phosphate DH. On the other hand, restoration of NAD for that enzyme has multiple possibilities. Three are outlined in Figure 2.7. The formation of lactate marks the end of the typical glycolytic pathway: NADH is oxidized by the action of lactate DH. In yeast cells, pyruvate is decarboxlated, and NADH is oxidized instead by alcohol DH, with the reduction of acetaldehyde to ethanol. The third possibility outlined is

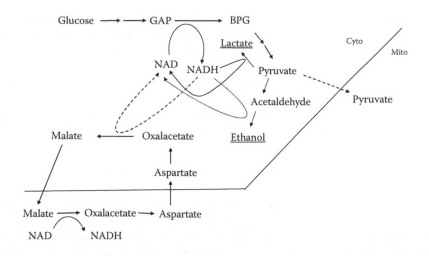

FIGURE 2.7 Balance of cofactors NAD and NADH. Glycolysis converts NAD to NADH at the glyceraldehyde phosphate DH step. Three routes of reoxidation of NADH to NAD are shown: lactate DH, alcohol DH, and malate DH of the malate/aspartate shuttle for moving reducing equivalents into the mitochondria.

that pyruvate enters the mitochondria to serve as a substrate for complete oxidation. In this case, the NADH is oxidized by malate DH, as part of the malate/aspartate shuttle to move reducing equivalents from cytosol to the mitochondria, as illustrated. The shuttle pathway can operate at the same time as either lactate DH or alcohol DH. Thus, multiple pathways are responsible for NAD/NADH redox balance.

2.13   INORGANIC PHOSPHATE AS A MOBILE COFACTOR

Like the other mobile cofactors, inorganic phosphate is also both a reactant and product from the reaction view, or a connector to multiple steps in the pathway view. However, rather than having a matching molecule, P_i appears alone as a mobile cofactor. While a stoichiometric balance to the reaction of ATP "hydrolysis" or synthesis, in fact a literal "ATP hydrolysis" does not exist in cells. It is convenient shorthand to name the Na pump the NaKATPase, and the Ca pump the CaATPase, but the phosphoryl group of ATP is initially transferred to an enzyme intermediate to facilitate the overall process. This is also true of the mitochondrial ATP synthetase (colloquially, $F_1F_0ATPase$) and other molecular motors. While cellular contents of Pi are relatively abundant (in the high millimolar range), it is nonetheless a mobile cofactor that requires regeneration like the others.

2.14   TOTAL VERSUS FREE CONCENTRATIONS

Modern methods for the analysis of metabolites have become increasingly sophisticated from a chemical standpoint. However, it is important to recognize that the total concentration may not represent the solution concentration. This is widely known for substances in the blood; for example, most are aware that lipid soluble compounds are largely bound to blood proteins. In addition, only about half of serum Ca^{2+} is free to react. The same situation applies to certain compounds in the cell. Of the five mobile cofactors involved in glycolysis, two—ADP and NADH—are known to be mostly bound to proteins in the cell cytosol, with only a small fraction free in water solution to react. This is a specific form of compartmentation, an issue discussed further in Chapter 4.

2.15 NUCLEOTIDES OTHER THAN ADENINE

While we are used to thinking about ATP and its other phosphorylated relatives ADP and AMP, there are also a variety of other nucleotides in cells. These are usually of interest to investigators interested in genetics as these nucleotides are needed for the formation of the information molecules DNA and RNA. However, it has long been known that the cell makes (admittedly less frequent) use of some other nucleotides in intermediary metabolism. In terms of the next most abundant nucleotide, guanine nucleotides (present at about one-tenth the concentration of adenine nucleotides [47]) find use in the eponymous "G-proteins," where typically the proteins are bound either to a GTP molecule (usually its active state) or GDP (usually its inactive state). Conversion between them is catalyzed *in situ*, corresponding to the inactivation of the intermediary signaling protein. Overall, the reaction catalyzed in this and in other reactions such as the gluconeogenic PEPCK [48] and the mitochondrial Krebs Cycle enzyme succinyl CoA synthetase [49], is characterized by the half-reaction

$$GTP \rightarrow GDP + Pi \qquad (2.15)$$

and the standard phosphorylation potential is the same as that for ATP hydrolysis. It is not clear why GTP is used rather than ATP, but it is likely a function of the different phosphorylation potentials. A rule of thumb that has emerged is that use of these more minor nucleotides in reactions corresponds to some cellular pathways. Thus, while glycolysis itself uses just ATP, glycogen metabolism uses UTP; as noted above, gluconeogenesis uses GTP, and lipid biosynthetic routes use CTP.

While some other aspects of glycogen metabolism will be explored in Chapter 6, it is important for us to consider that it is strongly connected to glycolysis in that in those tissues in which glycogen is in appreciable quantities (liver and muscle) it can be considered to be the pathway substrate of glycolysis.

In order to rephosphorylate nucleotides other than ATP, it is necessary for cells to make use of a commonly occurring reaction catalyzed by the NDP kinase [50]:

$$ATP + NDP \rightarrow ADP + NTP \qquad (2.16)$$

where N represents any non-adenine nucleotide. The reaction is needed to restore nucleotides other than ATP to their triphosphate state. The energy generating pathways of glycolysis as well as creatine phosphokinase and oxidative phosphorylation form strictly ATP.

The cellular equilibrium position of NDP kinase is uncertain [51]. The reaction clearly has an equilibrium constant close to 1 with any nucleotide diphosphate substrate (one side reacts a triphosphate with a diphosphate), Thus its standard free energy is zero:

$$\Delta G^\circ = -RT\ln K_{eq} = 0.$$
(2.17)

However, there is no reason to assume that this is also the case for the cellular free energy, and some reasons to suspect it may not be. First, the metabolic purpose of rephosphorylation of an NDP is clear: after its energy is expended, it is important to rephosphorylate the molecule for reuse. However, there is no known metabolic role for the reverse reaction. Assuming, for example, that G proteins use GTP precisely because it is *not* in near-equilibrium with the adenine nucleotide pool, then it would make sense to assume that the NDP kinase is mIRR. From the standpoint of glycolysis, we can say that it is connected in parallel with the rephosphorylation of the diphosphate forms of other nucleotides.

2.16   GLYCOLYSIS HAS CONNECTIONS

Beyond the cofactor balance issue, glycolytic intermediates connect to a large number of other pathways of intermediary metabolism. Glycolysis can connect to glycogen metabolism, the pentose phosphate shunt, and amino acid metabolism. From an energetic point of view, glycolysis can also be considered as a route that feeds the pathway that normally provides the bulk of cellular energy: mitochondrial oxidative phosphorylation.

REFERENCES

1. Newsholme, E. A., and Start, C. (1973) *Regulation in Metabolism*, Wiley, London.
2. Newsholme, E. A., and Leech, A. R. (1986) *Biochemistry for the Medical Sciences*, Wiley, New York.
3. Brown, G. K. (2000) Glucose transporters: Structure, function and consequences of deficiency. *J. Inherit. Metab. Dis.* 23, 237–246.
4. Chen, L. Q., Cheung, L. S., Feng, L., Tanner, W., and Frommer, W. B. (2015) Transport of sugars. *Annu. Rev. Biochem.* 84, 865–894.
5. Thorens, B., and Mueckler, M. (2010) Glucose transporters in the 21st century. *Am. J. Physiol. Endocrinol. Metab.* 298, E141–E145.
6. Rosa, J. C., and Cesar, M. C. (2016) Role of hexokinase and VDAC in neurological disorders. *Curr. Mol. Pharmacol.* 9, 320–231.
7. Inui, M., and Ishibashi, S. (1979) Functioning of mitochondria-bound hexokinase in rat brain in accordance with generation of ATP inside the organelle. *J. Biochem.* 85, 1151–1156.
8. Cornish-Bowden, A., and Cardenas, M. L. (1991) Hexokinase and 'glucokinase' in liver metabolism. *Trends Biochem. Sci.* 16, 281–282.
9. Schray, K. J, and Benkovic, S. J. (1978) Anomerization rates and enzyme specificity for biologically important sugars and sugar phosphates. *Acc. Chem. Res.* 11, 136–141.
10. Krebs, H. A. (1969) The role of equilibria in the regulation of metabolism. in *Current Topics in Cellular Regulation* (Horecker, B. L., and Stadtman, E. R. eds.), Academic Press, New York.

11. Krebs, H. A. (1975) The role of chemical equilibria in organ function. *Adv. Enzyme Regul.* 13, 449–472.
12. Guynn, R. W., Gelberg, H. J., and Veech, R. L. (1973) Equilibrium constants of the malate dehydrogenase citrate synthase citrate lyase and acetyl coenzyme A hydrolysis reactions under physiological conditions. *J. Biol. Chem.* 248, 6957–6965.
13. Minakami, S., Zuki, C., Saito, T., and Yoshikawa, H. (1965) Studies on erythrocyte glycolysis I. Determination of the glycolytic intermediates in human erythrocytes. *J. Biochem.* 58, 543–550.
14. Minakami, S., and Yoshikawa, H. (1966) Studies on erythrocyte glycolysis II. Free energy changes and rate limitings steps in erythrocyte glycolysis. *J. Biochem.* 59, 139–144.
15. Connett, R. J. (1985) In vivo glycolytic equilibria in dog gracilis muscle. *J. Biol. Chem.* 260, 3314–3320.
16. Huang, X., Holden, H. M., and Raushel, F. M. (2001) Channeling of substrates and intermediates in enzyme-catalyzed reactions. *Annu. Rev. Biochem.* 70, 149–180.
17. Ovadi, J., and Srere, P. A. (1996) Metabolic consequences of enzyme interactions. *Cell. Biochem. Funct.* 14, 249–258.
18. Stanley, W. C., and Connett, R. J. (1991) Regulation of muscle carbohydrate metabolism during exercise. *FASEB J.* 5, 2155–2159.
19. Soling, H. D., Bernhard, G., Kuhn, A., and Luck, H. J. (1977) Inhibition of phosphofructokinase by fructose 1 6-diphosphatase in mammalian systems: Protein–protein interaction or fructose 1 6-diphosphate trapping? *Arch. Biochem. Biophys.* 182, 563–572.
20. Ochs, R. S., Hanson, R. W., and Hall, J. (1985) *Metabolic Regulation*, Elsevier, Amsterdam.
21. Meyerhof, O., and Green, H. (1949) Synthetic action of phosphatase: Equilibria of biological esters. *J. Biol. Chem.* 178, 655–667.
22. Ingraham, L. L., and Pardee, A. B. (1967) Free energy and entropy in metabolism. in *Metabolic Pathways* (Greenberg, D. M. ed.), Academic Press, New York.
23. Hamori, E. (1975) Illustration of free energy changes in chemical reactions. *J. Chem. Educ.* 52, 370.
24. Hue, L., Blackmore, P. F., Strikama, H., Robinson-Steiner, A., and Exton, J. H. (1982) Regulation of fructose-2,6-P content in hepatocytes, perfused hearts, and perfused hindquarters. *J. Biol. Chem.* 257, 4308–4313.
25. Mor, I., Cheung, E. C., and Vousden, K. H. (2011) Control of glycolysis through regulation of PFK1: Old friends and recent additions. *Cold Spring Harbor. Symp. Quant. Biol.* 76, 211–216.
26. Zammit, V. A., and Newsholme, E. A. (1978) Properties of pyruvate kinase and phosphoenolpyruvate carboxykinase in relation to the direction and regulation of phosphoenolpyruvate metabolism in muscles of the frog and marine invertebrates. *Biochem. J.* 174, 979–987.
27. Patel, M. S., Nemeria, N. S., Furey, W., and Jordan, F. (2014) The pyruvate dehydrogenase complexes: Structure-based function and regulation. *J. Biol. Chem.* 289, 16615–16623.
28. Kanehisa, M., and Goto, S. (2000) KEGG: Kyoto encyclopedia of genes and genomes. *Nucleic Acids Res.* 28, 27–30.
29. Smith, A. J., Partridge, C. J., Asipu, A., Mair, L. A., Hunter, M., and Sivaprasadarao, A. (2006) Increased ATP-sensitive K^+ channel expression during acute glucose deprivation. *Biochem. Biophys. Res. Commun.* 348, 1123–1131.
30. Zhang, Q., Ramracheya, R., Lahmann, C., Tarasov, A., Bengtsson, M., Braha, O., Braun, M., Brereton, M., Collins, S., Galvanovskis, J., Gonzalez, A., Groschner, L. N., Rorsman, N. J., Salehi, A., Travers, M. E., Walker, J. N., Gloyn, A. L., Gribble, F., Johnson, P. R., Reimann, F., Ashcroft, F. M., and Rorsman, P. (2013) Role of KATP channels in glucose-regulated glucagon secretion and impaired counterregulation in type 2 diabetes. *Cell Metab.* 18, 871–882.

31. Rorsman, P., Ramracheya, R., Rorsman, N. J., and Zhang, Q. (2014) ATP-regulated potassium channels and voltage-gated calcium channels in pancreatic alpha and beta cells: Similar functions but reciprocal effects on secretion. *Diabetologia*. 57, 1749–1761.

32. Atkinson, D. E. (1968) The energy charge of the adenylate pool as a regulatory parameter. Interaction with feedback modifiers. *Biochemistry*. 7, 4030–4034.

33. Noat, G., Ricard, J., Borel, M., and Got, C. (1970) Kinetic study of yeast hexokinase. Inhibition of the reaction by magnesium and ATP. *Eur. J. Biochem.* 13, 347–363.

34. Viñuela, E., Salas, M. L., and Sols, A. (1963) End-product inhibition of yeast phosphofructokinase by ATP. *Biochem. Biophys. Res. Commun.* 12, 140–145.

35. Wood, T. (1968) The inhibition of pyruvate kinase by ATP. *Biochem. Biophys. Res. Commun.* 31, 779–785.

36. Hill, A. V. (1950) A challenge to biochemists. *Biochim. Biophys. Acta.* 4, 4–11.

37. Krause, U., and Wegener, G. (1996) Control of adenine nucleotide metabolism and glycolysis in vertebrate skeletal muscle during exercise. *Experientia*. 52, 396–403.

38. Powers, S. K., and Howley, E. T. (2009) *Exercise Physiology: Theory and Application to Fitness and Performance*, McGraw-Hill Higher Education, New York, NY.

39. Hargreaves, M., and Spriet, L. L. (2006) *Exercise Metabolism*, Human Kinetics, Champaign, IL.

40. Voet, D., and Voet, J. G. (2011) *Biochemistry*, John Wiley & Sons: Hoboken, NJ.

41. McArdle, W. D., Katch, F. I., and Katch, V. L. (1986) *Exercise Physiology: Energy, Nutrition, and Human Performance*, Lea & Febiger, Philadelphia.

42. Reeves, R. E., South, D. J., Blytt, H. J., and Warren, L. G. (1974) Pyrophosphate:D-fructose 6-phosphate 1-phosphotransferase. A new enzyme with the glycolytic function of 6-phosphofructokinase. *J. Biol. Chem.* 249, 7737–7741.

43. Veech, R. L., Cook, G. A., and King, M. T. (1980) Relationship of free cytoplasmic pyrophosphate to liver glucose content and total pyrophosphate to cytoplasmic phosphorylation potential. *FEBS Lett.* 117 Suppl, K65–K72.

44. Gitomer, W. L., and Veech, R. L. (1986) The accumulation of pyrophosphate by rat hepatocytes. *Toxicol. Ind. Health.* 2, 299–307.

45. Ronner, P., Friel, E., Czerniawski, K., and Frankle, S. (1999) Luminometric assays of ATP, phosphocreatine, and creatine for estimation of free ADP and free AMP. *Anal. Biochem.* 275, 208–216.

46. KEGG (2004) http://www.genome.ad.jp/kegg/kegg4.html

47. Bergmeyer, H. U. (1974) *Methods of Enzymatic Analysis*, Verlag, New York.

48. Jomain-Baum, M., and Schramm, V. L. (1978) Kinetic mechanism of phosphoenolpyruvate carboxykinase (GTP) from rat liver cytosol. Product inhibition, isotope exchange at equilibrium, and partial reactions. *J. Biol. Chem.* 253, 3648–3659.

49. Johnson, J. D., Muhonen, W. W., and Lambeth, D. O. (1998) Characterization of the ATP- and GTP-specific succinyl-CoA synthetases in pigeon. The enzymes incorporate the same alpha-subunit. *J. Biol. Chem.* 273, 27573–27579.

50. Moréra, S., Lascu, I., Dumas, C., LeBras, G., Briozzo, P., Véron, M., and Janin, J. (1994) Adenosine 5'-diphosphate binding and the active site of nucleoside diphosphate kinase. *Biochemistry*. 33, 459–467.

51. Cerpovicz, P. F., and Ochs, R. S. (1988) Method for determination of guanine nucleotide concentrations. *Fed. Proc.* 47, 6117.

ENDNOTES

    i. As noted in the reference cited for this point, the notion of assigning a Km value to glucokinase is not meaningful, as this enzyme has sigmoidal kinetics and thus has no Km. Strictly speaking, it is the $S_{0.5}$ value that should be used in this case to compare with the prevailing substrate concentration.

    ii. There are other effects of fatty acids, discussed in Chapter 6. Essentially, the formation of citrate requires that fatty acids enter the cell, and become converted to mitochondrial acetyl-CoA, and then citrate, which can be exported to the cytosol.

    iii. While this statement is correct, it is a tricky one. For one thing, every enzyme is altered to some extent by changes in substrate concentration. Consider too the case of the pyruvate dehydrogenase complex, which is activated by pyruvate. This would appear to be a clear violation of the principle, but it is not: it happens that pyruvate stimulates for an entirely different reason: it inhibits a kinase that itself inhibits the pyruvate dehydrogenase complex flux [27].

    iv. The notion that this sort of reaction might occur in mammals was proposed by Veech and colleagues [43], who suggested this based on an assay of pyrophosphate that showed very high concentrations in rat liver. However, this idea was discarded in later studies from that same group that indicated it was simply a measurement error [44].

3 Mitochondria and Energy Production

While glycolysis is traditionally the first pathway presented in biochemistry courses, a case can be made for the Krebs cycle as the metabolic hub for cellular function, with oxidative phosphorylation functioning to remove its reducing equivalents. This is because most cellular energy involves the mitochondria, and this organelle participates in virtually every significant metabolic route in the cell.

We can consider that the general function of mitochondria is to ferry electrons from two mobile cofactors—NADH and QH_2 (ubiquinone)[i]—into the machinery of the electron transport chain (or respiratory chain) to produce cellular ATP, mostly for consumption in the cytosol.

There are two connections between glycolytic and mitochondrial oxidation of glucose: pyruvate and NADH. Both must gain entry into the mitochondria across the double membrane barrier of this organelle. The first membrane encountered is the outer membrane, which is usually not considered a barrier for small molecules.

3.1 THE MITOCHONDRIAL OUTER MEMBRANE

The reason the outer membrane is not considered in mitochondrial studies stems from to the fact that various treatments that remove the outer membrane, such as digitonin, leave most functional properties intact (e.g., ref. [2]). There have been many studies of the porin protein that is normally thought of as a pore for small molecules (at least up to 5 kDa); as such, it does not influence transport of most metabolites. The protein in eukaryotic cells is also known as the voltage-dependent anion channel (VDAC) and has *in vitro* properties that would suggest it could regulate ATP exchange [3]. The notion that a channel that is essentially a large pore with no molecular selectivity expresses a voltage sensitivity is unusual. The electrical property is evident only when the protein is inserted into artificial membranes. Its voltage dependence shows a maximum conductivity at a zero potential [4]. Since zero volts is also the potential across the outer membrane, it is likely that voltage dependence is not a property related to metabolism. Porin is not likely involved in metabolism, but may play a role in apoptosis.

Porin binds not only ATP but enzymes such as hexokinase; these have already been suggested to be more significant in apoptosis through interactions with bcl2 proteins [3]. Thus, it is more likely that porin interactions with various proteins are involved in the orderly dismantling of mitochondria for the process of apoptosis. Even for this activity, the exact mixture of proteins involved is uncertain, as various proteins in the inner and outer membrane can substitute for one another in forming a complex that leads to cell destruction [5]. For the purposes of understanding metabolism, we will consider the outer membrane in the classical sense: as a pore.

3.2 EXCHANGES ACROSS THE INNER MITOCHONDRIAL MEMBRANE

The pathway for the complete oxidation of pyruvate begins with an inner membrane transporter, a symport with protons,[ii] for entry into the mitochondrial matrix [6]. Subsequently, pyruvate is metabolized by the pyruvate dehydrogenase complex to form acetyl CoA. The latter molecule is also the metabolic product of the ketogenic amino acids and most fatty acids (Figure 3.1). Thus this "active acetate" [7] is a C2 product that is the common endpoint of the major metabolic pathways. Its conversion to CO_2 requires the Krebs cycle, which we will examine below.

Meanwhile, glycolytic NADH can only indirectly enter mitochondria [8] using a shuttle system for the indirect transfer of those electrons (Figure 3.2; see also Figure 2.7). Metabolic shuttles have a common construction: isozymes of an enzymatic reaction are

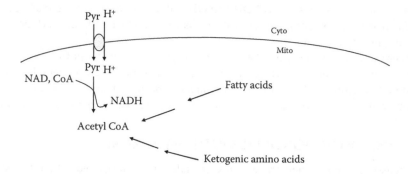

FIGURE 3.1 Mitochondrial acetyl CoA formation. The membrane shown dividing cytosol (cyto) from mitochondria (mito) is the mitochondrial inner membrane; the mitochondrial outer membrane is not a barrier to small molecules. Pyruvate (pyr), fatty acids, and ketogenic amino acids are precursors to the common intermediate acetyl CoA.

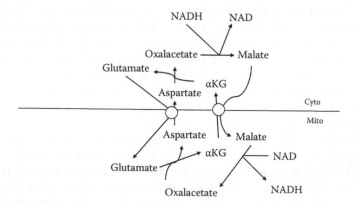

FIGURE 3.2 Malate–aspartate shuttle. This is the major route for moving reducing equivalents of cytosolic NADH into the mitochondria. The aspartate/glutamate exchange is electrogenic and mIRR under physiological conditions. αKG, α-ketoglutarate.

present in both the cytosolic and the mitochondrial space. These are near-equilibrium steps, run in opposing directions. The membrane separating these spaces—the mito-chondrial inner membrane—contains transporters which enable selective exchanges.

A major pathway for import of the reducing equivalents of NADH is the malate/aspartate shuttle (Figure 3.2). NADH is indirectly transported into the mitochondria by reducing oxaloacetate to malate in the malate DH reaction in the cytosol [9]. Next, malate can be transported into the mitochondria through an exchange with α-ketoglutarate.[iii] Once inside the mitochondria, mitochondrial malate DH enables the formation of NADH from NAD, and the effective transfer of reducing equivalents is complete. The shuttle pathway, however, is a cyclic pathway that requires further steps to achieve balance. The oxaloacetate formed as the product of mitochondrial malate DH is not permeable to the mitochondrial membrane (i.e., there is no transporter for the molecule). Thus, another indirect exchange occurs: oxaloacetate reacts with glutamate in the transami-nase reaction to form aspartate, which can exit the mitochondria through a transporter in exchange with glutamate. Once in the cytosol, the presence of a distinct isozyme of aspartate transaminase can enable regeneration of the oxaloacetate, and the overall shuttle components are balanced. Considered as a pathway, the import of NADH reduc-ing equivalents into the mitochondria through the malate/aspartate shuttle is irrevers-ible. The one metabolically irreversible step is the glutamate/aspartate exchanger, which is electrogenic [10,11]: the mitochondria exchanger exports aspartate^{-2} and imports glutamate^{-1}. As the mitochondria maintains a negative potential inside under physi-ological conditions, this process is metabolically irreversible.

A third pathway for entry of materials to supply mitochondrial energy is the oxidation of fatty acids. The route of long-chain fatty acid oxidation is outlined in Figure 3.3. Following cytosolic activation of the fatty acid to its CoA ester, the enzyme carnitine acyltransferase I (CAT I) catalyzes the formation of the carnitine ester. A transport protein exists for this acyl ester in the mitochondrial membrane in exchange for free carnitine. Inside the mitochondrial matrix, CAT II catalyzes

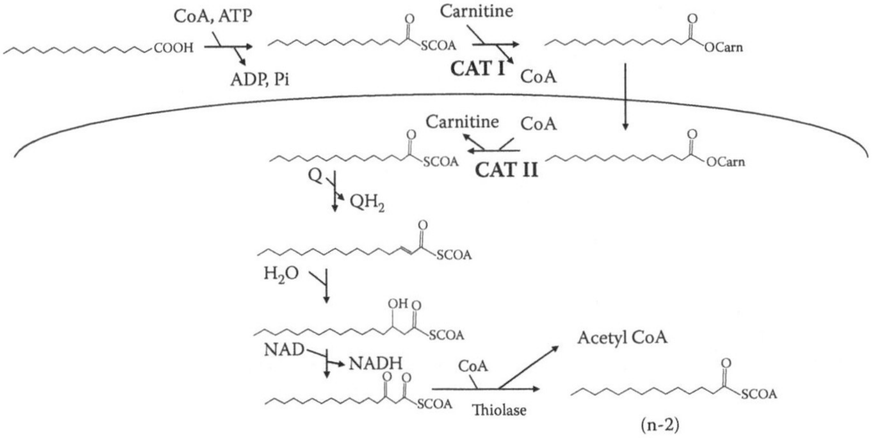

FIGURE 3.3 Mitochondrial fatty acid oxidation. CAT, carnitine acyltransferase; Q, QH$_2$ ubiquinone oxidized and reduced.

the regeneration of the acyl CoA. This portion of the pathway thus constitutes a second example of a shuttle, transferring an acyl CoA indirectly across the membrane. The ensuing steps in the mitochondrial matrix convert the acyl CoA to the two-carbon unit acetyl CoA. These are (i) oxidation of the acyl chain to form a double bond and the reduced cofactor QH_2, (ii) a hydration step to produce an alcohol from the double bond, (iii) oxidation of the alcohol to carbonyl, transferring electrons to NADH. Overall the process leads to acetyl CoA, a two-carbon shortened acyl-CoA. Repeating this process ultimately produces several molecules of QH_2, NADH, and acetyl CoA for each fatty acid molecule.

A separate means of producing mitochondrial acetyl CoA is the oxidation of amino acids. Only two amino acids—leucine and lysine—can completely form either acetyl CoA or the ketone body acetoacetate. These are known as *ketogenic* amino acids. Most amino acids form only an intermediate to gluconeogenesis, and are thus called *glucogenic* amino acids. There are some that are not so easily categorized: isoleucine, phenylalanine, tyrosine, and tryptophan can form both intermediates. The pathway for leucine conversion to acetyl CoA is shown in Figure 3.4. The transamination to α-ketoisocaproate takes place in extrahepatic tissues (largely muscle); the remained of the pathway in the mitochondria of liver cells [12]. After the extrahepatic transamination, the first step is oxidation of the short-chain ketoacid α-ketoisocaproate into the CoA ester isovaleryl CoA, using an enzyme complex analogous to the pyruvate dehydrogenase complex (PDC) that converts the ketoacid pyruvate to acetyl CoA. In turn, isovaleryl CoA is oxidized to β-methylcrotonyl CoA, with reducing equivalents ultimately transferred to the mobile carrier QH_2.[iv] Next, the molecule is carboxylated in a biotin-dependent step to β-methylglutaconyl CoA, converting an odd-numbered

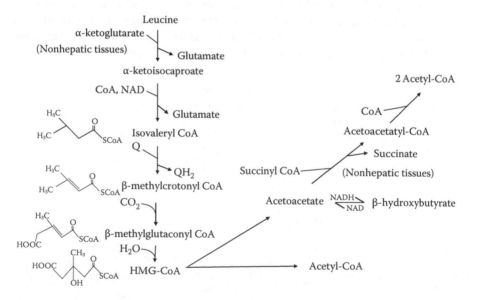

FIGURE 3.4 Leucine catabolism. The composite pathway shown illustrates first the transamination of leucine (in muscle) to α-ketoisocaproate, followed by hepatic mitochondrial steps to acetoacetate, and then extrahepatic utilization to acetyl CoA.

carbon into an even-numbered one. This is an example of apparent "carbon fixation"; however, it is not a stable incorporation as all the carbon is ultimately converted to acetyl-CoA, which leads to CO_2 formation by the Krebs cycle. The next reaction in the pathway is a hydratase-catalyzed reaction that leads to the formation of HMG-CoA (hydroxymethylglutaryl CoA). This is a point of convergence with liver ketogenesis.[v] For both lysine catabolism to HMG-CoA and ketone body synthesis, the next step is the same: lyase cleavage to acetyl CoA plus acetoacetate. The former is, of course, the convergence point for carbohydrate, lipid, and some amino acid metabolism. The latter, acetoacetate, is a ketone body that has no further metabolism in liver, but can be released for oxidation to acetyl CoA in the mitochondria of other tissues, such as muscle, heart, and brain. Thus, all major catabolic routes lead to acetyl CoA formation.

3.3 KREBS CYCLE OXIDIZES ACETYL CoA

The most unappreciated fact about the Krebs cycle is that it has just one substrate: acetyl CoA. To understand this, we must carefully distinguish between pathway intermediates on the one hand, and pathway substrates and reaction metabolites on the other. Every intermediate in the Krebs cycle—and each is a pathway intermediate—intersects with another pathway in mitochondrial metabolism. If we limit our consideration to the role of these metabolites in the Krebs cycle per se then the pathway substrate is acetyl CoA and the pathway product is CO_2. The pathway also generates three mobile cofactors: the electron-rich NADH and QH_2, and a single high-energy phosphate (GTP or ATP, depending on the cell).

Connecting the Krebs cycle to other reactions in metabolism highlights two features of the pathway that are consequences of its cyclic nature. First, any reaction supplying carbon to the cycle must be matched by other reactions that remove carbon at the same rate. If this were not the case, the intermediates would rapidly and drastically grow or shrink in concentration. The second is a related feature: the amount of intermediate supplied is *catalytic*.

Let us consider the reactions of the cycle (Figure 3.5). The first, the condensation of oxaloacetate with acetyl-CoA, catalyzed by citrate synthase, is a mIRR reaction. Like glycolysis, it is not the only such reaction in the pathway (Table 3.1), so that flux can be regulated at different points. The ensuing reaction, catalyzed by aconitase, converts citrate first to *cis*-aconitate and subsequently to isocitrate. While a chemically simple reaction—this is after all just a dehydration and rehydration—this step was historically the topic of a controversy, since it converts the symmetrical citrate into the molecule isocitrate which has two chiral centers. This was apparently resolved by Ogston [14] in 1948, who suggested a model in which this might happen by positioning the substrate into a decidedly chiral enzymatic environment. However, the original suggestion was speculative, and made in the complete absence of understanding enzyme mechanisms. In the following chapter, I elaborate this mechanism as an example of how prochirality can be understood. Note that the exposed intermediate of the reaction—*cis*-aconitate—is not always presented in displays of the Krebs cycle. I argue that it should be when considering the pathway from a metabolic view, since at least one reaction connects to *cis*-aconitate outside of the Krebs cycle (see Chapter 4).

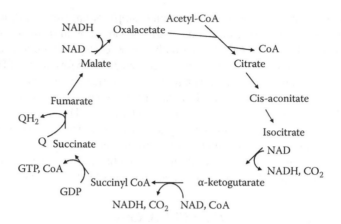

FIGURE 3.5   Krebs cycle.

TABLE 3.1
Krebs Cycle Enzymes and Their Equilibrium Status

Equilibrium	Metabolically Irreversible
Aconitase	Citrate synthase
Fumarase	α-Ketoglutarate DH
Malate DH	Succinate DH
Isocitrate DH	

Source: Krebs, H. A. 1969. The role of equilibria in the regulation of metabolism. In Bernard L. Horecker and Earl R. Stadtman (eds.), *Current
Topics in Cellular Regulation* (Academic Press: New York, London).

The next reaction is metabolically irreversible, and represents the first NADH formed by the cycle, as well as the first CO_2 released. However, some recent studies have suggested that flow into the Krebs cycle can be reversed at this reaction [15]. A possible resolution of this paradox lies in the fact that there are actually three isozymes of isocitrate DH [16–18]. One of these is the Krebs cycle enzyme shown; the other two use the NADP/NADPH pair as cofactors. Of those, there is a mitochondrial and a cytosolic form. The mitochondrial form could readily account for a flow from α-ketoglutarate to isocitrate. This would explain how the NAD-linked enzyme could catalyze a mIRR reaction, and yet backward flow of the pathway intermediates occurs. The two isozymes of isocitrate DH catalyze similar, but not identical, reactions. It also requires us to consider the distinctive roles of the two nucleotide pairs NAD/NADH and NADP/NADPH in mitochondria. Effectively, the two reactions together constitute a substrate cycle, and while neither is NEQ, the conversion between them is *pathway-reversible*.

The first pair is directly linked with extraction of energy and provision of electrons to the respiratory chain. The NADP/NADPH in mitochondria has a less well-defined role, however. NADPH is thought to be synthesized by the energy-linked transhydrogenase [19] that catalyzes the reaction

$$NADH + NADP \rightarrow NAD + NADPH \tag{3.1}$$

The reaction proceeds only in the forward direction, under conditions in which the membrane potential of the mitochondria is maintained, that is, under what we can assume are physiological conditions. Thus, the energy-linked transhydrogenase is mIRR. This keeps the redox ratios of the two nicotinamide systems in the mitochondrial matrix distinct, like the different ratios present in cytosol. This allows reactions to proceed in different directions so long as they use the distinctive cofactors.

Continuing with the Krebs cycle, the α-ketoglutarate DH step is another complex analogous to the PDC and the BCAA DH complex. Five dehydrogenase complexes are illustrated in Figure 3.6: the PDH, the α-KGDH, and the branched-chain dehydrogenase complexes operating on the three branched-chain amino acids. The similarity of the reactions includes not only the same mobile cofactors, but also bound cofactors within the complexes. The mechanism of reaction in all cases (Figure 3.7) requires attachment first to a thiamine bound to the first enzyme in the complex, E_1. This enzyme catalyzes decarboxylation as well as the reaction with E_2 to pass the shortened intermediate now attached as a thioester. The cofactor of E_2 is a covalently bound dithiol containing cofactor, lipoic acid, bound as an amide link to a lysine residue of E_2. E_2 catalyzes the release of the CoA ester, leaving E_2 in the disulfide form. This is oxidized by reaction of E_2 with the oxidized flavin FAD bound to E_3.

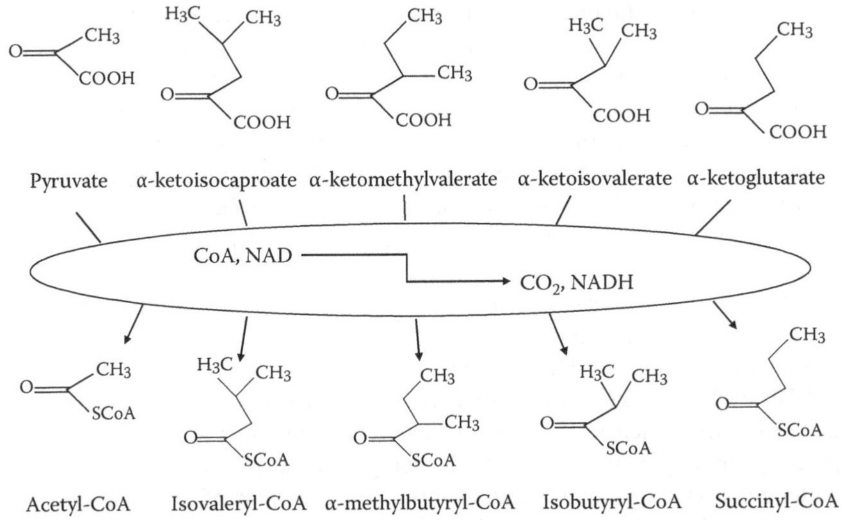

FIGURE 3.6 Dehydrogenase complexes. Acyl CoA substrates are shortened, CO_2 and NADH are produced, and a new CoA ester results from the similar complexes in which the substrate is passed to three consecutive enzymatic reactions.

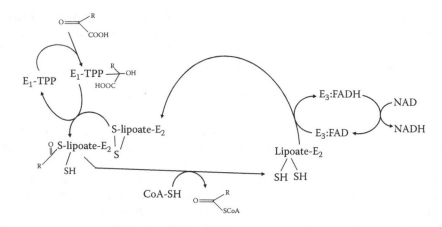

FIGURE 3.7 Mechanism of dehydrogenase complexes. In each case, the mechanism of action is similar. Aside from the three enzymes associated with the catalytic events of the complex, three prosthetic groups (TPP, thiamine pyrophosphate), lipoate, and FAD, as well as the mobile cofactors CoA and NAD/NADH participate. This is the "swinging arm" mechanism in which the central enzyme (E_2), which has a lysine residue bound to lipoate, sequentially reacts with E_1, CoA, and E_3.

This regenerates the original form of E_2, leaving E_3 bound to the reduced flavin, FADH. Finally, the original (oxidized) form of E_3 is restored by its reaction with NAD, producing the last product, the mobile cofactor NADH.

It is worth examining the mechanism of the dehydrogenase complexes, and not just because they are intrinsically interesting, and share similarity in chemical mechanisms with other metabolic reactions (such as the closely related glycine cleavage enzyme). Beyond this, they are clear representatives of the class of enzyme complexes, in which the enzymes are closely associated with each other at all times. From a metabolic perspective, this means none of the intermediates are released to the water solution (here, the mitochondrial matrix). While some investigators hold the view that this is the normal model for metabolic reactions [20], it is clearly a specialized set of reactions for a specific purpose: so that the intermediates are *not* connected to other pathways but instead routed to the final product only.

Just as proteins bound to each other cannot connect to outside reactions, bound cofactors like FAD cannot communicate between other soluble proteins. The notion of bound versus free cofactors was an idea introduced in Chapter 2. Note that the role of the free cofactor NADH is distinctive from that of FADH, for example. While the FADH cycle, with its oxidized state, remains attached to the enzyme throughout its catalysis, NADH leaves the PDC and can react with other sites, for example, the respiratory chain, glutamate DH, and the energy-linked transhydrogenase.

Note, too, how the enzyme mechanisms are microcosms of a cyclic pathway: the three enzymes form three connected cycles, regenerating the original enzyme form. Actually this is not a special feature of enzyme complexes. Every enzymatic reaction must form a cycle, regenerating the original enzyme form, or it could not otherwise be catalytic. It should thus be clear why the metabolic intermediates in a metabolic cycle are considered catalytic; adding a small amount of any of them will greatly accelerate the overall pathway.

Except for the α-KGDH of the Krebs cycle, the other dehydrogenase complexes are subject to phosphorylation/dephosphorylation, the result of two further activities in the complexes: a kinase (causing inactivation) and a phosphatase (causing activation). For the PDC, Ca^{2+} causes activation of the enzyme overall by stimulating the phosphatase [21,22]. For the α-KGDH, there is no covalent modification, although Ca^{2+} also causes activation by direct allosteric binding [23]. Beyond these modes of regulation, there is further control of the phosphorylation status of the PDC, as acetyl CoA inhibits the enzyme (activation of the kinase acting on E_1). Beyond this, there are no widely proven allosteric activations of enzymes of the Krebs cycle that are known to work both *in vitro* as well as in intact cells.

Next, the succinyl-CoA synthetase is the sole substrate-level phosphorylation reaction of mitochondria. Commonly the reaction converts GDP to GTP, although in some cases the nucleotide formed is ATP [24]. In the case of succinyl CoA synthetase, the reaction mechanism is known to involve the intermediate histidyl phosphate to transfer the phosphoryl to form the triphosphate. The mechanism is similar to the sequence that produces substrate level phosphorylation in the cytosol, the reactions glyceraldehyde phosphate DH and phosphoglycerate kinase in glycolysis. In early studies of mitochondria, the chemical isolation of an N-linked phosphoryl of histidyl phosphate was mistaken for the long-sought (but nonexistent) chemical intermediate for oxidative phosphorylation prior to the acceptance of the proton gradient [25].

Succinate DH catalyzes the subsequent reaction, a transfer of electrons from succinate to QH_2, with fumarate as product. The enzyme is a complex embedded in the mitochondrial inner membrane, and is also one of the five original complexes first isolated from the mitochondrial inner membrane, and named *Complex II*. The mobile cofactor pair Q/QH_2 shuttles electrons within the membrane, making it one of three different types of mobile cofactor. Another type, represented by cytochrome c, is restricted to the outer surface of the mitochondrial inner membrane. The third type includes all of the other mobile cofactors we have considered thus far, such as NAD and NADP. Succinate DH is clearly part of the Krebs cycle, but it is often viewed as being part of the respiratory chain as well. This is not correct. The idea seems natural because of the alternative name, Complex II, and because the other originally isolated complexes (I, III, IV, V) are involved in respiration. However, Complex II is not required for respiration at all. Unlike the other four complexes, it does not involve proton movement through the membrane. It is also the only one of the complexes entirely coded by the cell nucleus [26]. It is true that succinate DH generates QH_2 and this must be oxidized by the respiratory chain, but succinate DH is not *part of* the respiratory chain any more than any dehydrogenase that generates NADH (which of course needs the respiratory chain for oxidation) is an actual part of respiration. Often Complex II is included in diagrams of the electron flow for respiration, but when this is done, succinate DH represents a side branch. As a result, overall electron flow from NADH to O_2 does not depend on succinate DH, so this step of the Krebs cycle can be viewed as simply a source of electrons for the pathway originating from QH_2. In order for any pathway to proceed, all cofactors must be regenerated. Since the Q/QH_2 pair is a cofactor of the Krebs cycle, the only means of its turnover is the respiratory chain. In short, flow through the respiratory chain does not depend on the Krebs cycle, but flow through the Krebs cycle depends absolutely on flow through the respiratory chain.

In the subsequent step, fumarate is converted to malate by the fumarase reaction. This is a NEQ step, but also one which has intriguing stereochemistry, akin to the aconitase reaction; a further treatment is presented in Chapter 4. There is also a fumarase activity in the cytosol. It is not unusual to find cytosolic isozymes of mitochondrial enzymes. After all, that is the pattern for shuttle systems between the two metabolic spaces. What is unusual is that, for fumarase, it is the same protein in both metabolic spaces. A single nuclear gene product encodes fumarase, and upon its translation, only a portion of it is imported into the mitochondria [27,28]. There is some evidence that, apart from its metabolic uses, cytosolic fumarase is involved in DNA repair [29]. For the mitochondrial enzyme, however, it appears to strictly be a metabolic step of the Krebs cycle.

The malate formed is a substrate for the near-equilibrium malate DH step, which regenerates oxaloacetate that completes the Krebs cycle. The oxaloacetate thus produced at malate DH can condense with another molecule of acetyl-CoA and enter another round of this cyclic pathway. Oxidation of malate produces the mobile cofactor NADH, making it the last electron-rich mobile cofactor produced by the cycle.

3.4 PATHWAY SUBSTRATE AND PRODUCTS OF THE KREBS CYCLE

As emphasized above, the single-pathway substrate for the Krebs cycle is acetyl CoA. It is apparent from a consideration of the overall pathway (Figure 3.5) that it is the two carbons of acetyl CoA that are specifically the input carbons. The output carbons are the two fully oxidized CO_2 molecules. In addition, three NADH molecules are formed from NAD, and one QH_2 from Q. These reduced cofactors are intermediates in energy generation for the mitochondria, supplying electrons to the respiratory chain, and completing the full oxidation of the carbon moieties to CO_2 plus H_2O, as in any complete oxidation of organic molecules. We can now appreciate that these molecules can arise from any of the metabolic fuels: carbohydrate, fat, or protein (represented as amino acids). In a true pathway sense, it is only the CO_2 that can be considered a true pathway end-product, as the rest are intermediates in the overall pathway of energy formation.

There is a corollary to this statement that must be carefully considered as it is easy to err when considering energy extraction in the mitochondria. We might be tempted to suppose that there are other means by which we might produce energy. For example, it is clear that in isolated mitochondria experiments, the added substrates that lead to oxygen consumption and ATP formation are actually intermediates in mitochondrial metabolism. Thus, the commonly used combination of glutamate and malate is an excellent means of supporting respiration by isolated mitochondria. The route of this conversion (Figure 3.8) shows how NADH can be supplied and the mitochondrial transporters are balanced. While such experiments provide information concerning the operation of the mitochondria, they should not be confused with the *in vivo* operation of the cell. The system works because a large supply of precursor is artificially provided and over the course of the experiment a steady state

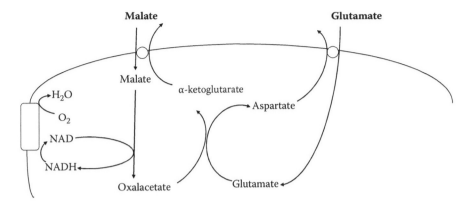

FIGURE 3.8 Isolated mitochondria and the substrate pair of malate and glutamate. Pathway for the utilization of added malate and glutamate in isolated mitochondrial preparations. It is a means of testing mitochondria activities *in vitro*, as it does not correspond to a physiological situation, but rate allows net utilization of the substrates and accumulation of α-ketoglutarate and aspartate.

is achieved. It is an isolated system that bypasses the cytosolic involvement but also removes the requirement that the Krebs cycle operate.

Closely related to this is the view that any intermediate of the Krebs cycle can connect with the respiratory chain. This would violate the principle of balance in this cycle pathway. For example, a view that succinate can accumulate and provide a key to mitochondrial metabolism is based on the Krebs cycle as a passive provider of succinate [30]!

Thus, the Krebs cycle, in common with fatty acid oxidation, and some amino acid oxidation pathways are generators of the energy-rich mobile cofactors NADH and QH₂. Our next consideration is how these cofactors are reoxidized.

3.5 MOBILE ELECTRON CARRIERS CONNECT CYTOSOLIC AND MITOCHONDRIAL CATABOLISM TO THE ELECTRON TRANSPORT CHAIN

The two energy-rich mobile cofactors within the mitochondria are NADH and QH₂. We have seen how each of these is formed within the mitochondrial matrix and from reactions in the cytosol prior to entering the Krebs cycle. For NADH, it is evident that shuttles are required to bring in the reducing equivalents. Within the mitochondria, it is not difficult to imagine that all sources of NADH converge into a common pool in the mitochondrial matrix.

This is also the case for QH₂, but the transport steps are not required, despite the fact that reactions producing this cofactor can occur within the cytosolic space or the mitochondrial space. As discussed above, the mobility of the Q/QH₂ cofactor pair is confined to the inner mitochondrial membrane lipid space. In order to see how transfer occurs, consider electrons from glycerol phosphate (Figure 3.9). This cytosolic substrate reacts at the cytosolic face of the inner membrane

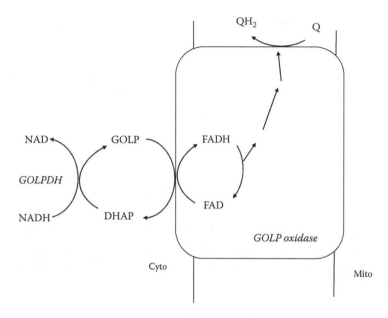

FIGURE 3.9 Cytosolic NADH to QH_2 in the glycerol phosphate shuttle. The pathway for converting cytosolic reducing equivalents into mitochondria QH_2 is shown to require the cytosolic GOLPDH (glycerol phosphate DH) and the mitochondrial embedded complex GOLP oxidase (glycerol phosphate oxidase).

embedded glycerol phosphate oxidase. Glycerol phosphate is converted to the oxidized product, dihydroxyacetone phosphate, by the glycerol phosphate oxidase complex, which donates the electrons to membrane QH_2. The immediate donation of electrons is to a bound FAD in the enzyme; this is subsequently oxidized and QH_2 formed.

Figure 3.10 shows five complexes embedded within the mitochondrial membrane that collect electrons into QH_2. Each complex spans the membrane and can be represented as similar to the GOLP oxidase of Figure 3.9. The name of each complex is written close to the side of the membrane that reacts with the reduced substrate. Aside from those already mentioned, the enzyme choline oxidase [31] is also mitochondrial, forming membrane QH_2 and matrix betaine (see Section 6.1.11 for full pathway). All of the reactions shown act in parallel. QH_2 is a pool collecting electrons from both sides of the mitochondrial inner membrane, passing them subsequently to the electron transport chain.

3.6 MITCHELL HYPOTHESIS

To this point in our examination of mitochondrial energy, we have stripped the electrons from the various substrates and funneled them into just two mobile cofactors: NADH and QH_2. NADH is the major input to the respiratory chain and it has long been known that as NADH is oxidized, O_2 is reduced to water, and ADP is converted to ATP. The linking of ATP formation with NADH oxidation is embedded

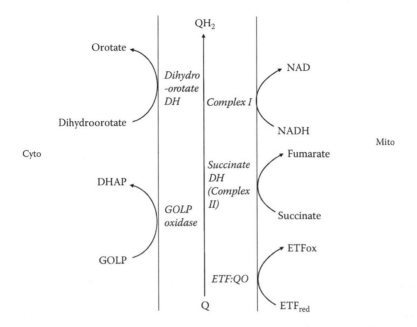

FIGURE 3.10 Mitochondrial complexes forming QH_2. Substrates from either cytosol or mitochondria can reduce Q to QH_2 to supply the respiratory chain. Each complex is indicated at the side of the membrane where it reacts with substrate, but all are membrane-spanning (as in Figure 3.9) and all produce QH_2.

in the name for the process, oxidative phosphorylation. The mechanism was a mystery for many years, finally resolved by the hypothesis of Peter Mitchell. Most investigators had believed that the formation of ATP was achieved in the same way as the substrate-level phosphorylation steps, as illustrated in glycolysis for the glyceraldehyde phosphate DH and phosphoglycerokinase reaction pair, and in the mitochondria for the succinyl CoA synthetase reaction. A formal acknowledgement of the Mitchell hypothesis by these same investigators and others prominent in mitochondrial energy production was published in 1977 [32].

Mitchell's idea was framed in terms of electrochemistry rather than mechanistic organic chemistry. The key point, illustrated in his monograph published in 1968 [33], was an analogy drawn from an electrochemical cell. The discussion below paraphrases the notions set out in that monograph. The electrochemical cell example of Figure 3.11 is for the redox reaction between Zn metal and $CuSO_4$:

$$Zn^0 + CuSO_4 \rightleftarrows Cu^0 + ZnSO_4 \qquad (3.2)$$

If this reaction is set up with Zn and Cu as solid bars suspended in sulfate solutions as shown in Figure 3.11a, the greater redox potential of Zn means that the initial direction is the oxidation of Zn in the left-hand solution and reduction of Cu^{2+} on the right, as indicated. In order to couple electron flow in these reactions but not allow them to mix directly, it is necessary to connect the metal bars with a wire. In addition,

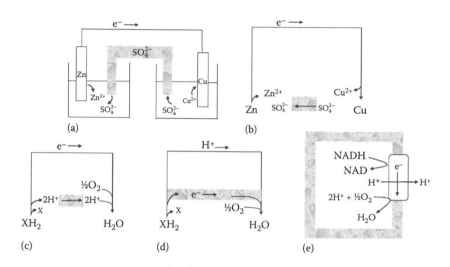

FIGURE 3.11 Proton motive force. (a) Electrochemical cell for Zn and Cu. (b) Abstract view of electrochemical cell. (c) Abstract view of electrochemical cell for substrate oxidation and oxygen reduction. (d) Abstract view exchanging roles of proton and electron in c. (e) Mitochondrial proton gradient production.

it is necessary to connect the solutions with a salt bridge, so that small amounts of counter ions can flow to complete the electric circuit. This is standard fare, but it is important for us to orient ourselves to electrochemistry in order to follow Mitchell's reasoning. What is represented here is the two distinct means by which electricity can flow: through a wire for the electrons and through a solution—including the salt bridge for its continuity—for ions. The driving force for the electrons is said to be an *electromotive force*. Removing the constraints of laboratory equipment in Figure 3.11b, we can clearly see the essence of the situation: Zn is oxidized on the left-hand side, electrons are driven by the electromotive force (here the difference in redox potential between Zn and Cu), and the electric charge is balanced by a net movement of SO_4^{2-}. For the mitochondrial system, we change the redox reactions slightly to indicate a substrate containing hydrogens and electrons, with XH_2 as donor and oxygen as acceptor. This is cast in a similar way to the electrochemical cell, as shown in Figure 3.11c. Notice the slightly distinctive change in completing the electrical circuit. In the case of the electrochemical cell, it is just the charges that are balanced, using sulfate ion migration. In the case of the proton balance through an imaginary salt bridge, they are in fact also products of the oxidation reaction and substrates of the reduction.

The genius of Mitchell was in conceptually reversing the roles of the electrons and the ions (in mitochondria, the ions are protons) as diagrammed in Figure 3.11d. If the electrons are driven by an electromotive force, then a proton circuit driving force should be called the **protonmotive force**. This flow is diagrammed in Figure 3.11e.

While the description of the electrochemical cell may seem elementary, and the history of the Mitchell hypothesis unimportant for modern times, in fact this central analogy explains mitochondria functions, and is not in need of modification fifty years later. The description is a broad outline: the movement of electrons requires the operation of multiple electron carriers enabled by the various redox potentials of the respiratory chain. Still, the analogy enables us to grasp immediately the overall working of the chain without becoming mired in detail.

Oxidative phosphorylation has now been extensively studied, and what was once known only as five embedded membrane complexes are now established as the molecular realization of the hypothesis we have just considered in theoretical terms [34]. The complexes are shown in Figure 3.12. Complexes I, III, and IV (Table 3.2) are components of the respiratory chain, and are fixed entities within the membrane. They are connected by mobile cofactors of three types, as shown in Figure 2.12. At the beginning and end of the sequence of electron carriers are water-soluble redox couples that react at the inner face of the mitochondrial membrane—NADH/NAD and O_2/H_2O. Thus, the first reaction feeding electrons in to the chain reduces Complex I (NADH oxidase); the last reaction donates the electrons to oxygen to form water at Complex IV (cytochrome c oxidase). The second type of mobile cofactor is the aforementioned QH_2/Q, which accepts electrons from Complex I, as well as Complex II and others, and donates them to Complex III (Q-c reductase). This pair exists entirely within the mitochondrial membrane. The third and final mobile cofactor is cytochrome c, which is reduced by Complex III and oxidized by Complex IV. The movement of this mobile cofactor is restricted to the outer surface of the inner mitochondrial membrane. This sliding is also noted in the membrane-bound G protein trimers of the plasma membrane that couple extracellular restricted hormones to intracellular signaling events (through activation of membrane-bound effector enzymes).

Complexes I, III, and IV also serve to transport protons from the mitochondrial matrix to the cytosol, which is linked to electron movement through the membrane.

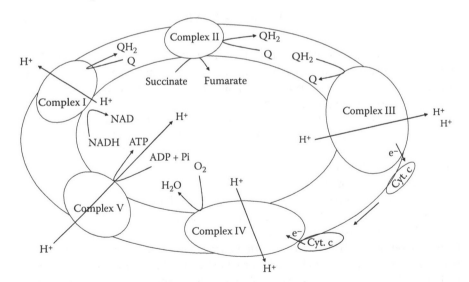

FIGURE 3.12 Complexes I–V of the mitochondrial inner membrane. The originally discovered five complexes and their activities are displayed. All but Complex II are involved in respiration, evident by their ability to transport protons.

TABLE 3.2

Five Mitochondrial Complexes in Energy Formation

Complex	Name	Mobile Cofactor Substrate	Mobile Cofactor Product	Bound Cofactors	Inhibitors
I	NADH oxidase	NADH, Q	NAD, QH_2	FAD/FADH Fe/S Protein	Rotenone Amytal
II	Succinate DH			FMN Fe/S Protein	Malonate Atpenins TTFA[a]
III	Q-c Reductase	QH_2, cyt. c (ox)	Q, cyt. c (red)	Fe/S, cyt. b	Antimycin A Stigmatellin
IV	Cyt. c Oxidase	cyt. c (red), O_2	cyt. c (ox), H_2O	Cu^{2+}, cyt. a	Azide Cyanide
V	F_1F_0 ATP Synthase	ADP, Pi	ATP		Oligomycin

[a] BAL, British anti-Lewisite (dimercaprol); TTFA, Thenoyltrifluoroacetone.

This *vectorial movement* is crucial to the mechanism of oxidative phosphorylation and results in the formation of a proton gradient. This forms a charge separation as well as a chemical separation as a result of that disparity, or a membrane potential and a chemical potential. The two components constitute the proton motive force, with the equation

$$\Delta p = \Delta pH + RT\Delta \Psi \tag{3.3}$$

which depicts the chemical and electrical nature of the proton gradient. While it is commonly shortened to a "pH gradient," this is a misrepresentation for mitochondria, as this ignores the membrane potential. In fact, 80% of the contribution to Δp is due to the membrane potential [35].

Complex V is the re-entry port of the proton, the ATP synthetase. In essence, this is a proton channel, with a conductance controlled by the need for ATP formation. This need is reflected in the concentration of cytosolic ADP. When that is increased, ADP enters the mitochondria through an exchange with ATP, catalyzed by the ATP/ADP exchange protein. Availability of ADP drives ATP formation by Complex V, which increases the proton conductance of the channel. Essentially, this is a drain on the Δp, so that during ADP driven ATP formation—the phosphorylation part of oxidative phosphorylation—the major component of Δp, the $\Delta \Psi$, decreases. The experimental observations behind these notions are an increase in oxygen consumption when ADP is added to mitochondria (provided with otherwise needed components, including Pi and oxidizable substrates), and a decrease in the measured $\Delta \Psi$ while ADP is present. If limiting amounts of ADP are added, then after all of the ADP is converted to ATP, the oxygen consumption resumes its lower rate and the $\Delta \Psi$ rises back to its resting level.[vi]

3.7   MITOCHONDRIAL STATES

In the days before the understanding of how the mitochondria operates, Britton Chance published a guide to experimental mitochondria based on oxygen consumption responses to various additions to the media [36]. These were designated as states 1 to 5, and included such experimental conditions as the absence of Pi or substrate. Only two of these operational states have survived, and they appear—almost mysteriously!—in current literature treatments. The two states currently used are state 4 and state 3. State 4 is the state of low oxygen consumption (and thus low but measurable proton conductance) in the presence of all components necessary to achieve ATP synthesis except for ADP. This was the "resting state" referred to above. State 3 has the most rapid rate of respiration, differing from state 3 only in having ADP present. The other states have not been found to be as experimentally useful so they have fallen out of vogue for the most part.

The ratio of state 3 to state 4 is an index of performance called the **respiratory control ratio**. The ratio depends in part on the quality of the preparation. Thus, less intact mitochondrial preparations have lower ratios. Generally, for example, freshly isolated muscle mitochondria are poorer than liver mitochondria, largely because the latter soft tissue is more experimentally amenable to the production of intact mitochondria. In addition, the choice of substrate also alters the ratio. One of the best combinations is glutamate/malate; this is usually higher than another popular choice, succinate/rotenone. The latter combination is important *in vitro* because in the absence of rotenone addition of exogenous succinate to mitochondria drives electrons in reverse through Complex I.

3.8   SEQUENCE OF ELECTRON CARRIERS

One of the earliest discoveries about mitochondria was the determination of the sequence of the electron carriers. The redox states of individual cytochromes is faithfully reported by measuring their distinctive absorbances [36]. Coupled with the ability to isolate the complexes and analyze them directly, the sequence of electron transfer was shown to be

$$NADH \rightarrow Complex\ I \rightarrow Q \rightarrow Complex\ III$$
$$\rightarrow cyt.\ c \rightarrow Complex\ IV \rightarrow O_2$$

(3.4)

In addition, substrates such as succinate (among others, as in Figure 3.10) directly transfers electrons to Q:

$$Succinate \rightarrow Q \rightarrow III$$

(3.5)

The transfer itself catalyzed by complex II. In part, this sequence was realized with the use of inhibitors that had enough selectivity to halt electron transfer at specific points. Thus, rotenone blocks Complex I, antimycin blocks Complex III, and cyanide

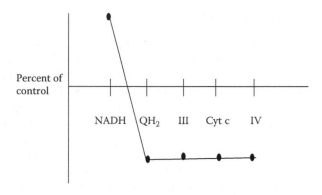

FIGURE 3.13 Crossover analysis of mitochondrial respiration. Redox states of the indicated intermediates are shown in the ordinate. The abscissa indicates, in percentage of control, the change in redox level of these intermediates upon the addition of a perturbant, in this case, rotenone. As indicated, it blocks between NADH and QH$_2$.

blocks Complex IV (Table 3.2). When the chain is blocked at any point, oxygen can still oxidize any complexes it is connected to so long as they are not subject to inhibition. For example, in the presence of rotenone, Complex IV, cyt. c, Complex III, and Q will all be oxidized, but NADH will stay reduced. The appearance of a crossover point between NADH and Q (Figure 3.13) demonstrates the action of rotenone on complex I. This idea was first suggested in 1958 [37] and later applied to metabolic pathways, as pointed out in Chapter 2.

3.9   PROTON GRADIENT CREATION, MEMBRANE POTENTIAL, AND UTILIZATION OF THE Δp

When electrons are passed through the complexes and mobile carriers, their traversal drives the formation of the proton gradient, with the attendant energy potential in the ΔpH and ΔΨ, as described above. The operation of just four complexes (I, III, IV, and V) is enough to explain the basic process of oxidative phosphorylation, driven by with the availability of mitochondrial matrix NADH and cytosolic ADP. Mechanisms for the operation of the three proton gradient-generating complexes— I, III, and IV—are described both in biochemical textbooks and more expansively in monographs, such as Nicholls and Ferguson [35]. Some aspects most relevant to metabolism are discussed here.

Complexes I and IV contribute to the proton gradient by a "pumping" mechanism. The mechanism is similar to GLUT type glucose transporters in that the protons being moved across the membrane bind at the matrix side in one conformation, and that binding site is flipped to face the cytosol in a second conformation. That second conformation is driven by the movement of electrons through the protein complex. Complex IV has an additional means of creating the gradient: its reaction with O$_2$ and protons occurs at the matrix side, so that protons are "annihilated" exclusively from that side. This is equivalent to the transport of protons across the membrane, even though there is no net movement.

The operation of Complex III is a "loop" mechanism, in which Q and QH_2 pass from one membrane face to the other, with an electron flow that directs one of the electrons through to cytochrome c, and the other back through an inner loop through the membrane composed of two b cytochromes, b_L and b_H, as described in the section for this complex below.

The overall operation is simply the movement of electrons through the membrane and of protons across it. The major utilizer of the energy is the F_1F_0ATP synthetase. This protein is a molecular machine, in which the proton movement causes a spinning of subunits that moves different components of the synthase into place to ultimately direct the formation of ATP from ADP and Pi.

3.9.1 COMPLEX I ESSENTIALS

The first complex is also the largest: 45 proteins in the mammalian form. Of those, 15 are closely analogous to the bacterial enzyme complex, and they are also the only ones of this complex synthesized by mitochondrial DNA. In general, the mitochondrial DNA codes for only the barest but most essential elements of mitochondrial function: portions of the respiratory chain. The core 15 proteins are also the only ones known to play a metabolic role and whose function is understood. The rest are unknown; it is hypothesized that they may play a role in complex assembly. If so, this role must be distinct from that of bacteria, in which assembly occurs in the absence of the extra proteins.

The complex is divided into three functional and spatial portions: the N, Q, and P regions (Figure 3.14). The whole complex in cross section appears to be shaped like an "L" in which one piece extends into the mitochondrial matrix, with the rest, arranged nearly perpendicular to the first, embedded in the membrane. The N region, which stands for NADH binding, is the jutting-out peninsula in which NADH transfers its hydride to FMN (the bound cofactor that is identical to FAD but missing the adenine moiety at the other end of the molecule). The transfer occurs in essentially a water phase, made clear by the location of the subunit, which is not embedded in membrane. The form of reduced FMN is the negatively charged FMNH⁻, the

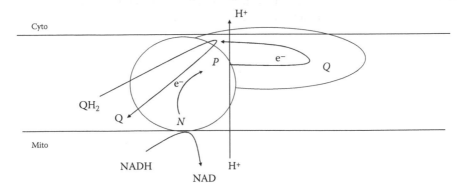

FIGURE 3.14 Complex I. The three regions, N, P, and Q, are indicated in the multisubunit complex.

result of hydride transfer. After accepting matrix space protons, the bound cofactor is $FMNH_2$. This next undergoes one-electron transfers to the Fe/S proteins. Thus, the flavin cofactor transduces a 2-electron transfer (from the obligate hydride transfer cofactor NADH) to 1-electron transfer (to the Fe/S proteins). In the Q region of Complex I, two sequential single-electron transfers result in the reduction of Q first to $Q^{·-}$ and then to the fully reduced form, QH_2, in which two protons from the mitochondrial matrix are incorporated.

Between accepting electrons from NADH and donating them to Q, electrons pass through the P region, in which they drive the translocation of protons from the matrix to the cytosol space. The P region contains transport proteins with strong homology to the membrane-bound Na/H exchange proteins. While catalyzing that exchange reaction would not provide a proton gradient, it would provide a chemical gradient of hydrogen ions, which would then participate in equilibria with water and cytosolic buffers. One proposal [38] is that the Na/H exchange occurs in the less-active or "deactive" state, the term used to describe the inactive state of Complex I that exists in the absence of substrate. Another intriguing possibility is that the complex undergoes a similar type of Q-cycle to that of complex III (see below).

The analogy to Na/H exchangers extends to the ability of amiloride[vii] to bind and inhibit the complex [40]. Amiloride contains a guanadino group, as shown in Figure 3.15. Also displayed in this figure is another guanadino-containing molecule, metformin, which also inhibits Complex I in isolated mitochondria and in permeabilized cells [41]. Supporting the notion that this may account for the inhibition, genetic substitution of a yeast homolog of Complex I which lacks proton translocation is not affected by metformin [42]. The fact that there is no metformin action at all in these constructs led those authors to conclude that Complex I is the actual target of metformin in intact cells. However, the *in vivo* action is entirely a different matter. The action of amiloride, for example, is accounted for as a Na/H inhibitor, not a Complex I inhibitor. While complex I inhibition is a popular hypothesis for metformin action, there are serious difficulties in accepting this mechanism, as I discuss in Chapter 10.

Complex I is the locus of most reactive oxygen species (ROS) generation in mitochondria. This represents essentially a competition between the respiratory chain and oxygen itself for the unpaired electrons present at the beginning and end of the sequence of reactions. It is unclear just how important ROS generation that is not

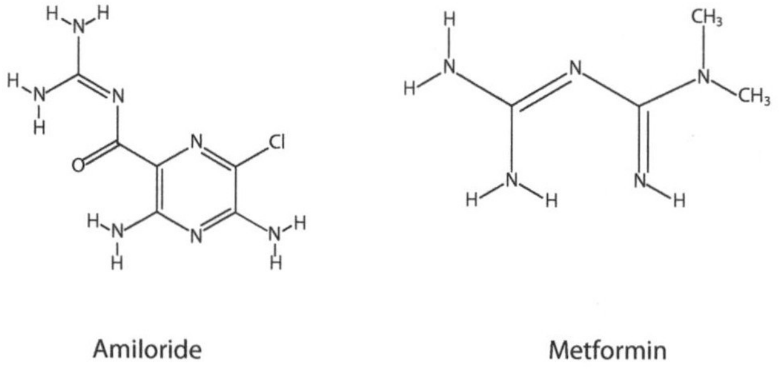

Amiloride   Metformin

FIGURE 3.15   Amiloride and metformin. Structural similarity between two compounds that inhibit Complex I *in vitro*.

tempered by the glutathione-linked systems is, although there is clearly a good deal of research interest in metabolic processes [43,44] beyond the usual role of such species in the nonspecific immune system [45]. In fact, an entire treatise [46] is devoted to an alternative metabolic regulation due to ROS rather than the principles described in the present work and earlier monographs on metabolic control. The scope of ROS interest is likely to be dampened by the finding that oxidative damage may be unrelated to longevity [47]. A more nuanced interpretation of the correlation between ROS production and lifespan was presented by Sanz et al. [48].

There are also suggestions that Complex I can be reversed *in vivo*. There is no doubt that this activity can be observed with *in vivo* preparations. Moreover, early studies had suggested that perhaps the entire span of electron flow from NADH at least to cytochrome c might be in near-equilibrium with the phosphorylation potential (ATP/ADP*Pi), and that as a result all steps in between were also at near-equilibrium [49]. This seems unlikely in the intact cell for several reasons. First, the earlier studies made assumptions about the assessment of ADP that have since been re-evaluated [50]. For example, most of the total ADP is bound to cytosolic proteins. Furthermore, conditions for reversal of complex I, such as a relatively oxidized NAD/NADH, are unlikely to prevail [35]. It would, for example, preclude any utilization of mitochondrial NADH.

3.9.2 COMPLEX III

The flow of electrons and protons through complex III is unique. The "loop mechanism" proposed by Peter Mitchell achieves the proton gradient by reaction of the cofactor ubiquinone (Q) at both membrane faces. An illustration of the flows through complex III is provided in Figure 3.16. To orient us to the distinct membrane faces of the mitochondrial inner membrane, I have used the more orthodox nomenclature of **N** for the matrix side space and **P** for the cytosolic facing space. These refer to the charge developed after the proton translocation; negative to the inside and positive to the outside. Two separate flows are indicated to elucidate proton and electron movement.

Starting at (1), the solid trace takes a QH_2 produced from the pool of this reduced cofactor. For linear flow through the respiratory chain, this would arise from complex I; it could of course come from any of the non-proton translocating sources, like Complex II, or GOLP oxidase (see Figure 3.10). At the **P** side, QH_2 donates one electron to an Fe/S protein, which connects to cytochrome c_1 and then to the mobile cofactor cytochrome c (as shown in the figure inset). At the same time, protons are released at the **P** side, and the radical ion $Q^{\cdot-}$ is formed, where the charged portion faces the water phase. From here, another electron is passed to cytochrome b_L, and the resulting fully oxidized Q is now fully mobile within the lipid portion of the membrane. Two separate flows now occur: the electron from b_L is passed to b_H, and Q can cross the membrane to the **N** face. The two streams now join up, and the electron from cytochrome b_H is taken up by Q to form $Q^{\cdot-}$, this time with the charged portion facing the **N** side.

To complete the picture, we must now take a second molecule of QH_2 and start from (2), following the dotted pathway. The path of this QH_2 is the same as the first up until the formation of $Q^{\cdot-}$ at the P face. From this radical ion, an electron is donated to b_L, leaving Q. At this juncture, to construct a complete pathway follow the electron

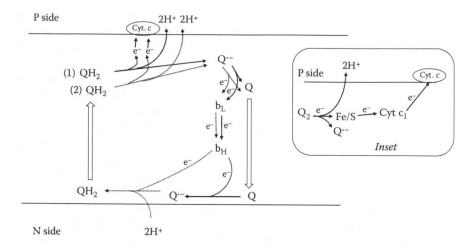

FIGURE 3.16 Q cycle: Complex III operation. The pathway is broken down into two parts. (1) leaves the radical ion species $Q^{\cdot-}$ at the N side (negatively charged, matrix space). (2) completes the pathway by further reducing $Q^{\cdot-}$ at the N side, accepting protons from the N side, forming QH_2 that can move through the membrane to the P side (positively charged, cytosolic space).

pathway indicated by the dotted arrows. The electron from b_L is transferred b_H. From here, the electron adds to the $Q^{\cdot-}$ at the **N** face, protons are added from the same face, and QH_2 results. This balances the pathway, at least in the steady state.

A satisfying experimental confirmation is the finding of two pools of $Q^{\cdot-}$, obtained by electron paramagnetic resonance [51], due to the distinct electrical environment of the two otherwise identical species. The other famous fact about the Q cycle is that it explains a paradox of "reductant induced oxidation" [52] that was the impetus for the suggestion of the pathway in first place.

This paradox can be explained by considering the action of the inhibitor antimycin. Its site of action is drawn in an abbreviated diagram of the Q cycle shown in Figure 3.17. This inhibitor blocks at the site of electron flow to form Q^- at the **N** face, and so prevents electron flow through the b cytochromes. Note, however, that electron flow is still possible from QH_2 to cytochrome c. If b cytochromes were part of a linear sequence like the other complexes, it is hard to imagine how addition of antimycin could cause their oxidation. It is clear, however, that they are not part of a linear electron flow. Instead, they span the membrane and provide a back flux of electrons through the membrane. Also indicated are the points of interaction of other inhibitors of complex III—stigmatellin and BAL (British anti-Lewisite [53])—which *do* stop all flow of electrons from the mobile cofactor QH_2 to cytochrome c, and thus halt both further electron flow outside the complex as well as flow to the b cytochromes. Thus, these inhibitors produce no anomalous results.

The point of the story, aside from a history of how an unusual mechanism was resolved, is to show that not all inhibition analysis is as simple as blocking a linear flow. Beyond this point, there is an analogy to two other events: the flow of carbon in the Krebs cycle and the pathway of galactose catabolism in liver. In the Krebs cycle,

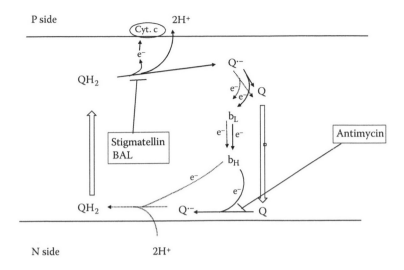

FIGURE 3.17 Inhibitors of Complex III. The site of inhibition of three inhibitors, including BAL (British anti-Lewisite) within this complex that explains unusual crossover data resulting from antimycin.

the carbons that are incorporated into citrate from acetyl CoA are not the carbons released in that particular cycle as CO_2; it is only the later turns of the cycle where that carbon is released as CO_2. In the steady state, we can say that the two carbons are released as CO_2, but not from the first turn of the cycle. The second example is the catabolism of galactose.

In the galactose pathway (Figure 3.18) liver-specific galactokinase catalyzes galactose-1-P formation, followed by reaction with UDP-glucose to form UDP-galactose plus glucose-1-P [54]. This uridylyltransferase reaction produces glucose-1-P directly, but this does not represent net formation of a pathway product. Consider the next step: UDP-galactose is converted to UDP glucose by the action of an epimerase which uses NAD first to oxidize the alcohol at the 4-position of the sugar, and then uses the NADH formed to reduce the ketone group, regenerating NAD. The effect is to convert one epimer to the other by the intermediacy of a keto group (Figure 3.18). In this unusual mechanism, NAD is itself cycled.

In all these cases, it is clearly necessary to traverse the scheme more than once to establish the actual pathway of net conversion in a steady state. Thus, the Q cycle is not unique, but rather an example of a pathway construction that requires a repeated sequence of steps to achieve a net pathway flow.

Electrons from cytochrome c are donated exclusively to Complex IV (see the next section). From what we have seen of all of the other mobile factors, there are multiple connections. We might expect cytochrome c, like Q, to have many inputs. However, the only one known is Complex III. It is true that, *in vitro*, it is possible to use an artificial electron donor to supply electrons to cyt c, such as tetramethylethylenediamine (TEMED) in assays of Complex IV [55]. Cytochrome c does play a non-metabolic role in the initiation of apoptosis [56]. It is possible that a future discovery of a reductant from the cytosol could input electrons to the mitochondria at this stage.

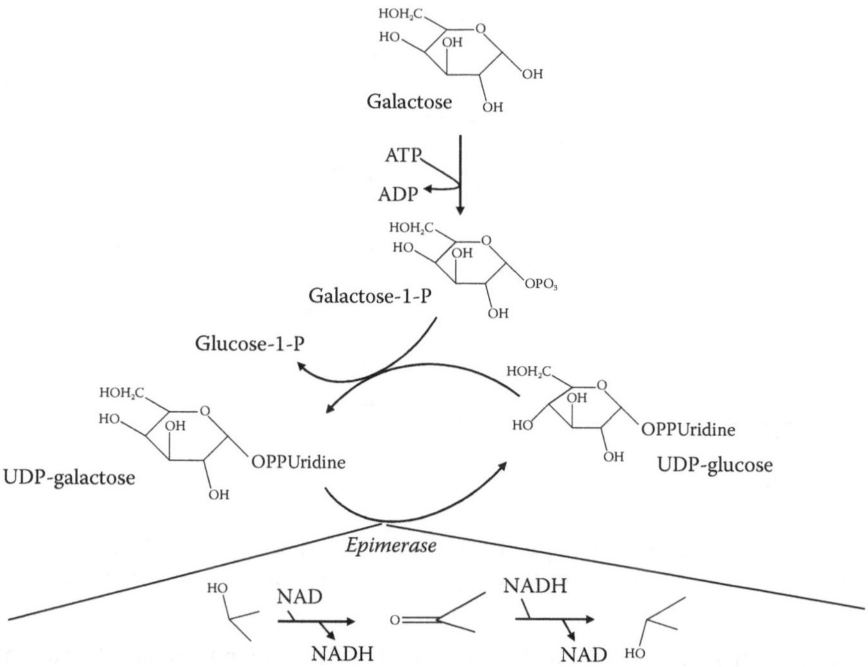

FIGURE 3.18 Galactose catabolism. Galactose conversion to glucose-1-P takes place in liver cytosol using enzymes unique to that tissue.

3.9.3 COMPLEX IV AND ITS CONNECTIONS TO MOBILE COFACTORS

Complex IV, like all complexes, is an island in a membrane sea connected to other complexes by mobile cofactors. In this case, it is the movement of cytochrome c along the P side surface of the membrane that donates electrons. A series of cytochromes accepts and relays them, along with two Cu proteins. Like Complex I, the protons move across the membrane in response to the redox activity, and like Complex I it is not exactly clear how electron and proton movements are engaged. However, the input and output ends of Complex IV are significant for understanding metabolic features as they can interact with other electron carriers apart from cytochrome c and oxygen, respectively.

As a method for donating electrons in an experimental setting, the pair of agents ascorbate and TEMED and are typically used to bypass the upstream portions of the respiratory chain [57]. Electrons can also be drawn off of cytochrome c by the use of FeCN; it would seem likely that it is also possible to draw off electrons to other acceptors.

In this context it is of interest in considering that the substrate for Complex IV, while normally having no alternative pathways in mammalian mitochondria, could have some in the case of exogenous molecules. One such likely agent is the molecule used to assess cell viability, MTT, a tetrazolium salt that is reduced by mitochondria and forms an insoluble formazan. The latter is dissolved in organic solvent and measured spectroscopically as an estimate of cell integrity. There are papers discussing the pros and cons of this technique [58–60], but the measurement itself is not actually one of cell integrity but rather of electron flow from Complex I to III to cytochrome c to MTT [58]. The argument is that this will reflect cell viability in so much as mitochondrial electron flow is preserved in intact cells, but not in compromised cells. While this is true, and often there is a close correspondence between a vital dye method and MTT reduction, they can easily diverge. In fact, we have demonstrated that different cell viability results ensue from three separate methods of analyzing this parameter: by total cellular ATP [61], by MTT, or by trypan blue, a vital dye [62]. Thus, tetrazolium salt methods such as MTT actually measure a portion of the respiratory chain to cytochrome c, and need not always correlate to cell viability.

3.10   UNCOUPLING

A strong feature of the Mitchell hypothesis is that it explains the otherwise perplexing issue of why so many different chemical structures exhibit the phenomenon of uncoupling. Operationally, an uncoupler allows electron flow to proceed despite little or no ATP formation. Compounds acting as uncouplers have little in common structurally apart from the properties of being a weak acid (or base) and relatively lipophilic in both the charged and uncharged forms. The simple activities of being a protonophore and a charged molecule that can cross the membrane by spreading out its charge are consistent with the proton gradient hypothesis, but impossible to fit into a chemical intermediate hypothesis. For the Mitchell hypothesis, the uncoupler provides a shortcut to proton re-entry into the mitochondria: the protonophore itself. For a classical treatment of a number of uncouplers and their properties, see Heytler [63]. In that article the ability of fatty acids to uncouple is also noted, with the caveat that they exert this ability only in the absence of media protein; thus their ability to exert this action in intact cells is unlikely.

3.10.1   Uncouplers and Weight Loss

The use of uncouplers as a weight-loss drug has a long history [64], including its use as a diet drug in the early twentieth century. It is still possible to purchase dinitrophenol (DNP) from the internet, despite well-known dangers of uncouplers *in vivo*, including blindness and death.

More recently, a reformulation of this early uncoupler has been reported that releases DNP to the circulation slowly [65]. In the rat, no toxicity was observed, and the investigators thus suggest this compound may be used as a weight-loss agent that would thereby reduce the incidence of associated diseases (i.e., obesity, diabetes, hypertension as part of the collection of diseases known as metabolic syndrome). There is little doubt that, apart from the toxicity of DNP, it is an excellent weight-loss agent: oxidation of substrates is not converted to energy but rather dissipated as heat.

3.10.2   Protein Uncouplers

It has long been known that uncoupling occurs in a physiological context, at least for certain animals such as hibernating species. The source of this activity is the uncoupling proteins, associated with an adipocyte variant rich in mitochondria, called the brown fat. These have long been issues of research study [66,67]. The recent discovery of brown fat in humans has initiated an avalanche of research on this topic (e.g., refs. [68–71]). It is possible that clinical manipulation of the expression of proteins involved (the UNC proteins) may achieve the same end as exogenous uncouplers, but in a more controlled way.

Mechanistically, the proteins provide a proton re-entry site rather than serve directly as carriers like the classical small molecule uncouplers. The proteins enable for example a heat rise for hibernation and potentially an enormous diet niche. For a good (brief) discussion of the uncoupler proteins, along with a summary of the controversy of uncoupler proteins discovered by homology, see Nicholls and Ferguson [35].

3.11 CONTROL OF ATP PRODUCTION BY MITOCHONDRIA: BALANCE

The control of ATP formation by mitochondria is surprisingly an issue that is not entirely resolved. As an operational hypothesis consistent with most investigations related to overall cell metabolism, the free concentration of cytosolic ADP reflects cellular energy demand and thus drives ATP formation. It is true that there are ATP utilizing steps in the mitochondrial matrix, such as reactions in the formation of urea in liver (carbamoyl phosphate synthetase), and hepatic gluconeogenesis at least in part (mitochondrial PEPCK). However, the vast majority of cell ATP consumption and thus ADP formation is cytosolic.

Two other analyses should be considered. First, the exchange of cytosolic ADP for matrix ATP is electrogenic, and, in functional mitochondria, this is invariably metabolically irreversible. One experimental approach in intact cells was conducted by the Krebs lab [72]. In this experiment, the idea was to use the "spare capacity" titration approach (also used by Rognstad [73]). The concept is that a metabolic pathway is measured in an intact cell preparation and titrated with a selective inhibitor. A plot of inhibitor concentration versus pathway product should show a lag phase if the target is not limiting in amount, but a straight line through the origin if it is. For example, gluconeogenesis from lactate shows an expected lag [73]. However, in the experiments of Krebs and colleagues [72], evidence suggesting control by the adenine nucleotide translocase resulted from an inhibitor titration study like the Rognstad experiment just discussed. In their study of gluconeogenesis and urea synthesis from various substrates such as lactate/ammonia and alanine, no reserve capacity was observed with inhibitors of the mitochondrial adenine nucleotide exchanger (atractyloside and carboxyatractyloside). The experiments showed an inhibition plot with no lag at all. This confounded the investigators. They recognized that the amount of ATP the mitochondria must produce must be determined by metabolic need, yet this experiment would suggest instead that a kinetic limitation of the exchanger is rate limiting for mitochondrial energy production. The clash between the experimental result and the interpretation was not resolved.

The second approach was to analyze the pathway using a variant of *sensitivity analysis*, an engineering concept applied to metabolic pathways. As discussed in Chapter 9, the method essentially provides a measure of rate limitation by an expansion of the relatively simple inhibitor titration just considered. By analysis of the same problem, Groen et al. [74] concluded that the adenine nucleotide exchanger as well as Complex IV were control points for isolated mitochondria. This follows the spirit of the analysis, which suggests that there is not just one rate limiting step, but many; there must be a sharing of the control. Since the approach of these investigators required isolated mitochondria rather than intact cells, it is not possible to project what they might mean. Thus, definitive control of mitochondrial respiration is not available. For our purposes, control by cytosolic ADP must fill in as an expression of cellular demand. In a similar way, control by Ca^{2+} follows the system principle that the cell cytosol exerts control over the mitochondria, rather than the other way around (see Chapter 7).

3.12   CURIOUS CASE OF THE RHO⁰ CELL

A more extreme example of the operation of mitochondrial pathways is found in studies of the laboratory creation known as rho⁰ cells. These are cells in which mitochondrial DNA is mutated, usually by treatment with ethidium bromide. A cell-selection process ensures, and some cells survive when supplemented with pyruvate and uridine [75]. A study of the mechanism for cell survival showed that the cells maintain a modest but definite membrane potential, despite the fact that the respiratory complexes are not functional [76]. In particular, the surviving cells maintain an adenine nucleotide exchange and an ATP synthetase, but the mitochondria do not actually synthesize ATP. The proposal was that both the exchange process (running in reverse of the physiological direction) and the F_1ATPase (that is, synthetase running in reverse) contribute to the membrane potential. While not explicitly diagrammed in that work, a model consistent with the evidence is drawn as Figure 3.19.

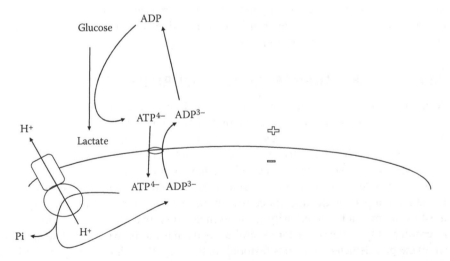

FIGURE 3.19   Energy metabolism in rho⁰ cells. This engineered cell type has mitochondrial damage; the ability to maintain a mitochondrial membrane potential uses ATP formation from glycolysis, and reverse flows through the ATP/ADP translocase and the F_1F_oATPase.

The authors suggested that the glycolytic pathway contributes the ATP to drive the adenine nucleotide exchange (the authors calculated that less than 10% of the cellular ATP would be needed for the maximum membrane potential the cells could achieve). The ATP enters the mitochondria in exchange for ADP *via* the adenine nucleotide translocase, which they show to be essential for the membrane potential (inhibited by the selective agents atractyloside and bonkrekic acid). The activity of the F_1F_oATPase is needed to regenerate ADP. The reversal of the proton through the F_o portion of the complex is responsible for the Δp. A smaller contribution is also possible due to the reversed exchange of adenine nucleotides, as both are fully charged species. It was proposed that the resulting membrane potential is essential for survival of these cells, at least in part for the import of cell nucleus encoded proteins into the mitochondrial matrix.

While the rho^0 cells may be useful in understanding certain extremes of mitochondrial behavior, their metabolic significance can be overextended. For example, Chen et al. [77] suggested that inhibiting a component of the normal ATP synthetase might be a therapeutic option as it restored ATP and membrane potential in a cell line, pointing to the example of the rho^0 cells. However, this is not actually restoring cellular function, but creating a rho^0 type mutant that is growth-competent but not really an amelioration of cell function. Similarly, Wheaton et al. [42] suggested that the ATP synthetase might be able to reverse directions in cells, based on a finding that oligomycin led to a diminished membrane potential. Thus, they proposed that normal cells could behave on occasion like rho^0 cells. This was necessary because the authors were proposing that metformin inhibits complex I, but somehow in a different way than established complex I inhibitors like rotenone. In particular, a membrane potential is required for metformin action. Perhaps cells would operate like the model of Figure 3.19 after metformin is introduced. This seems unlikely; the ATP synthetase only acts in reverse in rho^0 cells because it is a mutant form; in fact, in those cells it is insensitive to oligomycin [76]. There is no evidence that intact cells can reverse the direction of electron flow despite the fact that this property can be elicited in isolated mitochondrial preparations.

3.13   WHOLE-CELL MEASUREMENTS OF RESPIRATION

As a final point, we will consider a newer method for the measurement of respiration in intact cells. While oxygen utilization by isolated cells can be readily measured by a Clark-type electrode for cell suspensions (e.g., ref. [78]), analyzing cells adhering to a plate, as in cell culture, has usually been problematical. The recent introduction of the Seahorse oxygen analysis has led to a blossoming of analyses [79–81].

The principle of the instrument is unique: rather than measure the oxygen content in the bulk, a plate with adherent cells contains wells from which a probe displaces most of the media to temporarily form an incubation medium volume of just a few microliters. O_2 is measured for several seconds and the probe is lifted so that the small oxygen-depleted volume is restored. In this way, the volume of oxygen utilized during the measurement is a significant fraction of the total, and yet the cells are not subject to anoxia.

The instrument is sold by just one company, and they have also proposed a sequence of additions that can be made during the assay; ports are available for different solutions. In Figure 3.20, the record of three separate additions is displayed, with the suggested interpretation for differences in oxygen consumption[viii] noted in the rectangles of the drawing. In the absence of additions, the respiration is basal. Upon the addition of oligomycin, which should block the F_o portion (proton pore) of the F_1F_oATP synthetase, respiration declines. The difference in oxygen consumption before and after oligomycin is taken as "ATP production." Next, the uncoupler FCCP is added and respiration reaches its maximal value. The last-calculated difference, between maximal respiration and basal respiration is labeled as "spare capacity." Last, rotenone plus antimycin is added and the mitochondrial electron flow halted. The difference between this rate and the true zero is taken as "non-mitochondrial respiration." The distinction between oxygen consumption after oligomycin and after respiratory chain inhibition is taken as "proton leak."

These are fairly reasonable interpretations and make use of compounds that can be added to whole cells. Of course, the basal respiration is a variable that is dependent on the cell situation at the beginning of the experiment and is independent of the respiration observed after oligomycin. The measurements are somewhat indirect, as it is not possible to measure the states discussed earlier in the chapter; most compounds such as ADP are impermeable to the plasma membrane. Despite its limitations (e.g., ref. [82]), the method provides an insight into an intact cell system in culture that is distinct from isolated mitochondria.

With a few pathways described, we have material to work with to consider the issue of how they are connected. This issue has already been approached, both by discussing pathway intermediate connections (series) and mobile cofactor connections (parallel). We will now turn to a more elaborate discussion of those connections.

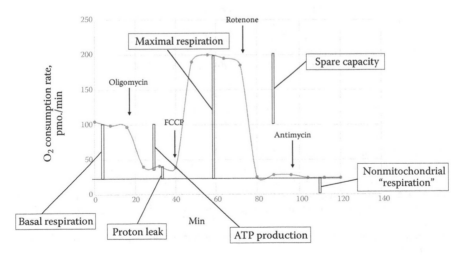

FIGURE 3.20 Respiration in cell cultures by oxygen-sensitive dye measurement instrument.

REFERENCES

1. Cammack, R. (1987) FADH$_2$ as a 'product' of the citric acid cycle. *TIBS*. Oct., 377–378.
2. Greenawalt, J. W. (1979) Survey and update of outer and inner mitochondrial membrane separation. *Methods Enzymol.* 55, 88–98.
3. Clemencon, B., Fine, M., and Hediger, M. A. (2016) Conservation of the oligomeric state of native VDAC1 in detergent micelles. *Biochimie.* 127, 163–172.
4. Hille, B. (2001) Calcium dynamics, epithelial transport, and intercellular coupling. In *Ion Channels of Excitable Membrane*, Sinauer Associates Inc., Sunderland, MA.
5. Izzo, V., Bravo-San Pedro, J. M., Sica, V., Kroemer, G., and Galluzzi, L. (2016) Mitochondrial permeability transition: New findings and persisting uncertainties. *Trends Cell Biol.*
6. Lin, R. Y., Vera, J. C., Chagantis, R. S. K., and Golde, D. W. (1998) Human monocarboxylate transporter 2 (MCT2) is a high affinity pyruvate transporter. *J. Biol. Chem.* 273, 28959–28965.
7. Kresge, N.,Simoni, R. D., and Hill, R. L. (2005) Fritz Lipmann and the discovery of coenzyme A. *J. Biol. Chem.* 280, e18.
8. Lehninger, A. L. (1951) Phosphorylation coupled to oxidation of dihydrodiphosphopyridine nucleotide. *J. Biol. Chem.* 190, 345–359.
9. Mehlman, M. A., Walter, P., and Lardy, H. A. (1967) Paths of carbon in gluconeogenesis and lipogenesis. VII. The synthesis of precursors for gluconeogenesis from pyruvate and bicarbonate by rat kidney mitochondria. *J. Biol. Chem.* 242, 4594–4602.
10. Lanoue, K. F., Meijer, A. J., and Brouwer, A. (1974) Evidence for electrogenic aspartate transport in rat liver mitochondria. *Arch. Biochem. Biophys.* 161, 544–550.
11. LaNoue, K. F., and Williamson, J. R. (1971) Interrelationships between malate-aspartate shuttle and citric acid cycle in rat heart mitochondria. *Metab. Clin. Exp.* 20, 119–140.
12. Shimomura, Y., Honda, T., Shiraki, M., Murakami, T., Sato, J., Kobayashi, H., Mawatari, K., Obayashi, M., and Harris, R. A. (2006) Branched-chain amino acid catabolism in exercise and liver disease. *J. Nutr.* 136, 250s–253s.
13. Bertrand, C., Dumoulin, R., Divry, P., Mathieu, M., and Vianey-Saban, C. (1992) Purification of electron transfer flavoprotein from pig liver mitochondria and its application to the diagnosis of deficiencies of acyl-CoA dehydrogenases in human fibroblasts. *Clin. Chim. Acta.* 210, 75–91.
14. Ogston, A. G. (1948) Interpretation of experiments on metabolic processes, using isotopic tracer elements. *Nature.* 162, 963.
15. Teicher, B. A., Linehan, W. M., and Helman, L. J. (2012) Targeting cancer metabolism. *Clin. Cancer Res.* 18, 5537–5545.
16. Smith, C. M., and Plaut, G. W. E. (1979) Activities of NAD-specific and NADP-specific isocitrate dehydrogenases in rat-liver mitochondria. Studies with D-threo-alpha-methylisocitrate. *Eur. J. Biochem.* 97, 283–295.
17. Sies, H., Akerboom, T. P. M., and Tager, J. M. (1977) Mitochondrial and cytosolic NADPH systems and isocitrate dehydrogenase indicator metabolites during ureogenesis from ammonia in isolated rat hepatocytes. *Eur. J. Biochem.* 72, 301–307.
18. Cupp, J. R., and McAlister-Henn, L. (1993) Kinetic analysis of NAD(+)-isocitrate dehydrogenase with altered isocitrate binding sites: Contribution of IDH1 and IDH2 subunits to regulation and catalysis. *Biochemistry.* 32, 9323–9328.
19. Moyle, J., and Mitchell, P. (2014) The proton-translocating nicotinamide–adenine dinucleotide (phosphate) transhydrogenase of rat liver mitochondria. *Biochem. J.* 132, 571–585.
20. Srere, P. A. (1987) Complexes of sequential metabolic enzymes. *Annu. Rev. Biochem.* 56, 89–124.
21. Hems, D. A., McCormack, J. G., and Denton, R. M. (1978) Activation of pyruvate dehydrogenase in the perfused rat liver by vasopressin. *Biochem. J.* 176, 627–629.

22. Roche, T. E., and Cate, R. L. (1977) Purification of porcine liver pyruvate dehydrogenase complex and characterization of its catalytic and regulatory properties. *Arch. Biochem. Biophys.* 183, 664–677.

23. Panov, A. V., and Scaduto, R. C. Jr. (1996) Substrate specific effects of calcium on metabolism of rat heart mitochondria. *Am. J. Physiol.* 270, H1398–H1406.

24. Johnson, J. D., Muhonen, W. W., and Lambeth, D. O. (1998) Characterization of the ATP- and GTP-specific succinyl-CoA synthetases in pigeon. The enzymes incorporate the same alpha-subunit. *J. Biol. Chem.* 273, 27573–27579.

25. Boyer, P. D., Deluca, M., Ebner, K. E., Hultquist, D. E., and Peter, J. B. (1962) Identification of phosphohistidine in digests from a probable intermediate of oxidative phosphorylation. *J. Biol. Chem.* 237, PC3306–PC3308.

26. Cardaci, S., Zheng, L., MacKay, G., van den Broek, N. J., MacKenzie, E. D., Nixon, C., Stevenson, D., Tumanov, S., Bulusu, V., Kamphorst, J. J., Vazquez, A., Fleming, S., Schiavi, F., Kalna, G., Blyth, K., Strathdee, D., and Gottlieb, E. (2015) Pyruvate carboxylation enables growth of SDH-deficient cells by supporting aspartate biosynthesis. *Nat. Cell Biol.* 17, 1317–1326.

27. Yogev, O., Yogev, O., Singer, E., Shaulian, E., Goldberg, M., Fox, T. D., and Pines, O. (2010) Fumarase: A mitochondrial metabolic enzyme and a cytosolic/nuclear component of the DNA damage response. *PLoS Biol.* 8, e1000328.

28. Yogev, O., Naamati, A., and Pines, O. (2011) Fumarase: A paradigm of dual targeting and dual localized functions. *FEBS J.* 278, 4230–4242.

29. Bardella, C., Olivero, M., Lorenzato, A., Geuna, M., Adam, J., O'Flaherty, L., Rustin, P., Tomlinson, I., Pollard, P. J., and Di Renzo, M. F. (2012) Cells lacking the fumarase tumor suppressor are protected from apoptosis through a hypoxia-inducible factor-independent, AMPK-dependent mechanism. *Molec. Cell. Biol.* 32, 3081–3094.

30. Pell, V. R., Chouchani, E. T., Frezza, C., Murphy, M. P., and Krieg, T. (2016) Succinate metabolism: A new therapeutic target for myocardial reperfusion injury. *Cardiovasc. Res.* 111, 134–141.

31. Ueland, P. M. (2011) Choline and betaine in health and disease. *J. Inherit. Metab. Dis.* 34, 3–15.

32. Boyer, P. D., Chance, B., Ernster, L., Mitchell, P., Racker, E., and Slater, E. C. (1977) Oxidative phosphorylation and photophosphorylation. *Annu. Rev. Biochem.* 46, 955–966.

33. Mitchell, P. (1968) *Chemiosmotic Coupling and Energy Transduction*, Glynn Research Laboratories, Cornwall.

34. Racker, E. (1976) *A New Look at Mechanisms in Bioenergetics*, Academic Press, New York.

35. Nicholls, D. G., and Ferguson, S. J. (2013) Bioenergetics. in, 1 online resource (xiv, 419 pp.).

36. Chance, B., and Williams, G. R. (1955) Respiratory enzymes in oxidative phosphorylation: III. The steady state. *J. Biol. Chem.* 217, 409–428.

37. Chance, B., Holmes, W., Higgins, J., and Connelly, C. M. (1958) Localization of interaction sites in multi-component transfer systems: Theorems derived from analogues. *Nature.* 182, 1190–1193.

38. Roberts, P. G., and Hirst, J. (2012) The deactive form of respiratory complex I from mammalian mitochondria is a Na+/H+ antiporter. *J. Biol. Chem.* 287, 34743–34751.

39. Blazer-Yost, B. L., Liu, X., and Helman, S. I. (1998) Hormonal regulation of ENaCs: Insulin and aldosterone. *Am. J. Physiol.* 274, C1373–C1379.

40. Murai, M., Murakami, S., Ito, T., and Miyoshi, H. (2015) Amilorides bind to the quinone binding pocket of bovine mitochondrial complex I. *Biochemistry.*

41. Bridges, H. R., Jones, A. J., Pollak, M. N., and Hirst, J. (2014) Effects of metformin and other biguanides on oxidative phosphorylation in mitochondria. *Biochem. J.* 462, 475–487.

42. Wheaton, W. W., Weinberg, S. E., Hamanaka, R. B., Soberanes, S., Sullivan, L. B., Anso, E., Glasauer, A., Dufour, E., Mutlu, G. M., Budigner, G. S., and Chandel, N. S. (2014) Metformin inhibits mitochondrial complex I of cancer cells to reduce tumorigenesis. *Elife.* 3, e02242.

43. Bedard, K., and Krause, K. H. (2007) The NOX family of ROS-generating NADPH oxidases: Physiology and pathophysiology. *Physiol. Rev* 87, 245–313.

44. Katz, A. (2007) Modulation of glucose transport in skeletal muscle by reactive oxygen species. *J. Appl. Physiol.* 102, 1671–1676.

45. Guzik, T. J., Korbut, R., and Adamek-Guzik, T. (2003) Nitric oxide and superoxide in inflammation and immune regulation. *J. Physiol. Pharmacol.* 54, 469–487.

46. Chandel, N. S. (2015) *Navigating Metabolism*, Cold Spring Harbor Laboratory Press, Cold Spring Harbor, New York.

47. Perez, V. I., Bokov, A., Van Remmen, H., Mele, J., Ran, Q., Ikeno, Y., and Richardson, A. (2009) Is the oxidative stress theory of aging dead? *Biochim. Biophys. Acta.* 1790, 1005–1014.

48. Sanz, A., Fernandez-Ayala, D. J., Stefanatos, R. K., and Jacobs, H. T. (2010) Mitochondrial ROS production correlates with, but does not directly regulate lifespan in Drosophila. *Aging (Albany NY).* 2, 200–223.

49. Wilson, D. F., Stubbs, M., Veech, R. L., Erecinska, M., and Krebs, H. A. (1974) Equilibrium relations between the oxidation–reduction reactions and the adenosine triphophate synthesis in suspension of isolated liver cells. *Biochem. J.* 140, 57–64.

50. Veech, R. L., Lawson, J. W. R., Cornell, N. W., and Krebs, H. A. (1979) Cytosolic phosphorylation potential. *J. Biol. Chem.* 254, 6538–6547.

51. Zhang, H., Osyczka, A., Dutton, P. L., and Moser, C. C. (2007) Exposing the complex III Qo semiquinone radical. *Biochim. Biophys. Acta.* 1767, 883–887.

52. Slater, E. C. (1981) *Chemiosmotic Proton Circuits in Biological Membranes: In Honor of Peter Mitchell*, Addison-Wesley, Advanced Book Program/World Science Division, Reading, MA.

53. Slater, E. C. (1983) The Q cycle, an ubiquitous mechanism of electron transfer. *Trends Biochem. Sci.* 8, 239–242.

54. Cohn, R. M., and Segal, S. (1973) Galactose metabolism and its regulation. *Metab. Clin. Exp.* 22, 627–642.

55. Bucher, J. R., and Penniall, R. (1975) The subunit composition of beef heart cytochrome c oxidase. *FEBS Lett.* 60, 180–184.

56. Sedlak, T. W., and Snyder, S. H. (2006) Messenger molecules and cell death: Therapeutic implications. *JAMA.* 295, 81–89.

57. Gnaiger, E., and Kuznetsov, A. V. (2002) Mitochondrial respiration at low levels of oxygen and cytochrome c. *Biochem. Soc. Trans.* 30, 252–258.

58. Berridge, M. V., and Tan, A. S. (1993) Characterization of the cellular reduction of 3-(4,5-dimethylthiazol-2-yl)-2,5-diphenyltetrazolium bromide (MTT): Subcellular localization, substrate dependence, and involvement of mitochondrial electron transport in MTT reduction. *Arch. Biochem. Biophys.* 303, 474–482.

59. Berridge, M. V., Herst, P. M., and Tan, A. S. (2005) Tetrazolium dyes as tools in cell biology: New insights into their cellular reduction. *Biotechnol. Annu. Rev.* 11, 127–152.

60. Calabro, A. R., Konsoula, R., and Barile, F. A. (2008) Evaluation of in vitro cytotoxicity and paracellular permeability of intact monolayers with mouse embryonic stem cells'. *Toxicol. In Vitro.* 22, 1273–1284.

61. Cornell, N. W. (1983) Evaluation of hepatocyte quality: Cell integrity and metabolic rates. in *Isolation, Characterization, and Use of Hepatocytes* (Harris, R. A., and Cornell, N. W. eds.), Elsevier Biomedical, New York.

62. Vytla, V. S., and Ochs, R. S. (2013) Metformin increases mitochondrial energy formation in L6 muscle cell cultures. *J. Biol. Chem.* 288, 20369–20377.

63. Heytler, P. G. (1979) Uncouplers of oxidative phosphorylation. *Methods Enzymol.* 55, 462–42.

64. Parascandola, J. (1974) Dinitrophenol and bioenergetics: An historical perspective. *Molec. Cell. Biochem.* 5, 69–77.

65. Perry, R. J., Zhang, D., Zhang, X. M., Boyer, J. L., and Shulman, G. I. (2015) Controlled-release mitochondrial protonophore reverses diabetes and steatohepatitis in rats. *Science.* 347, 1253–1256.
66. Lee, S., Hamilton, J., Trammell, T., Horwitz, B., and Pappone, P. (1994) Adrenergic modulation of intracellular pH in isolated brown fat cells from hamster and rat. *AJP.* 267, C349–C356.
67. Czech, M. P., Lawrence, J. C., and Lynn, W. S. (1974) Hexose transport in isolated brown fat cells. A model system for investigating insulin action on membrane transport. *J. Biol. Chem.* 249, 5421–5427.
68. Ramseyer, V. D., and Granneman, J. G. (2016) Adrenergic regulation of cellular plasticity in brown, beige/brite and white adipose tissues. *Adipocyte.* 5, 119–129.
69. Lin, J. Z., and Farmer, S. R. (2016) Morphogenetics in brown, beige and white fat development. *Adipocyte.* 5, 130–135.
70. Hibi, M., Oishi, S., Matsushita, M., Yoneshiro, T., Yamaguchi, T., Usui, C., Yasunaga, K., Katsuragi, Y., Kubota, K., Tanaka, S., and Saito, M. (2016) Brown adipose tissue is involved in diet-induced thermogenesis and whole-body fat utilization in healthy humans. *Int. J. Obes. (Lond.).*
71. Zhang, Y., Xie, C., Wang, H., Foss, R., Clare, M., George, E. V., Li, S., Katz, A., Cheng, H., Ding, Y., Tang, D., Reeves, W. H., and Yang, L. J. (2016) Irisin exerts dual effects on browning and adipogenesis of human white adipocytes. *Am. J. Physiol. Endocrinol. Metab.* ajpendo.00094.2016.
72. Stubbs, M., Vignais, P. V., and Krebs, H. A. (1978) Is the adenine nucleotide translocator rate-limiting for oxidative phosphorylation? *Biochem. J.* 172, 333–342.
73. Rognstad, R. (1979) Rate-limiting steps in metabolic pathways. *J. Biol. Chem.* 254, 1875–1878.
74. Groen, A. K., Wanders, R. J. A., Westerhoff, H. V., Van Der Meer, R., and Tager, J. M. (1982) Quantification of the contribution of various steps to the control of mitochondrial respiration. *J. Biol. Chem.* 257, 2754–2757.
75. King, M. P., and Attardi, G. (1989) Human cells lacking mtDNA: Repopulation with exogenous mitochondria by complementation. *Science.* 246, 500–503.
76. Appleby, R. D., Porteous, W. K., Hughes, G., James, A. M., Shannon, D., Wei, Y. H., and Murphy, M. P. (1999) Quantitation and origin of the mitochondrial membrane potential in human cells lacking mitochondrial DNA. *Eur. J. Biochem.* 262, 108–116.
77. Chen, W. W., Birsoy, K., Mihaylova, M. M., Snitkin, H., Stasinski, I., Yucel, B., Bayraktar, E. C., Carette, J. E., Clish, C. B., Brummelkamp, T. R., Sabatini, D. D., and Sabatini, D. M. (2014) Inhibition of ATPIF1 ameliorates severe mitochondrial respiratory chain dysfunction in mammalian cells. *Cell Rep.* 7, 27–34.
78. Ochs, R. S., and Harris, R. A. (1986) Mechanism for the oleate stimulation of gluconeogenesis from dihydroxyacetone by hepatocytes from fasted rats. *Biochim. Biophys. Acta.* 886, 40–47.
79. Technologies, Aligent. 2016. Publications with Seahorse XF Data. http://www.agilent.com/publications-database/.
80. Jiang, L., Shestov, A. A., Swain, P., Yang, C., Parker, S. J., Wang, Q. A., Terada, L. S., Adams, N. D., McCabe, M. T., Pietrak, B., Schmidt, S., Metallo, C. M., Dranka, B. P., Schwartz, B., and DeBerardinis, R. J. (2016) Reductive carboxylation supports redox homeostasis during anchorage-independent growth. *Nature.* 532, 255–258.
81. Xu, M., Xiao, Y., Yin, J., Hou, W., Yu, X., Shen, L., Liu, F., Wei, L., and Jia, W. (2014) Berberine promotes glucose consumption independently of AMP-activated protein kinase activation. *PLoS One.* 9, e103702.
82. Mookerjee, S. A., Goncalves, R. L., Gerencser, A. A., Nicholls, D. G., and Brand, M. D. (2015) The contributions of respiration and glycolysis to extracellular acid production. *Biochim. Biophys. Acta.* 1847, 171–181.

ENDNOTES

i. Many sources refer to flavin cofactors such as FAD as mobile, either explicitly or implicitly, suggesting, for example, that the mitochondrial respiratory chain collects electrons in part through FAD, or by showing FAD as a cofactor in a manner identical to NAD. However, this is an old textbook error [1]; FAD is a tightly bound prosthetic group that does not leave the enzyme during catalysis and must be converted to its original form at the end of the catalytic cycle; it is the opposite of a mobile cofactor.

ii. Equivalently, the transporter is an antiport with hydroxide ions.

iii. There is also the possibility of exchanging malate with inorganic phosphate, although the overall balance of the pathway is no longer preserved. In a sense, this is mere "paper biochemistry." Experimental studies to actually determine which pathway is traversed are not available.

iv. Isovaleryl CoA oxidation is more complicated than indicated. While it is true, as shown, that reducing equivalents are ultimately transferred to form QH_2, the proximal acceptor for the electrons is the electron transferring flavoprotein [13]; this process is described in the context of long-chain fatty acid oxidation in Chapter 6.

v. Note how metabolic compartmentation prevents the intersection of cytosolic HMG-CoA and this pool. In the cytosol, this intermediate is the substrate of HMG-CoA reductase, the key enzyme of cholesterol and branched chain lipid biosynthesis. However, this is completely separate from the oxidative pathways of ketone bodies and of leucine catabolism as the pools don't mix. In fact, no CoA-containing intermediate can cross the mitochondrial membrane.

vi. Much of this can be found in monographs of oxidative phosphorylation, including that of Racker [34], and more extensively in Nicholls and Ferguson [35].

vii. Amiloride is an established diuretic compound that blocks the Na/H exchange activity in the distal tubule of kidney. These channels are known as ENaC, or epithelial sodium channels [39].

viii. D. Nicholls, who is an author of the monograph cited in this chapter [35], is a consultant for the company. While the company name – Seahorse – should bring to mind the action of the bobbing probe, just like the ocean animal, according to a company representative, this was not the name origin. Instead it was just assigned from inside the company.

4 Principles of Pathway Connectivity

Having examined the pathways of glycolysis, the Krebs cycle, and mitochondrial respiration, it is already evident that a simple linear connection of one pathway to another is the fundamental means of connecting metabolic pathways. We might call this join between pathways a "head-to-tail" type. In this chapter, I examine other types of connections that exist between metabolic pathways, offering examples of three distinct types. The first of these is the connection of gluconeogenesis to glycolysis. These pathways overlap to a large extent; that is, they share a majority of enzymatic steps, with shared reactions running in opposite directions. The second is a sharing of just one intermediate, the case of connection of glycolysis with glycogen metabolism. Even for this example, we will find that the situation may involve more extensive overlap than is apparent at first sight. A further example of this type involves the Krebs cycle enzyme aconitase, and draws in issues of stereochemistry and enzyme mechanism. The third connectivity is the parallel connections of mobile cofactors, a topic introduced in prior chapters. I conclude by considering metabolic compartmentation, mobile cofactors, and transport across molecules.

4.1 EXTENSIVE OVERLAP: GLYCOLYSIS AND GLUCONEOGENESIS

The first type of connection is one in which two pathways share most of the same enzymes in common. A clear example of this is the liver pathways of glycolysis and gluconeogenesis (Figure 4.1). The reactions in the direction of the top of the diagram down represent glycolysis. The gluconeogenic pathway is shown only in sketch outline, omitting movement between cell spaces (transporters) and mobile cofactors. What remains clear is that most of the reactions are the same, but utilize the near-equilibrium steps to produce pathways in opposing directions.

As a result, it is not possible to run both of these pathways at the same time. The liver must be either glycolytic or gluconeogenic, but not both. It is possible, of course, for separate reactions that connect the same points to run simultaneously, and they clearly do. Thus, the different routes connecting three segments:

glucose/glucose-6-P
fructose-6-P/fructose bisphosphate
PEP/pyruvate

are connected by mIRR enzymes that, running simultaneously, produce three small cycles. Any of the enzymes can therefore regulate flux of either pathway. As the operation of the cycle itself in two of the cases serve to effectively just split ATP to ADP and Pi, they expend energy and were originally called "futile cycles" (see Chapter 6).

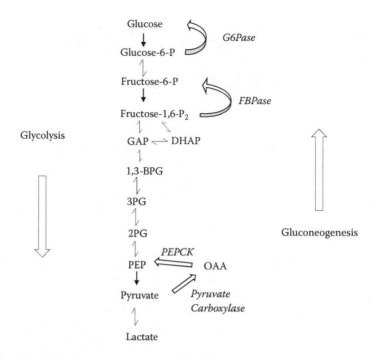

FIGURE 4.1 Glycolysis and gluconeogenesis. The opposing pathways share most of the steps at NEQ enzymes.

Aside from the fact that it is structurally not possible to run both of these pathways simultaneously, it is also physiologically satisfying. For example, in the fasted state of liver, it is important for the flow to be towards glucose as the liver must provide this substrate to key organs such as brain. In the fed state, glycolysis can provide the pathway of fatty acid synthesis with pyruvate to form fatty acids (Figure 4.2).

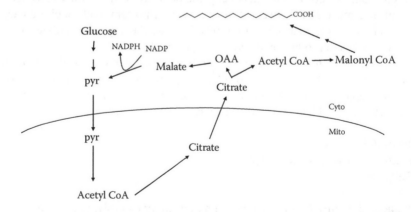

FIGURE 4.2 Glycolysis and lipogenesis. These pathways share few intermediates; they intersect at pyr (pyruvate) as part of the citrate shuttle for the lipogenic production of cytosolic acetyl CoA.

Cytosolic pyruvate is the sole connection between glycolysis and fatty acid bio-synthesis, as evident from Figure 4.2. Both glycolysis and the citrate export cycle of fatty acid biosynthesis produce pyruvate.

If we follow the physiological direction argument invoked for glycolysis/gluconeogenesis and extend it to glycolysis/fatty acid synthesis, then we might expect that the latter pair would always be active at the same time, that is, in the fed state. This expectation is borne out, but in this case the lack of overlap of these pathways means nothing prevents them from diverging in pathway direction. In fact, there is an interesting situation in which gluconeogenesis and fatty acid synthesis are simultaneously active. That is to say, there is a simultaneous pathway for gluconeogenesis/fatty acid biosynthesis.

This occurs in the case of type II diabetes, and is considered a "paradox" because it violates the feeding/fasting paradigm [1]. What it really says is that under these conditions, gluconeogenesis is properly reflecting insulin insensitivity (and is thus over-active), but that fatty acid synthesis is not suppressed. A recent study has proposed a mechanism to account for this paradox: the enzyme complex TORC1, which is a regulator of transcriptional events, is connected not only to insulin signaling but also to direct nutrient stimulation, such as amino acids [2]. The latter serve to increase transcription of lipogenic enzymes and thus produce fatty acids in the face of an effective insulin deficit which should by itself suppress lipogenesis.

Admittedly, this is a pathological condition, but it illustrates the distinction in pathway connections that are structurally enchained (glycolysis/gluconeogenesis) and simply linked by a substrate–product combination (glycolysis/fatty acid biosynthesis).

4.2   INTERSECTING PATHWAYS: GLYCOGEN METABOLISM

This type of connection involves an intermediate of one pathway that intersects with another. A classical connection of this sort is that between glycolysis and glycogen metabolism (Figure 4.3). This figure does not account for all of glycogen

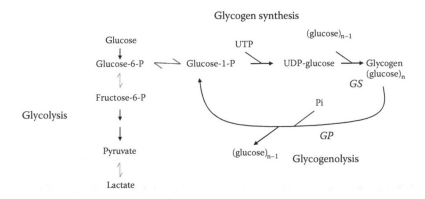

FIGURE 4.3   Glycolysis and glycogen metabolism. Glycogen metabolism branches from glycolysis at glucose-6-P.

metabolism, as branching and debranching, for example, are not shown. It does include the key regulated enzymes of this pathway, GS (glycogen synthase) and GP (glycogen phosphorylase). It appears from the diagram that just one intermediate of glycolysis is common to both pathways: glucose-6-P. However, to construct a complete pathway for either synthesis or degradation of glycogen, it is necessary to include other reactions. For example, in liver, the supply of glucose-6-P may come from either glucose (during states of high blood glucose, as in feeding), or through the gluconeogenic pathway in postprandial states. The latter is known as the *indirect pathway* [3,4].

A variation on the connection between glycolysis and glycogen metabolism was proposed that may be of importance for fast muscle [5]. Glycolysis may be considered to begin with either the entry of glucose into cells or the breakdown of glycogen. In this study, it was suggested that, for rapid contraction, the pathway starts with glycogen. Glycolysis-from-glycogen is a pathway that can occur in the resting phase. In the active state of contraction, glycolysis-from-glycogen can produce more ATP than the usually glycolytic route. The situation is illustrated in Figure 4.4. In panel (a) glycolysis-from-glucose produces the expected net two ATPs. In panel (b) the storage of glucose as glycogen requires an input of three ATPs; this is the resting state. Finally, in panel (c) active contraction produces three ATPs as the lactate is formed from glycogen. This separation of glycogen formation via external glucose from resting-state glycogen synthesis thus permits more ATP formation in the active state, but requires more ATP utilization for the storage of glycogen. Clearly, some glycolysis-from-glucose or aerobic metabolism must take place for maintenance of cellular function.

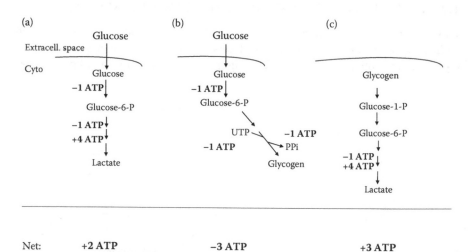

FIGURE 4.4 Energy balance in glycolysis with different pathway substrates. The net ATP formation is two from extracellular glucose (first panel), but three from glycogen (last panel). The extra ATP formation is the result of glycogen formation at rest from extracellular glucose (middle panel), which requires input of three ATPs.

4.3   STEREOCHEMISTRY AND METABOLIC CONNECTIONS

In the early history of the Krebs cycle [6] there was a controversy concerning the notion that citrate could lead to isocitrate formation, as the precursor is symmetrical, while the product is chiral (Figure 4.5). While Krebs had proposed citrate as the first intermediate in 1937, this stereochemical anomaly prevented acceptance of the idea. The situation changed suddenly in 1948. In that year, Ogston [7] proposed a "three-point landing" of a molecule having only three distinct substituents attached to a carbon center might still position itself in an enzyme to produce just one stereoisomer. This was rapidly accepted, followed by the acceptance of the Krebs cycle itself. Much later, a modification was introduced by Koshland [8], who proposed a fourth binding site might be of importance if multiple prochiral centers were involved. This was extended further to cases of even greater numbers.[i]

The Ogston hypothesis, however, remains the foundational idea. The paper itself [7] is a short communication that points out possible interpretation errors of radiolabeling studies to determine possible enzyme mechanisms. The two examples were serine hydroxymethyltransferase and aconitase. The first was a critique of a paper by Shemin [9], who had ruled out the possible intermediate amino malonic acid, as it was symmetrical and thus could not explain the labeling results. Ogston suggested it could make a three-point landing on the enzyme surface, placing the two carboxyl groups in distinct environments (Figure 4.6). While this is an insightful point, it was made without knowing that this happened not to be the actual intermediate of the reaction. Not only that, this reaction does not actually produce a chiral molecule: the product is glycine! So, while this is an interesting example of a theoretical three-point landing, it cannot be taken as evidence for the mechanism of prochirality.

The other example, aconitase, was not represented as a figure in the Ogston paper. If it were there would be a clear difficulty. Citrate, the substrate shown in Figure 4.7, has two identical substituents at C5, but they are C-H groups. These cannot attach to an enzyme surface, as C-H has essentially no polarity (the electronegativities being essentially identical). However, this is only a part of the problem. The reaction proceeds through the intermediate *cis*-aconitate, for which the enzyme is named (Figure 4.7). Both the reaction of citrate to *cis*-aconitate, and *cis*-aconitate to isocitrate, are catalyzed by aconitase. Since neither citrate nor *cis*-aconitate are chiral, it is clear that citrate is not even a prochiral molecule! That designation applies only to

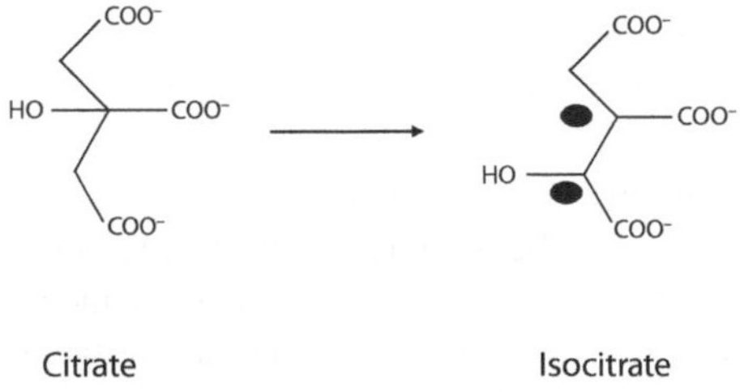

Citrate   Isocitrate

FIGURE 4.5   Chiral product formation in aconitase. The product isocitrate has two chiral centers indicated by dots; there are none in citrate.

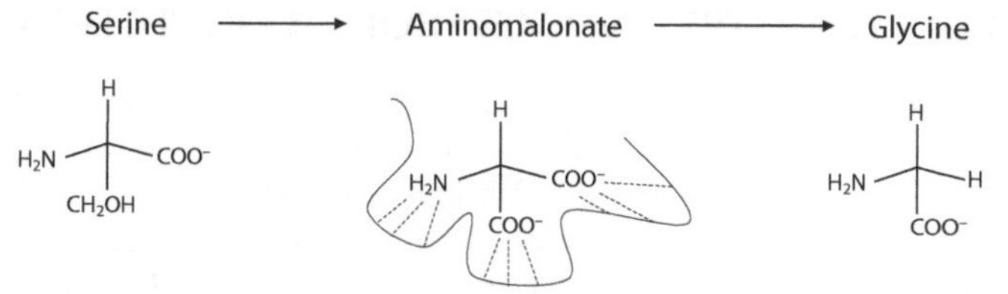

FIGURE 4.6 Proposed intermediate in serine hydroxymethyltransferase. The intermediate proposed has three attachment sites according to the Ogston hypothesis. However, later studies showed that this is not the actual intermediate for this enzyme.

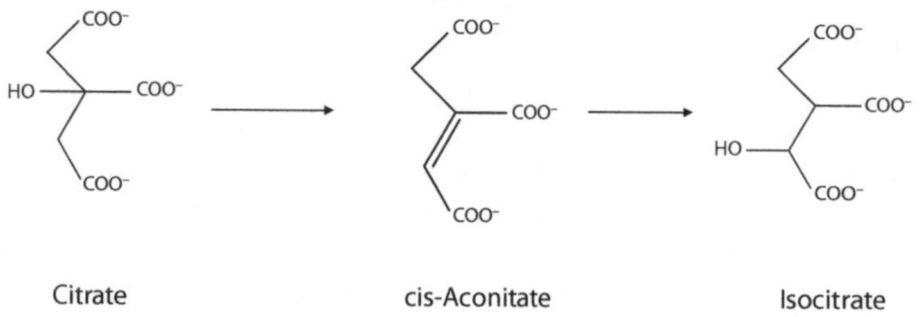

Citrate   cis-Aconitate   Isocitrate

FIGURE 4.7   Aconitase. The enzyme catalyzes two sequential reactions: first to *cis*-aconitate, and then to isocitrate.

cis-aconitate. However, once we consider *cis*-aconitate as the precursor, we can no longer apply the Ogston hypothesis having two identical groups out of four attached to a carbon, since there are only three bonds to C5 in *cis*-aconitate. That is to say, the actual prochiral molecule here is sp^2 hybridized and not sp^3 hybridized. In fact, in every case of prochiralty of enzymology, *all* of them are sp^2 hybridized.

A new model for prochirality is required. We have proposed one using *in silico* modeling studies[ii] of aconitase and a similar, but simpler enzymatic reaction catalyzed by fumarase. In the latter case (Figure 4.8) hydration of the double bond requires fixation of the substrate on the enzyme, and positioning of the water so that just one stereoisomer of malate is formed. From this, we propose three requirements for enzymatic prochiral formation:

1. The substrate has a geometrical feature of asymmetry.
2. The substrate has a fixed attachment to the enzyme.
3. The enzyme has catalytic features positioned to selectively attack the substrate, or positioned binding sites for other molecules that do.

The first of these is a feature of the substrate. In the case of fumarase, this is the trans-orientation of the double bond. The second is a feature of both substrate and enzyme: the fumarate carboxyl groups each selectively attach to several amino acids in the catalytic pocket as shown. The third is a feature of the enzyme, here the histidine base assisted water attack on the Si face, with a proton abstracted from a serine residue on the Re face. The result of these three properties is that only the S-malate can be the product of fumarase.

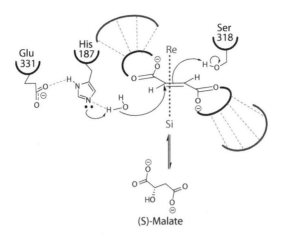

(S)-Malate

FIGURE 4.8 Mechanism of fumarase. Water attacks the double bond of fumarate from the si face and pi electrons of the double bond attach to the hydrogen of Ser318. The dashed lines show the attachment of fumarate to the enzyme. Just two groups of attachments exist: the two carboxyl groups.

Applying these three criteria to the aconitase reaction, we must recall that the rules concern only the intermediate aconitate, and not the substrate or product. This is simply a restatement of the fact that only aconitate is prochiral. Nonetheless, there is a curious asymmetry to the reaction when it is run in the forward direction as opposed to the reverse. In cells, aconitase is NEQ and in fact does catalyze the over-all reaction in both directions.

Starting with the forward direction, the first catalytic event is the conversion of citrate to *cis*-aconitate. After the dehydration step, but before the release of *cis*-aconitate, we say that the enzyme binds *cis*-aconitate in the "citrate mode." This is the first transfor-mation of Figure 4.9. Notice that if this complex were subject to water addition, it could only form citrate once again, by the reverse direction. It can never form iso-citrate in this form. Instead, it must undergo a 180° rotation of *cis*-aconitate [10,11]. This now places the double bond in a position to add water to form isocitrate.

The enzyme-*cis*-aconitate form that exists after 180° rotation is known as the "isocitrate mode" of binding. Thus, while the chemical transformation is similar to fumarase, it is clear that a new feature is present in aconitase that somehow requires a 180° rotation of *cis*-aconitate in order to change it from the citrate mode to the isoci-trate mode. This event is known as a "180° flip." How this flip occurs is controversial.

4.3.1 "Flip" and Metabolic Insight

There are two ideas for how the flip might be accomplished. One is that *cis*-aconitate remains on the enzyme surface during the course of the reaction, rotating 180° in place. The second is that the intermediate aconitate dissociates from the enzyme in one binding form and then rebinds in the other.

Consistent with the first mechanism is a study in 1967 that showed that labelled proton removed in forming the double bond intermediate was retained in the product [12].

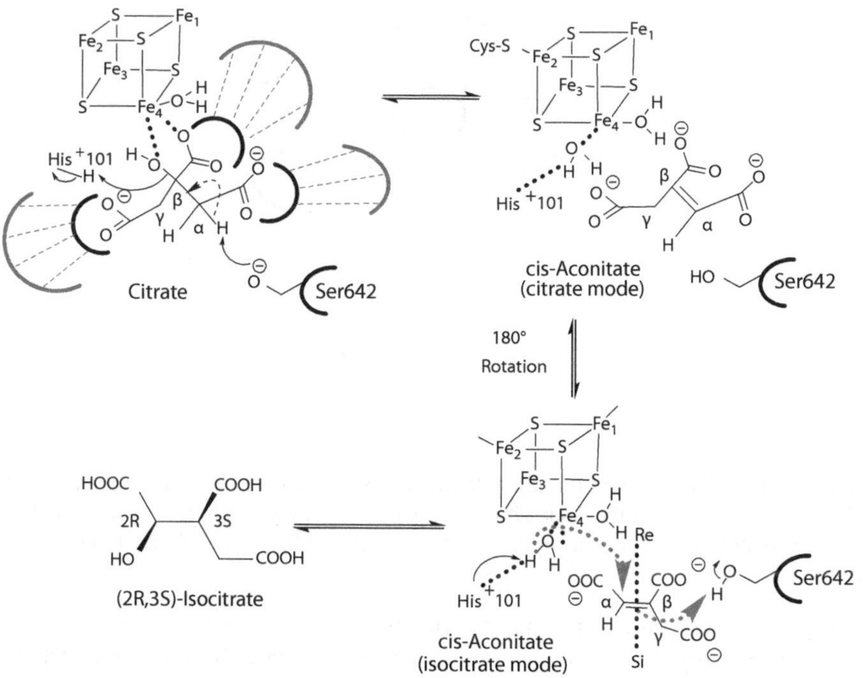

FIGURE 4.9 Mechanism of aconitase. Citrate forms *cis*-aconitate by dehydration as shown, bound to the enzyme in the citrate mode. The intermediate must rotate 180° to form the same complex, but in the isocitrate mode. Hydration leads to the chiral isocitrate molecule.

The idea is best visualized by reducing the overall aconitase reaction to its simplest form: the elements of water in citrate are removed to form a *cis*-double bond intermediate (aconitate), which is then rehydrated to isocitrate. Since the experiments showed that the label was retained, rather than exchanged with solvent, it seemed reasonable to suggest that the intermediate somehow remained affixed to the enzyme surface and rotated in place by 180°. However, it is now known that the residue that removes this proton in the enzymatic reaction is from serine, and there is no reason to believe that this serine can readily exchange with water. Thus, it is not easy to prove the case for this mechanism. Beinert et al. [13] propose essentially the second mechanism, suggesting that a solution aconitate molecule displaces the existing bound intermediate, but in the opposite (180°) orientation.

While we have pushed somewhat deeply into the details of enzyme mechanism, there is a reason for doing so. We can apply metabolic information to help resolve this controversy. At least one enzymatic reaction is known to use *cis*-aconitate as substrate: *cis*-aconitate decarboxylase. This enzyme catalyzes the decarboxylation of aconitate to produce itaconoate (Figure 4.10) [14]. Measurable pools of *cis*-aconitate are found in both bacteria [15] and mammalian mitochondria [16,17]. These findings demonstrate that *cis*-aconitate is a branch point connecting two metabolic pathways, and thus cannot remain bound to its enzyme over the course of the two reactions, but rather must be free in solution. The isotopic data suggesting

cis-Aconitate   Itaconoate

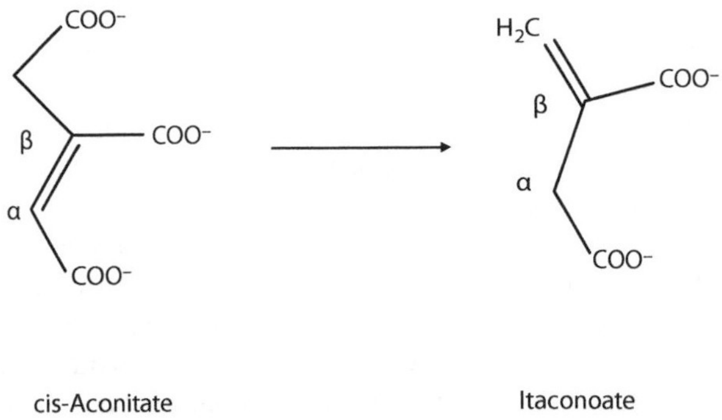

FIGURE 4.10 Itacanoate produced from *cis*-Aconitate. This is a separate reaction from the first product of aconitase. The existence of this intermediate shows that *cis*-aconitate must not remain bound to aconitase during its catalytic cycle.

otherwise must therefore have an alternative explanation, but it is unlikely to apply to the situation in intact cells.

There is a similar conflict between *in vitro* enzymology results and metabolic pathways in the case of xanthine oxidase. This enzyme also catalyzes two sequential reactions (Figure 4.11). According to a kinetic analysis of the enzyme *in vitro*, the enzyme first oxidizes hypoxanthine, and the product does not leave the enzyme surface, but rather rotates 180° while fixed to the enzyme surface [18]. While the overall mechanism of the enzymatic catalysis remains controversial [19], there have been no challenges to the idea that xanthine exists only as a bound intermediate in the xanthine oxidase reaction. However, as also shown in Figure 4.11, xanthine oxidase is the end stage of the pathway for purine degradation. Adenine nucleotides are

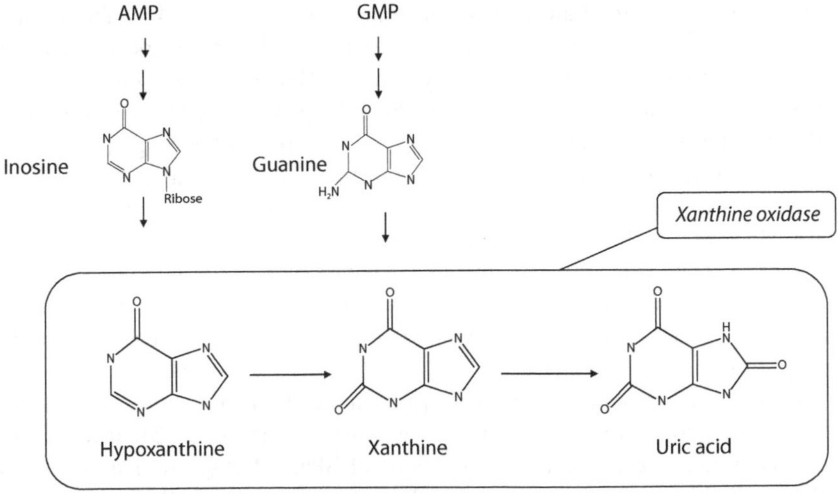

FIGURE 4.11 End stages of purine degradation. The breakdown of adenine and guanine nucleotides lead to two substrates of xanthine oxidase. Thus, xanthine has a connection to guanine and must be a free intermediate in the two reactions catalyzed by xanthine oxidase.

degraded to inosine and then to hypoxanthine. Guanine nucleotides are degraded to guanine and then xanthine, the intermediate of xanthine oxidase. Thus, in order to connect the degradation of guanine nucleotides to their ultimate fate – uric acid – xanthine oxidase must use its namesake, xanthine, as a substrate. This would be impossible if xanthine only exists as a bound intermediate. The argument is essentially identical to that offered for aconitase.

The distinction between enzyme clusters and individual steps is a simple and functional one. For both aconitase and xanthine oxidase, the intermediate does not stay bound to the enzyme but rather dissociates in solution, so in this case it is distinctive from situations discussed previously like palmitate synthase, in which reactions are catalyzed by separate enzymes without (mammalian) or with (bacterial) dissociation into solution. Thus, the real meaning of enzyme complexes, and its rarity, becomes clear. When reactions need to connect to other reactions, enzymes catalyze separate reactions. When an assembly needs to be completed while preventing any side reactions, a complex is needed.

4.4   COMPARTMENTATION

Compartmentation is a broad hypothesis holding that molecules are not free to diffuse in a defined cell water space. It is often advanced not as a specific mechanism but rather an expedient hypothesis to resolve a paradox. One type of compartmentation hypothesis was introduced in Chapter 1: the idea that molecules might somehow cluster in the cytosol in apparent defiance of diffusion. As discussed there, it clashes with the fact that most metabolites are found at a concentration close the Km values of the enzymes for which they are a substrate [20]. We revisit the issue here from the standpoint of metabolic connections.

As specific case, compartmentation was raised to explain the discordant results of measuring an intermediate in liver gluconeogenesis (Figure 4.1). The intermediate oxaloacetate in that pathway is the substrate for the rate-limiting step in the pathway, phosphoenol pyruvate carboxykinase (PEPCK). It was known that the concentration of oxaloacetate in liver is about 10 μM, whereas measurements of the Km for the enzyme were much higher, in the millimolar range [21–23]. The anomaly was rationalized by positing a compartmentation of oxaloacetate, suggesting it was in much greater concentration near the enzyme surface than in the bulk cytosol. However, as the later studies showed, this was simply an error in measurement. Oxaloacetate is chemically unstable in solution, especially in the presence of Mn^{2+}. This metal is routinely included in the enzymatic measurements. Once this is taken into consideration, the Km value and the properly measured intermediate concentrations match.

A second example of protein clusters having a metabolic role concerned the interaction of a glycolytic and gluconeogenic enzyme. Soling et al. [24] investigated a proposed physical interaction between PFK and FBPase. The question to resolve was whether, as had been previously suggested, the enzymes physically interact so that, for example, FBPase inhibits PFK, or whether it is merely the fact that FBPase, by removing fructose bisphosphate, decreases its *in vitro* activity. The latter was shown to be the case, based upon a number of manipulations, and no evidence for any protein-protein interaction was evident.

A recent examination of compartmentation from a global standpoint, based upon estimates of diffusion and enzymatic activity, demonstrated that many metabolites, such as ATP, lactate, and pyruvate, show behavior of a homogenous distribution within cell cytosol [25]. Nevertheless, in that same study, the authors suggested that two metabolites might in fact be localized: H^+ and Ca^{2+}. Yet, H^+ can travel at speeds greater than diffusion in a water solution (e.g. ref. [26]). The situation with Ca^{2+} is separate and discussed in detail in Chapter 7. It can be pointed out here that ATP, known to be uniformly distributed, as the authors also assert, can bind Ca^{2+} and contribute to a rapid and uniform cellular distribution [27].

Glycolysis exists entirely within the cytosolic space, occurs in virtually all cell types, and essentially every intermediate connects to another cellular pathway. This can be appreciated by examining a metabolic chart (e.g., ref. [28]). Thus, it is difficult to accept arguments that any of these intermediates can be compartmented as they must be shared with a wide number of separate pathways. In fact, all of the analytical tools used in studying the response of metabolic flows both in laboratory and clinical investigation depend on this fact.

There is nonetheless the belief that molecules within a single reactive water space like the cytosol are inhomogeneous. As evidence, electron micrographs of cells have been offered along with elaborate drawings reconstructing molecules within cells, depicting what is called "molecular crowding" [29]. Thus, proponents suggest that small molecules and large cannot really be evenly distributed but rather cluster in various portions of the cytosol. In fact, recent proposals have gone even further, suggesting that as a rule most enzymes bind other enzymes. This is the result of a method of analysis, the "two-hybrid cross," a genetic method to evaluate potential protein-protein interactions. By this standard, virtually all enzymes stick to other enzymes. Databases of the results have been accumulated (e.g. ref. [30]). The value of this information lies more in the area of long-term signaling, where multiple clusters are important for forming an active complex. It is only in its extension to metabolic interactions where the concept finds contradiction with established function. This is most evident when interconnected pathways are considered.

Proponents of enzyme complexes operating in metabolic pathways on a large scale argue that this would enhance metabolic efficiency [29,31]. The basis of this argument is theoretical calculation, buttressed by the idea that metabolites can more easily pass from one enzyme surface to the other, rather than having to visit the water space.

There are cases of enzyme complexes that we have already seen, such as the pyruvate dehydrogenase complex. However, the principle that emerges from others is that metabolic connections require release of intermediary metabolites, as was the case with aconitase and xanthine oxidase. Consider the mammalian fatty acid synthase enzyme. This contains seven reactions, all on the surface of a continuous protein. While we might take that for high efficiency, in many bacteria, this occurs as seven separate proteins [32]. It is difficult to argue that the mammalian form is somehow more streamlined or more efficient than bacteria; the latter are hard to beat in evolving the ultimate in efficiency, consider the organism's entire life span can be on the order of 30 minutes or even less.

Some compartmentation arguments combine unequal distribution of large and small molecules. For example, it has been known for some time that upon extraction of tissues, much hexokinase is found in the mitochondrial fraction [33]. The conclusion is not the prosaic one that it sticks under *in vitro* conditions upon cell rupture. Instead, it is widely concluded that it must be close to the source of ATP, that is, mitochondria. This explanation is, of course, itself *ad hoc*, and cannot account for the fact that there are hundreds of other kinases; they too would need ATP. Moreover, there is no evidence in favor of ATP gradients nor of a paucity of ATP for distant kinases.

More recently, an explanation has emerged that could explain the *in vitro* result in the context of a separate biological event: apoptosis [34]. It is now evident that in this process, a complex can form between proteins of the mitochondrial outer membrane, the inner membrane, and a prolyl isomerase of the matrix space. While different proteins from the cytosol can participate in this complex and successfully trigger apoptosis, one of them is hexokinase. It is known to bind to porin, the protein that forms a channel in the mitochondrial outer membrane and allows molecules up to several thousand in molecular weight entry into the inner membrane space. For apoptosis, this complex leads to the opening of the outer membrane not to small molecules, but to large ones, in particular, cytochrome c, which can participate in further steps of apoptosis with cytosolic proteases. Thus, the hexokinase binding is a reflection of a biological event, just not a metabolic one.[iii]

True compartmentation exists in the cell in terms of membrane delineated water spaces. The two metabolically most important of these are the cytosol and mitochondrial matrix. In most cases, this distinction has been ignored. It was evaluated in hepatocytes, making use of rapid separation techniques for these spaces [36]. For the most part, the results suggest that measured total concentrations of metabolites are approximately equal to cytosolic concentrations. This is due primarily to the much larger volume of the cytosol, about an order of magnitude greater than the mitochondrial matrix. Then too there is the fact that many exchange proteins present in the mitochondrial membrane are at near-equilibrium, so that large disparities in concentration are few.

4.5   DIFFUSION

Closely related to the notion of compartmentation as an explanation of metabolic anomalies is the idea that diffusion limitations exist to prevent mixing. This idea is an implicit part of the image of a "crowded" cytosol in large and small molecules are unable to distribute themselves evenly throughout a water space such as the cytosol. This is therefore an unstated assertion that the principle of diffusion does not apply within cells. Since we would be forced to accept this notion if we are to accept this compartmentation argument, it may be worthwhile to consider some principles of diffusion itself [37].

Diffusion was developed in the mid-nineteenth century by Fick, based upon heat convection ideas formulated by Fourier. Fick developed two laws that he experimentally verified, and these have been adopted ever since, along with some further developments to explain the forces behind movement of particles at random by Brown and Einstein. Fick's first law relates flux of material in a fixed time to a concentration gradient. To obtain the rate of change of material at a particular point, which is germane to the notion of apparent inhomogeneity at spots in the cytosol, Fick's second law is applicable. This relates flux to time changes in concentration. The equations themselves are not important even for proponents of compartmentation, but one feature is: Fick's second law is a non-steady-state law. Upon reflection, it is obvious that this is the case, since a change in concentration with time during the process is anathema to the very notion of a steady state.

Where does this leave us? If we employ a compartmentation argument for metabolism, we must accept a model that involves non-steady-state events, which rules out most known models for the analysis of events themselves. Equilibrium binding constants are out, as are kinetic constants such as Vmax and Km values, as well as inhibition constants. Thus, for any sort of metabolic analysis, the assumption of non-steady-state, which is inexorably tied to compartmentation, leaves us without any of the tools that we routinely use to quantify metabolic relationships.

4.6 METABOLITE BINDING

An issue that bears some resemblance to compartmentation is the finding that, in some cases, the measured, total values of metabolic intermediates is not the same as the solution concentration. It has been known for decades that several metabolites are largely bound—mostly to proteins—rather than free in solution [38]. Well-established examples are: ADP, AMP, NADH, FBP, and Ca^{2+}. There is no evidence that most metabolites are extensively bound, probably because for the most part metabolites interact with few specific binding sites. Thus, consider the intermediates of glycolysis. Just one of these—FBP—is likely to be appreciably bound and that is due to the low level of this molecule and thus its binding to the enzymes PFK and FBPase [39].

In more recent studies, the distinction between free and total concentrations is not commonly considered. For example, a survey of methods for measurements of large numbers of metabolites does not include this distinction [40]. In that study, most of the methodology detects only total values (e.g., mass spectroscopy, HPLC), although one of the methods, NMR, does measure strictly free metabolites (when attached, molecules are not free to tumble in solution, broadening their resonance lines and rendering them invisible).

As an example of using the NMR to measure metabolites, one study estimated the concentration of free ADP by utilizing the near-equilibrium property of two reactions in muscle: the creatine phosphokinase reaction, and the adenylate kinase reaction. With the assumption of a constant total creatine pool (the measurement used was ^{31}P-nmr, which will not therefore measure creatine), ADP was calculated to be in the tens of micromolar concentration [41]. It is also possible to make this estimate by direct measurement of total ATP, creatine, and creatine phosphate by tissue extraction and enzymatic endpoint analysis. Using the same equilibrium values (and near-equilibrium assumption), both the values of free ADP and AMP were determined, and found to lie in tens of micromolar and micromolar range, respectively [42]. A separate experiment has been performed that confirms the cellular values of free ADP. In this case, isolated mitochondria were incubated with various ratios of ATP/ADP*Pi, established by titration with purified preparations of mitochondrial complex V, which acts *in vitro* as a simple ATPase. These titrations revealed that the same concentration range found in free ADP measurement experiment fell in the titratable range of oxygen consumption by mitochondria [43,44].

4.7 CONSTANCY OF CELLULAR ATP

As discussed in Chapter 2, ATP concentration is little changed in cells during active metabolism. In one sense that may seem surprising since ATP must donate its phosphoryl group in hundreds of reactions. Thus, its actual concentration cannot be entirely unchanged, or it could never participate in high-energy phosphate donations[iv].

Let us consider how this comes about. Consider the reaction of ATP to ADP as the general phosphate donation reaction to provide energy for the cell:

$$ATP + acceptor \rightarrow ADP + acceptor\text{-}P \qquad (4.1)$$

where the notation -P is taken as the phosphoryl group, and a generalized acceptor is indicated, commonly a hydroxyl group. The drop in ATP according to this equation must equal the rise in ADP. The key issue is the absolute decrease in the ATP concentration and the absolute increase in ADP. The concentrations of ATP in cells is very high: between 2 and 5 mM. However, free concentration of ADP, as indicated in the prior section, is two orders of magnitude less, on the order of tens of micromolar. Thus, the decrease in ATP concentration is on the order of 1%. As pointed out recently, the Km values for cellular kinases are usually far below the concentration of ATP; this provides a separate explanation for why ATP concentrations do not regulate reactions in which they participate [45].

It is also the case that, as a mobile cofactor, ADP must be utilized in the steady state at the same rate at which it is generated. The pathways for replenishing ATP are glycolysis and oxidative phosphorylation. In some cells, such as muscle and nerve, there is the additional use of the phosphocreatine kinase reaction, but the principle is the same. Since the amount of breakdown of ATP is about 1%, there is no significant change in ATP concentration, either in terms of the ability of the cell to respond to changes or the ability of the analyst to detect it.

Nonetheless, in some experimental settings it is possible to observe large drops in the concentration of ATP. Positioned against the overwhelming amount of data from other studies in cells, tissues and *in vivo*, changes in ATP concentration are not to be taken as an important control element. A full treatment of the situation is available in a monograph published in 1983 [46].

The third major adenine nucleotide is AMP, which is also mostly a bound nucleotide. Its free concentration is even lower than ADP. The nucleotides enjoy a near-equilibrium through the adenylate kinase reactions:

$$ATP + AMP \rightleftharpoons 2ADP \qquad (4.2)$$

Considering the fact that this equilibrium constant is approximately 1, the near-equilibrium reaction provides an estimate of AMP that shows it to be about an order of magnitude below that of ADP [42]. This provides AMP at a concentration at which it can modulate certain reactions in the cell, such as the now-widely studied AMP-activated kinase, and the earlier discovered allosteric actions of glycogen phosphorylase and phosphofructokinase. This reaction also provides a simple explanation for how a rise in ADP concentration can lead to a rise in AMP without a change in ATP. The reason for the lack of change in ATP concentration is the same as before: the relative levels of free nucleotide mean that large percentage changes in ADP and AMP lead to negligible percentage changes in ATP.

4.8   ENERGY CHARGE

A phrase invented by Atkinson to account for overall cellular regulation is the "energy charge" [47]. The idea was an early attempt to rationalize changes in the adenine nucleotides and establish a means for regulation of the many enzymes that used them as substrate, and might be modified by them. The concept was simple: like a battery, cell energy would be high if there was more ATP and lower if there was more ADP and AMP. Since ADP would still contain (potentially) some bond energy, it was also counted as a potential donor. The original ratio was set as

$$\text{Energy charge} = ([\text{ATP}] + \tfrac{1}{2}[\text{ADP}]) / ([\text{ATP}] + [\text{ADP}] + [\text{AMP}]) \qquad (4.3)$$

This could theoretically run from zero to 1, and the number would be a readout of energy. Subsequent studies of energy charge, however, invariably found that measures of these nucleotides always led to a value of about 0.8, irrespective of cellular state (e.g., ref. [48]). Part of the problem is likely due to the fact that total, rather than free values, of ADP and AMP are typically used for this ratio. In fact, viewed from the standpoint of the actual changes that take place in free adenine nucleotides, since

$$[\text{ATP}] \gg [\text{ADP}] > [\text{AMP}] \qquad (4.4)$$

The numerator and denominator of Equation 4.3 should be essentially just [ATP], so the energy charge should always be very close to 1.

Curiously, the revival of energy charge has been identified with a new ratio of [AMP]/[ATP], corresponding to the allosteric actions of the two nucleotides on the AMPK enzyme in *in vitro* assays [49]. Putting aside the fact that the name was usurped for a new ratio, the new energy charge also uses total rather than free concentrations; in this case, just values for AMP.

As an example of citation of the energy charge, a review of nucleotide metabolism focusing on the AMP deaminase [27] the authors suggested that this enzyme acts to "stabilize the energy charge" in muscle as part of the purine nucleotide cycle (see Chapter 10). Thus, some believe this is a ratio that *needs* to be constant, although, like the energy charge itself, the idea is not connected to a broad view of metabolism. It is time to once again retire the "energy charge."

4.9   REDOX COFACTORS: NAD, NADH, NADP, AND NADPH

Cells have multiple redox states. The global ratios of NAD/NADH and NADP/NADPH are maintained at very different potentials in the cell. Their standard redox potentials are the same, reflecting identical chemical structure in the nicotinamide ring. In cells, having two separate nicotinamide redox potentials allows reactions with very different redox potentials, such as the NAD-linked lactate DH and the NADP-linked glucose-6-P DH, to proceed simultaneously. The mitochondrial matrix maintains a different NAD/NADH as well as a different NADP/NADPH ratio.

Beyond these, there are redox states not considered in further detail here, such as the Q/QH$_2$ that exists entirely inside the inner mitochondrial membrane, and the various reactive oxygen species. The ubiquinone pair is involved in collecting reducing equivalents from various donors, as discussed in the prior chapter, and delivering them to Complex III. Beyond this mobile cofactor role, a possible role of Q and QH$_2$ in the control of respiration is not well established. There are studies of this cofactor, such as the correlation of levels to disease states [50,51], and how it may enable a "superstructure" of multiple membrane complexes [52]. As discussed in the context of compartmentation, a true complex would vitiate the ability of Q/QH$_2$ to serve as a mobile cofactor. Studies of reactive oxygen species invariably involve Q, as free radical donation to oxygen at sites connected to Q is the origin of the radicals. While this is of importance to conditions of oxidative stress, it is not considered further in the present work.

In the cytosol, the ratio of NAD/NADH is about 100,000 times more oxidized than the NADP/NADPH [53]. Broadly speaking, the NAD pair extracts electrons from molecules that are oxidized to provide cellular energy, while NADPH donates electrons for biosynthesis. This generalization explains, for example, why glycolysis uses NAD whereas lipid biosynthesis uses NADPH. Moreover, it easily explains an otherwise-puzzling problem of how two nearly identical molecules with identical standard redox potentials can participate in reactions in opposing directions in the same reactive space.

Mitochondria maintain NAD/NADH distinct from NADP/NADPH through the energy-linked transhydrogenase. In this space, two enzymes are known to use both nucleotide pairs: the isocitrate DH, and glutamate DH. The means of doing so, however, is distinct. As discussed in the prior chapter, isocitrate DH is a mIRR reaction. There are two separate isozymes of isocitrate DH. One uses NAD/NADH and is involved in the Krebs cycle. The other uses NADP/NADPH and is involved in the reverse reaction, the formation of isocitrate. The other reaction, glutamate DH, is known to be a NEQ reaction, although this is established only for the NAD/NADH pair. It is known that the enzyme *in vitro* can utilize the NADP cofactor [54], but it is not clear how this impacts cellular metabolism.

4.10   IS NAD A REGULATOR?

4.10.1   SIRT1

An enzyme discovered in yeast that leads to gene silencing was named *sir* (silent information regulator). Its activity is that of a protein deacetylase. However, unlike other known deacetylases, it transfers the acetyl group to NAD (Figure 4.12) rather

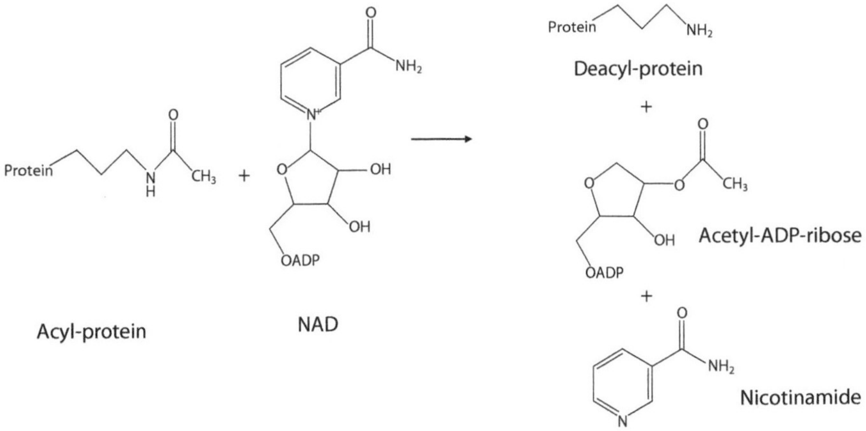

FIGURE 4.12 SIRT1 reaction. NAD rather than water is used in protein deacetylation catalyzed by SIRT1.

than water (i.e., in place of hydrolysis). A mammalian analog, known as SIRT1, is widely believed to be connected to the NAD/NADH redox state since NAD is a substrate of the SIRT1 reaction. It has already been pointed out that this is likely a misconception [55], based on the very high concentration of NAD. It is thus unlikely that it can be substantially changed in concentration in order to be considered a regulator.

The reasons behind the interest in NAD as a SIRT1 regulator stem in part from the fact that it seems like the cell is using important resources – NAD – for a task that could be accomplished simply by water hydrolysis. This was the argument advanced in favor of NAD as regulator by Walsh [56]. It was also suggested the hydrolysis is "thermodynamically favored."

It is useful to consider these views, in part because the idea that the involvement of NAD as a substrate in SIRT1 is related to the cellular NAD/NADH redox state is a common one [57–59]. If true, it would mean that SIRT1 is either controlled by the cellular cytosolic NAD/NADH, or that the cellular redox state is controlled by SIRT1.

While the NAD/NADH ratio varies depending upon the reduction level of the cellular substrates, there will always be a very large concentration of NAD; in fact, it is of similar order of magnitude to ATP. We know that ATP is used in other roles than that of mobile cofactor. For example, it is used to generate cyclic AMP. There are no theories that adenylate cyclase is regulated by ATP, but the same logic would suggest that ATP is a regulator of cAMP formation!

Another view might be that the ratio of NAD/NADH is communicating to the deacetylase, and thus a cellular communication is revealed by the mere presence of NAD as a substrate for the enzyme. However, on the face of it, this argument cannot be correct, as the ratio largely changes as a result of NADH changes. SIRT1 has only NAD as its substrate. Not only that, the ratio of NAD/NADH is part of a large number of near-equilibrium enzymes, and that ratio can change very rapidly. As stated above, the purpose is really to move electrons from one reaction to the next, extracting them from substrates as they are oxidized. The rapidity with which this ratio can change far outstrips the rapidity with which a protein can be deacetylated, so that this communication too seems unlikely.

There is another issue in the thought experiment trying to account for NAD as a substrate when other similar enzymes merely transfer the acetyl group to water. That is the comment that simple hydrolysis is thermodynamically favorable [56]. This does not take into account how thermodynamics applies to metabolic pathways. The fact is that every reaction of a pathway that proceeds in cells is thermodynamically favorable. If it were otherwise, the reactions would simply not proceed.

We are left with an interesting mystery: why indeed should the cell use an enzyme like SIRT1 when it could use a simple one? I admit to having no answers either, although I could also come up with a few speculations. For example, perhaps the use of NAD as part of the reaction produced a reaction that has a more favorable equilibrium position than simple hydrolysis for that reaction under metabolic conditions. While this is also a speculation, it does have the virtue of not contradicting established tenets of metabolism.

4.10.2   PARP

A second suggestion that NAD could be a regulator is in for reaction known as PARP, or the polyadenylation reaction that incorporates adenine-ribose into proteins during DNA repair. In fact, PARP and SIRT are considered new pathways for cellular regulation by NAD [60]. In the case of PARP, however, the situation is the opposite of that of SIRT: in this case, a substantial depletion of NAD can occur, leading to cell death [61]. This is not a means of cellular regulation by NAD or response to the NAD/NADH ratio, but rather an avenue to necrosis. A variety of conditions apart from those involving NAD alter the activity state of PARP [62].

In an experimental test of the NAD hypothesis, increasing the formation of NAD has been advanced, both by provision of the biosynthetic precursor nicotinamide, as well as by genetic insertion of the gene nmat (nicotinamide mononucleotide adenylyltransferase) [63]. However, these manipulations are themselves complicated. For example, nicotinamide itself inhibits PARP [64]. The genetic manipulation does produce intriguing results, but this also should be considered cautiously. Another name for this enzyme (indicating its other role) is visfatin [65], a hormone released by fat cells having widespread metabolic effects. This may complicate the interpretation of this genetic addition. As another approach, a tissue-specific induction of nmat in muscle such that NAD increases 50% did not alter mitochondrial metabolism [66].

4.10.3   LACTATE DH AS REGULATORY ENZYME

An old notion of lactate DH as a regulatory enzyme was a textbook speculation. The idea was that isoforms in heart and liver might enable the enzyme to run in different directions in those tissues. As the enzyme is NEQ in both cell types, this directionality idea has no basis. While it has faded from most textbooks, the idea that lactate DH might be regulatory has reappeared. In a recent article examining the contribution of glycolysis and mitochondrial glucose metabolism, lactate DH was genetically ablated and the consequences examined [67]. However, the study did not consider that, while this has a dramatic and interesting effect on the metabolism of the cell, the reason for it is not simply that glycolysis is diminished and oxidative phosphorylation turned on. Rather, the inability to form lactate means that the redox shuttle becomes the only way of removing cytosolic NADH for turnover at glyceraldehyde phosphate DH. This is one of the fundamental pathway connections explored in this chapter. Perhaps not surprisingly, the investigators concluded that glyceraldehyde phosphate DH is the key regulatory enzyme of glycolysis.

4.11 TRANSPORT

A final category of connections to consider are the membrane transporters. As these are effectively enzymes catalyzing a space translocation, I discuss this more extensively in the following chapter on kinetics. Here transporters are viewed globally for how they join pathways.

We have already seen that what we call a pathway spans membranes. Even glycolysis may be considered to require a plasma membrane transporter to admit the pathway substrate glucose and another to allow exit of the pathway products lactate and pyruvate. Gluconeogenesis spans multiple membranes: mitochondria, endoplasmic reticulum, and plasma membrane.

Transporters influence the pathway by separating portions between spaces that allow limited connections (i.e., across a semipermeable membrane). For example, with separate redox states, the mitochondrial portion of gluconeogenesis communicates with a distinct NAD/NADH ratio from the cytosolic portion. This means that cofactor balance, while still required, communicates with a different redox poise. Another way to think of it is that all of the mitochondrial intermediates are compartmented from the cytosolic ones, so that they share only the limited number of metabolites that have transporters.

A further consideration is the issue of whether the transport step is itself NEQ or mIRR. For glucose transport across the plasma membrane, this could be NEQ (as in liver and a few other tissues) or mIRR (as in most tissues). As discussed in the following chapter, in most cases studied, mitochondrial transport is NEQ. One of the cases of mIRR transport is the glutamate/aspartate exchanger, which exports mitochondrial aspartate. This exchanger is important in many cells as part of the malate-aspartate shuttle (Figure 3.2) to indirectly move NADH into the mitochondria, and for gluconeogenesis to indirectly move oxaloacetate into the cytosol. The exchange is electrogenic. Aspartate emerges from the mitochondria with a charge of -1, coupled to glutamate entry with a charge of zero (Figure 4.13). It is interesting that one study discussed this exchanger as "near-equilibrium" [68]. The context of that classification, however, was that the exchange was that it was in near-equilibrium with the potential itself. This is different from the use of NEQ in a metabolic context. In the latter case, the reaction is mIRR, since it is never reversed physiologically as the membrane potential is never collapsed. Even under conditions of lowered membrane potential (say, an increase in an uncoupling protein), an actual reversal of the transporter would require a positive membrane potential.

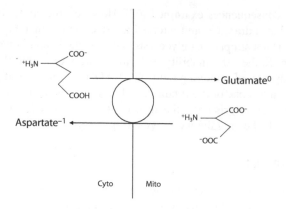

FIGURE 4.13 Glutamate/aspartate exchanger. The exchanger, present in the mitochondrial inner membrane, exports the -1 charged aspartate and imports the zero-charged glutamate.

REFERENCES

1. Brown, M. S., and Goldstein, J. L. (2008) Selective versus total insulin resistance: A pathogenic paradox. *Cell Metab.* 7, 95–96.
2. Feng, D., Youn, D. Y., Zhao, X., Gao, Y., Quinn, W. J., 3rd, Xiaoli, A. M., Sun, Y., Birnbaum, M. J., Pessin, J. E., and Yang, F. (2015) mTORC1 down-regulates cyclin-dependent kinase 8 (CDK8) and cyclin C (CycC). *PloS one* 10, e0126240.
3. Johnson, J. L., and Bagby, G. J. (1988) Gluconeogenic pathway in liver and muscle glycogen synthesis after exercise. *J. Appl. Physiol.* 64, 1591–1599.
4. Newgard, C. B., Hirsch, L. J., Foster, D. W., and McGarry, J. D. (1983) Studies on the mechanism by which exogenous glucose is converted into liver glycogen in the rat. A direct or an indirect pathway? *J. Biol. Chem.* 258, 8046–8052.
5. Shulman, R. G., and Rothman, D. L. (2001) The "glycogen shunt" in exercising muscle: A role for glycogen in muscle energetics and fatigue. *Proc. Natl. Acad. Sci.* 98, 457–461.
6. Ord, M. G., and Stocken, L. A. (1998) Aspects of carbohydrate, oxidation, electron transfer, and oxidative phosphorylation. in *Early Adventures in Biochemistry* (Ord, M. G., and Stocken, L. A. eds.), JAI. p. 79.
7. Ogston, A. G. (1948) Interpretation of experiments on metabolic processes, using isotopic tracer elements. *Nature* 162, 963.
8. Mesecar, A. D., and Koshland, D. E. (2000) Structural biology: A new model for protein stereospecificity. *Nature* 403, 614–615.
9. Shemin, D. (1946) Conversion of e-serine to glycine. *J. Biol. Chem.* 162, 297–307.
10. Lloyd, S. J., Lauble, H., Prasad, G. S., and Stout, C. D. (1999) The mechanism of aconitase: 1.8 A resolution crystal structure of the S642a:citrate complex. *Protein Sci.* 8, 2655–2662.
11. Schloss, J. V., Emptage, M. H., and Cleland, W. W. (1984) pH profiles and isotope effects for aconitases from Saccharomycopsis lipolytica, beef heart, and beef liver. alpha-Methyl-cis-aconitate and threo-Ds-alpha-methylisocitrate as substrates. *Biochemistry (Mosc).* 23, 4572–4580.
12. Rose, I. A., and O'Connell, E. L. (1967) Mechanism of aconitase action. I. The hydrogen transfer reaction. *J. Biol. Chem.* 242, 1870–1879.
13. Beinert, H., Kennedy, M. C., and Stout, C. D. (1996) Aconitase as iron–sulfur protein, enzyme, and iron-regulatory protein. *Chem. Rev.* 96, 2335–2374.

14. Strelko, C. L., Lu, W., Dufort, F. J., Seyfried, T. N., Chiles, T. C., Rabinowitz, J. D., and Roberts, M. F. (2011) Itaconic acid is a mammalian metabolite induced during macrophage activation. *J. Am. Chem. Soc.* 133, 16386–16389.

15. Sasikaran, J., Ziemski, M., Zadora, P. K., Fleig, A., and Berg, I. A. (2014) Bacterial itaconate degradation promotes pathogenicity. *Nat. Chem. Biol.* 10, 371–377.

16. Michelucci, A., Cordes, T., Ghelfi, J., Pailot, A., Reiling, N., Goldmann, O., Binz, T., Wegner, A., Tallam, A., Rausell, A., Buttini, M., Linster, C. L., Medina, E., Balling, R., and Hiller, K. (2013) Immune-responsive gene 1 protein links metabolism to immunity by catalyzing itaconic acid production. *Proc. Natl. Acad. Sci. U. S. A.* 110, 7820–7825.

17. Tang, M., Liu, B. J., Wang, S. Q., Xu, Y., Han, P., Li, P. C., Wang, Z. J., Song, N. H., Zhang, W., and Yin, C. J. (2014) The role of mitochondrial aconitate (ACO2) in human sperm motility. *Syst. Biol. Reprod. Med.* 60, 251–256.

18. Cao, H., Hall, J., and Hille, R. (2014) Substrate orientation and specificity in xanthine oxidase: Crystal structures of the enzyme in complex with indole-3-acetaldehyde and guanine. *Biochemistry (Mosc).* 53, 533–541.

19. Nishino, T., and Okamoto, K. (2015) Mechanistic insights into xanthine oxidoreductase from development studies of candidate drugs to treat hyperuricemia and gout. *J. Biol. Inorg. Chem.* 20, 195–207.

20. Cleland, W. W. (1970) Steady state kinetics. in *The Enzymes* (Boyer, P. D. ed.), Academic Press, New York. pp. 1–65.

21. Ballard, F. J. (1970) Kinetic studies with cytosol and mitochondrial phosphoenol-pyruvate carboxykinases. *Biochem. J.* 120, 809–814.

22. Ochs, R. S., Hanson, R. W., and Hall, J. (1985) *Metabolic Regulation*, Elsevier, Amsterdam.

23. Siess, E. A., Brocks, D. G., and Wieland, O. H. (1976) A sensitive and simple method for the study of oxaloacetate compartmentation in isolated hepatocytes. *FEBS Lett.* 70, 51–55.

24. Soling, H. D., Bernhard, G., Kuhn, A., and Luck, H. J. (1977) Inhibition of phosphofructokinase by fructose 1 6-diphosphatase in mammalian systems: Protein-protein interaction or fructose 1 6-diphosphate trapping? *Arch. Biochem. Biophys.* 182, 563–572.

25. Barros, L. F., and Martinez, C. (2007) An enquiry into metabolite domains. *Biophys. J.* 92, 3878–3884.

26. Mai, B. K., Park, K., Duong, M. P., and Kim, Y. (2013) Proton transfer dependence on hydrogen-bonding of solvent to the water wire: A theoretical study. *J. Phys. Chem. B* 117, 307–315.

27. Zabielska, M. A., Borkowski, T., Słominska, E. M., and Smolenski, R. T.(2015) Inhibition of AMP deaminase as therapeutic target in cardiovascular pathology. *Pharmacol. Rep.* 67, 682–688.

28. (2017) Roche Biochemical Pathways. Available at http://web.expasy.org/pathways.

29. Ellis, R. J. (2001) Macromolecular crowding: Obvious but underappreciated. *Trends Biochem. Sci.* 26, 597–604.

30. Ceol, A., Chatr Aryamontri, A., Licata, L., Peluso, D., Briganti, L., Perfetto, L., Castagnoli, L., and Cesareni, G. (2010) MINT, the molecular interaction database: 2009 update. *Nucleic Acids Res.* 38, D532–539.

31. Srere, P. A. (1987) Complexes of sequential metabolic enzymes. *Annu. Rev. Biochem.* 56, 89–124.

32. Hardie, D. G., and McCarthy, A. D. (1985) Fatty acid synthase: Probing the structure of a multifunctional protein by limited proteolysis. *Biochem. Soc. Trans.* 13, 297–299.

33. Rose, I. A., and Warms, J. V. (1967) Mitochondrial hexokinase. Release, rebinding, and location. *J. Biol. Chem.* 242, 1635–1645.

34. Izzo, V., Bravo-San Pedro, J. M., Sica, V., Kroemer, G., and Galluzzi, L. (2016) Mitochondrial permeability transition: New findings and persisting uncertainties. *Trends Cell Biol.*

35. Hille, B. (2001) Voltage-gated calcium channels. in *Ion Channels of Excitable Membrane*, Sinauer Associates Inc., Sunderland, MA. pp. 124–126.

36. Zuurendonk, P. F., Akerboom, T. P. M., and Tager, J. M. (1976) Metabolite distribution in isolated rat-liver cells and equilibrium relationships of mitochondrial and cytosolic dehydrogenases. in *Use of Isolated Liver Cells and Kidney Tubules in Metabolic Studies*, Elsevier, New York. pp. 17–27.

37. Cussler, E. L. (2009) *Diffusion: Mass Transfer in Fluid Systems*, 3rd ed., Cambridge University Press, Cambridge; New York.

38. Ottaway, J. H. (1979) Sequestration of metabolites: Insights into metabolic control. *Biochem. Soc. Trans.* 7, 1161–1167.

39. Veech, R. L., Raijman, L., Dalziel, K., and Krebs, H. A. (1969) Disequilibrium in the triose phosphate isomerase system in rat liver. *Biochem. J.* 115, 837–842.

40. Wishart, D. S. (2007) Current progress in computational metabolomics. *Brief Bioinform.* 8, 279–293.

41. Hancock, C. R., Brault, J. J., Wiseman, R. W., Terjung, R. L., and Meyer, R. A. (2005) 31P-NMR observation of free ADP during fatiguing, repetitive contractions of murine skeletal muscle lacking AK11. *AJP—Cell Physiol.* 288, C1298–C1304.

42. Ronner, P., Friel, E., Czerniawski, K., and Frankle, S. (1999) Luminometric assays of ATP, phosphocreatine, and creatine for estimation of free ADP and free AMP. *Anal. Biochem.* 275, 208–216.

43. Davis, E. J., and Davis-Van Thienen, W. I. A. (1984) Rate control of phosphorylation-coupled respiration by rat liver mitochondria. *Arch. Biochem. Biophys.* 233, 573–581.

44. Jong, Y. S. A., and Davis, E. J. (1983) Reconstruction of steady state in cell-free systems. Interactions between glycolysis and mitochondrial metabolism: Regulation of the redox and phosphorylation states. *Arch. Biochem. Biophys.* 222, 179–191.

45. Kinnaird, A., Zhao, S., Wellen, K. E., and Michelakis, E. D. (2016) Metabolic control of epigenetics in cancer. *Nat. Rev. Cancer* 16, 694–707.

46. Bridger, W. A., and Henderson, J. F. (1983) *Cell ATP*, Wiley, New York.

47. Atkinson, D. E. (1977) *Cellular Energy Metabolism and Its Regulation*, Academic Press, New York.

48. Mayr, G. W., and Thieleczek, R. (1991) Masses of inositol phosphates in resting and tetanically stimulated vertebrate skeletal muscles. *Biochem. J.* 280 (Pt 3), 631–640.

49. Hardie, D. G., and Hawley, S. A. (2001) AMP-activated protein kinase: The energy charge hypothesis revisited. *Bioessays* 23, 1112–1119.

50. Sharma, A., Fonarow, G. C., Butler, J., Ezekowitz, J. A., and Felker, G. M. (2016) Coenzyme Q10 and heart failure: A state-of-the-art review. *Circulation. Heart Fail.* 9, e002639.

51. Fischer, A., Onur, S., Niklowitz, P., Menke, T., Laudes, M., and Doring, F. (2016) Coenzyme Q10 redox state predicts the concentration of c-reactive protein in a large caucasian cohort. *Biofactors* 42, 268–276.

52. Alcazar-Fabra, M., Navas, P., and Brea-Calvo, G. (2016) Coenzyme Q biosynthesis and its role in the respiratory chain structure. *Biochim. Biophys. Acta* 1857, 1073–1078.

53. Reinke, L. A., Thurman, R. G., and Kauffman, F. C. (1979) Oxidation-reduction state of free NADP during mixed-function oxidation in perfused rat livers. Evaluation of the assumptions of near equilibrium by comparisons of surface fluorescence changes and calculated NADP:NADPH ratios. *Biochem. Pharm.* 28, 2381–2387.

54. Hoek, J. B., Ernster, L., De Haan, E. J., and Tager, J. M. (1974) The nicotinamide nucleotide specificity of glutamate dehydrogenase in intact rat-liver mitochondria. *Biochim. Biophys. Acta* 333, 546–559.

55. Nicholls, D. G., and Ferguson, S. J. (2013) *Bioenergetics*, pp. 1 online resource (xiv 491 pages).

56. Walsh, C. (2006) *Posttranslational Modification of Proteins: Expanding Nature's Inventory*, Roberts and Co. Publishers, Englewood, CO.

57. McGettrick, A. F., and O'Neill, L. A. J. (2013) How metabolism generates signals during innate immunity and inflammation. *J. Biol. Chem.* 288, 22893–22898.

58. Marcu, R., Wiczer, B. M., Neeley, C. K., and Hawkins, B. J. (2014) Mitochondrial matrix Ca^{2+} accumulation regulates cytosolic NAD+/NADH metabolism, protein acetylation and sirtuins expression. *Mol. Cell. Biol.*

59. Nogueiras, R., Habegger, K. M., Chaudhary, N., Finan, B., Banks, A. S., Dietrich, M. O., Horvath, T. L., Sinclair, D. A., Pfluger, P. T., and Tschöp, M. H. (2012) Sirtuin 1 and Sirtuin 3: Physiological modulators of metabolism. *Physiol. Rev.* 92, 1479–1514.

60. Faraone-Mennella, M. R. (2015) A new facet of ADP-ribosylation reactions: SIRTs and PARPs interplay. *Frontiers Biosci.* 20, 458–473.

61. Moroni, F. (2008) Poly(ADP-ribose)polymerase 1 (PARP-1) and postischemic brain damage. *Curr. Opin. Pharmacol.* 8, 96–103.

62. Grossmann, M. E., Nkhata, K. J., Mizuno, N. K., Ray, A., and Cleary, M. P. (2008) Effects of adiponectin on breast cancer cell growth and signaling. *Br. J. Cancer* 98, 370–379

63. Williams, P. A., Harder, J. M., Foxworth, N. E., Cochran, K. E., Philip, V. M., Porciatti, V., Smithies, O., and John, S. W. (2017) Vitamin B3 modulates mitochondrial vulnerability and prevents glaucoma in aged mice. *Science* 355, 756–760.

64. Bai, P., Cantó, C., Oudart, H., Brunyánszki, A., Cen, Y., Thomas, C., Yamamoto, H., Huber, A., Kiss, B., Houtkooper, Riekelt H., Schoonjans, K., Schreiber, V., Sauve, Anthony A., Menissier-de Murcia, J., and Auwerx, J. (2011) PARP-1 Inhibition Increases Mitochondrial Metabolism through SIRT1 Activation. *Cell Metab.* 13, 461–468.

65. Chang, Y. H., Chang, D. M., Lin, K. C., Shin, S. J., and Lee, Y. J. (2011) Visfatin in overweight/obesity, type 2 diabetes mellitus, insulin resistance, metabolic syndrome and cardiovascular diseases: A meta-analysis and systemic review. *Diabetes. Metab. Res. Rev.* 27, 515–527.

66. Frederick, D. W., Davis, J. G., Davila, A., Jr., Agarwal, B., Michan, S., Puchowicz, M. A., Nakamaru-Ogiso, E., and Baur, J. A. (2014) Increasing NAD synthesis in muscle via nicotinamide phosphoribosyltransferase is not sufficient to promote oxidative metabolism. *J. Biol. Chem.*

67. Peng, M., Yin, N., Chhangawala, S., Xu, K., Leslie, C. S., and Li, M. O. (2016) Aerobic glycolysis promotes T helper 1 cell differentiation through an epigenetic mechanism. *Science* 354, 481–484.

68. Kauppinen, R. A., Hiltunen, J. K., and Hassinen, I. E. (1983) Mitochondrial membrane potential, transmembrane difference in the NAD+ redox potential and the equilibrium of the glutamate-aspartate translocase in the isolated perfused rat heart. *Biochim. Biophys. Acta (BBA)—Bioenerg.* 725, 425–433.

69. Sundaresan, V., and Abrol, R. (2005) Biological chiral recognition: The substrate's perspective. *Chirality* 17, 530–539.

70. Testa, B., Vistoli, G., Pedretti, A., and Caldwell, J. (2013) Organic stereochemistry. *Helvetica Chimica Acta* 96, 747–798.

ENDNOTES

i. This was further extended by Sundaresan and Abrol [69] to even more binding points if there are more stereocenters to consider: an N + 2 binding sites depending on the number of chiral centers N that are produced. More subtleties of prochirality based on both the Ogston model and Koshland extension have been published and have significance largely for pharmacological matters [70].

ii. Ochs, R.S. and Tanaji, T.T. 2017. Unpublished observations.

iii. A more common recent name for porin is VDAC, or voltage-dependent anion channel. This name suggests that ions might have difficulty crossing the outer membrane, which is unexpected, since even molecules with molecular weights in the thousands can pass! In fact, this is not an ion channel in cells, but *in vitro* it can conduct anions, and opens at 0 mV [35]. This is convenient, as there is no voltage across this membrane under any known circumstances.

iv. Here I have reverted to the colloquialism of "phosphate donation" rather than the more strictly correct "phosphoryl transfer." The fact is, the product is a phosphorylated compound, and there is little doubt about what the reaction is. A distinct, more egregious error is to call the removal reactions "hydrolysis" when in fact very few reactions in cells are really hydrolysis reactions. The latter are more common in extracellular enzymatic reactions, as no other cofactors are available.

5 Enzymes and Their Inhibition

In prior chapters, some aspects of enzymology have already been encountered, including discussions of Km and Vmax values. Here, I present a treatment of enzyme kinetics focusing on inhibition, which expands the usual treatment that involves Vmax, Km, and double-reciprocal forms, using some analysis concepts derived from enzymologists (e.g., refs. [1,2]). One concept adopted from the latter is the introduction of V/K in place of Km for the analysis of enzyme inhibition. Without this substitution, a clear understanding of enzyme inhibition is not possible.

It has been pointed out that an understanding of the mode of inhibition has value in understanding the effects on a pathway. For example, Berridge discussed the effects of Li^+ as an uncompetitive inositol phosphatase inhibitor compared to competitive inhibition as the former has an action that cannot be overcome by substrate accumulation [3]. As a second example, the use of transition state analysis allows understanding of enzyme inhibitors that are not of the competitive type [4].

5.1 KINETIC ANALYSIS OF ENZYMES IN A METABOLIC CONTEXT

When measuring enzyme activity *in vitro* for the purpose of a kinetic analysis, it is necessary to use an unvarying portion of the time course for enzyme action (Figure 5.1). By universal agreement, the initial portion of the reaction progress curve is taken for the velocity measure, extrapolated to the condition of zero substrate. This ensures a velocity uncontaminated by product formation, and provides an unambiguous measure of enzyme activity. It should also be recognized that these measurements are performed in isolation, and that the resulting velocity is irreversible. Thus, even enzymes that catalyze near-equilibrium reactions in cells are irreversible in kinetic analysis. In this measurement system, enzymes may be subject to regulators that are not relevant in intact cells. Even enzymes known to be control points in cells, such as phosphofructokinase, exhibit properties under these conditions that may not be relevant in an intact cell. For example, interactions with other proteins such as fructose bisphosphatase, as indicated in the prior chapter [5], or of other regulators (such as its substrate ATP), have no correlate in cellular physiology.

As an example of an *in vitro* result involving a near-equilibrium enzyme, aldolase has been suggested to be regulated by actin [6]. In the cited report, the term "housekeeping enzyme" was used to indicate that actin is not itself involved in regulation. While this is likely the case, the terminology is unfortunate, not simply because it is colloquial[i] but because even enzymes normally present in large amounts sufficient to achieve near-equilibrium can change in amount under different circumstances [7,8].

It is the metabolically irreversible steps that are of consequence for control of metabolic reactions, and also the important drug targets, largely because they are

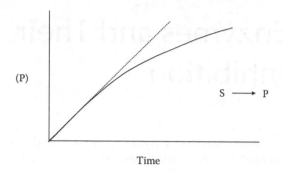

FIGURE 5.1   Reaction progress curve. Enzymatic (or nonenzymatic) time course of product formation.

the ones most sensitive to external conditions that regulate pathway flux. Of course, it is possible to inhibit any step of a pathway so long as enough inhibitor is present; the problem is that the large amounts of inhibitor thus required may exert additional effects on other steps. With these provisos, let us consider how to think about enzyme inhibition.

5.2   REVERSIBLE INHIBITION

The definition of a reversible inhibitor is that such a molecule exists in equilibrium with a *form* of the enzyme. During the catalytic cycle, an enzyme exists initially in its *free* form, followed by others, such as the enzyme-substrate form. In the simplest enzyme mechanism there are just two forms: E and ES,[ii] in the simplest mechanism:

$$E + S \underset{k_r}{\overset{k_f}{\rightleftharpoons}} ES \xrightarrow{k_{cat}} E + P \tag{5.1}$$

Equation 5.1 is a model of enzyme action, and represents a steady-state, as evident by the single arrow leading to product formation. The corresponding Michaelis–Menten equation

$$v_0 = Vmax[S]/(Km + [S]) \tag{5.2}$$

uses the steady state assumption that [ES] is constant with time. The two kinetic constants are

$$K_m = (k_r + k_{cat})/k_f \tag{5.3}$$

and

$$Vmax = k_{cat}[E]_{tot} \tag{5.4}$$

In previous sections in this work and in most biochemical interpretations, the Km is simplified to equal the equilibrium (dissociation) constant for ES:

$$Km = K_s = k_r / k_f \tag{5.5}$$

Notice that this assumes that the rate constant-producing product (k_{cat}) is much less than the rate constant leading to dissociation of ES, or

$$k_r \gg k_{cat} \tag{5.6}$$

This assumption is usually true, although it is not universally the case (e.g., ref. [9]). The assumption allows the comparison of Km values with existing substrate concentrations as mentioned previously.

However, using this assumption in the analysis of enzyme inhibition is a mistake. The change in behavior of Km in the different forms of enzyme inhibition *cannot* be analyzed as if this kinetic constant is the same as an equilibrium dissociation constant. Why not? Because inhibition forms, depending on whether the inhibitor binds the free enzyme, the bound enzyme, or both can cause an increase, a decrease, or no effect on the Km value. In none of these instances is the affinity of the enzyme for its substrate the correct interpretation.

In just one case, the change in Km seems to fit with the dissociation constant interpretation: competitive inhibition. Here, the inhibitor will cause an increase in the apparent Km, and it is commonly (and incorrectly) concluded that the affinity of the substrate has decreased because of the inhibitor. It is true that the relative reactivity of the enzyme with substrate has decreased, but that is not the same thing. What follows is a means of analysis for reversible enzyme inhibition that will enable a clear interpretation of the events, and allows an understanding of all forms of reversible inhibition, competitive or otherwise.

5.2.1 Competitive Inhibition

A competitive inhibitor (**I**) establishes an equilibrium with the free form of the enzyme, so that the equation

$$EI = E + I \tag{5.7}$$

is added to Equation 5.1. This has a dissociation constant written as K_i. The overall mechanism (Figure 5.2) shows that both the substrate and the free enzyme bind the inhibitor **I**. As [S] increases, less of the total enzyme is in the **E** form. Since **I** binds exclusively with **E**, inhibition is diminished. From this it is apparent that inhibition is greatest when substrate concentration is low.

A natural and unambiguous analysis can be made by taking the prior description and converting it into statements about the extremes (limits) of the situation. That is, as S $\rightarrow$ 0, inhibition is maximal; as S $\rightarrow$ infinity, inhibition is zero. To apply this method of analysis, we must apply limits to the graphical form of the

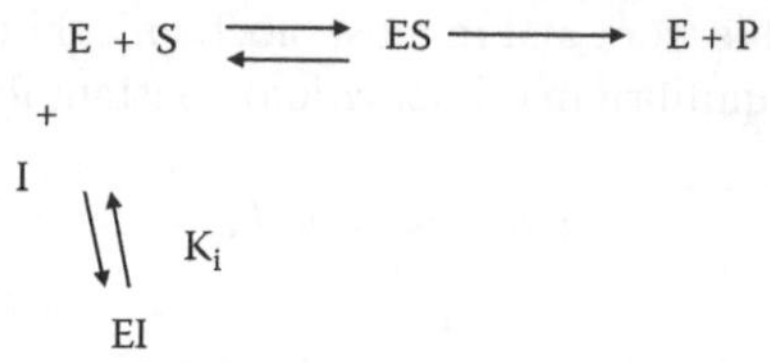

$$E + S \rightleftharpoons ES \longrightarrow E + P$$

$$+$$

$$I$$

$$\Updownarrow K_i$$

$$EI$$

FIGURE 5.2 Mechanism for competitive inhibition. The inhibitor binds exclusively to the free enzyme form, E.

Michaelis-Menten equation and show how these limits behave in the presence of the inhibitor. This is illustrated as Figure 5.3. The graph of velocity versus substrate concentration shows the expected curve (rectangular hyperbola), with the extrapolation to infinite (saturating) substrate, Vmax. Also shown is the other, corresponding limit: the extrapolation of velocity to the condition of zero substrate. It is important not to confuse the zero in the "v_0" with the zero in the extrapolated slope of Figure 5.3. The first is the time extrapolation of the temporal progress curve at one fixed substrate concentration; these velocities are collected and used as points to construct saturation curves such as Figure 5.3. The approach of substrate concentration to zero approximates a straight line. The slope of the asymptote line is the ratio Vmax/Km.

The pair of values—Vmax and Vmax/Km—is often abbreviated to just V and V/K. These values change with variations in substrate and inhibitor concentrations, and they are crucial for examining rate behavior. Notice how Km itself has been excluded. It is true that the saturation curve is also uniquely defined if we fix Vmax and Km. However, the Km does not lead to a meaningful interpretation if we stay with a steady state rather an equilibrium assumption. That is, if we are no longer able to consider Km a dissociation constant, then it is no longer useful for our analysis. That is the key advantage of using V and V/K.

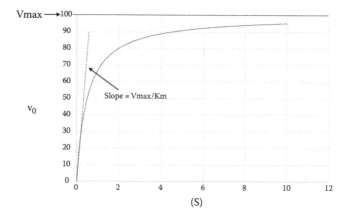

FIGURE 5.3 Asymptotes of the velocity–substrate profile. Vmax (V) is the horizontal asymptote as substrate approaches infinity. Vmax/Km (V/K) is the slope of the asymptote as substrate approaches zero.

The type of inhibition is readily evident from the asymptotes that produce V and V/K as they affect the direct plots of the Michaelis–Menten equation in the presence of inhibitors. Inhibition leads to a decrease in the V intercept, the V/K slope, or both.

In the competitive case (Figure 5.4) it is evident that the value of V is unchanged by the presence of the inhibitor, but the V/K is decreased. The V term is the familiar maximum velocity. Note that it also contains the enzyme concentration since

$$V = k_{cat} [E]_{total} \tag{5.8}$$

From a metabolic perspective, this means that an increase in enzyme concentration, perhaps by increased transcription and/or protein synthesis, will selectively increase this parameter. The second term is V/K, the slope of the asymptote line

$$v_0 = (V/K) * [S] \tag{5.9}$$

is less familiar, because it is less widely used. It is a formula for a straight line, which means that as substrate concentration approaches zero, velocity is a linear function of substrate concentration, and the slope can be interpreted as a first order rate constant. In fact, the true meaning of competitive inhibition emerges at this point: it is the inhibition that causes a decrease in this rate constant V/K as its sole action. This does not diminish our intuitive understanding of how the inhibitor competes with substrate, nor the important idea that molecules that chemically resemble the substrate but are not transformed themselves are often competitive inhibitors.

While analysis using V and V/K provides a straightforward analysis of competitive inhibition, it is also true that if we adhere to the usual means of analyzing Km, we would get a result that seems satisfying as well. After all, if Km were an equilibrium constant, the effect of a competitive inhibitor is to raise the measured Km, which means dissociation is smaller, and we would conclude the affinity of enzyme for substrate is poorer. However, this is not actually what is taking place. In order to

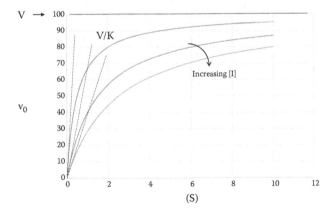

FIGURE 5.4 Competitive inhibition decreases V/K. In the presence of inhibitor I, the slope V/K decreases, but V is unchanged.

truly understand why it is not the way to analyze enzyme inhibition, it is simplest to consider forms other than competitive.

5.2.2 ANTICOMPETITIVE (UNCOMPETITIVE) INHIBITION

The first problem with uncompetitive inhibition is that it is not intuitive. How do two entities "uncompete?" The only obvious feature of this form of inhibition is that it is distinct from competitive. We do have a formal definition: it is the formation of a dead-end complex between inhibitor and the ES complex (Figure 5.5). Before we proceed with the analysis of this inhibition and how it affects the enzyme velocity, it is important to achieve a non-mathematical understanding of just what sort of inhibition this is. Essentially, an uncompetitive inhibitor is one that binds the enzyme at steps subsequent to the substrate binding to free enzyme. One analogy is that of a con game. For example, an entrepreneur might be tempted to engage in illegal activities, say a drug deal. If she accepts it, she moves from being a free unencumbered citizen to an outlaw engaged in a felony (a bound complex). It is only in the latter form that inhibition (arrest by authorities) is possible. As a second example, imagine a bear has discovered a hive with an opening so narrow that he must open his hand to fit into it to reach the honey. Yet, once he has grabbed a fist full of honey he cannot withdraw it. In each case, the free form is not subject to inhibition but the bound form is.

Turning from the whimsical to the analytical, we can use the same analysis developed for competitive inhibition to analyze uncompetitive inhibition. We need only add the equation

$$ESI \rightleftarrows ES + I \tag{5.10}$$

characterized by the equilibrium dissociation constant K_i to the Michaelis–Menten equation. The resulting equation, solved for v_0, can be plotted against substrate concentration for various concentrations of inhibitor, as shown in Figure 5.6. In this case, the behavior of the two constants is the reverse of the competitive case. Here, the inhibitor lowers the V, but does not affect the V/K. Since this is the mirror image of the effect of a competitive inhibitor, it is sometimes referred to as "anti-competitive" inhibition.

With these parameters we can interpret the actions of the anticompetitive inhibitor. Notice that the V is depressed, which means the k_{cat} is decreased. From the mechanism diagram of Figure 5.5, it is evident why this is the case: the inhibitor is removing ES, which is the direct precursor to product formation. The enzyme is shuttled between

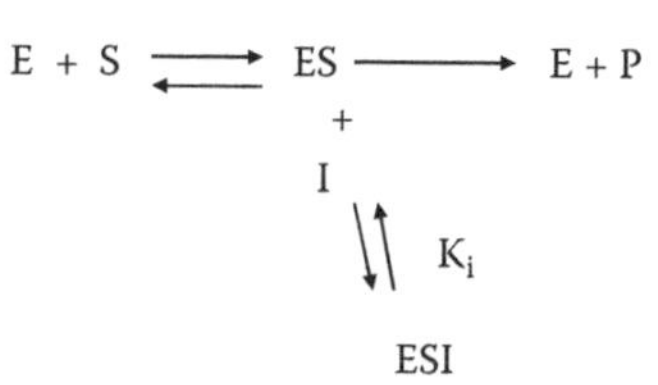

$$E + S \rightleftarrows ES \longrightarrow E + P$$

$$\begin{array}{c} + \\ I \end{array}$$

$$\Big\updownarrow K_i$$

$$ESI$$

FIGURE 5.5 Mechanism for anticompetitive inhibition. The inhibitor binds exclusively to the bound enzyme form, ES.

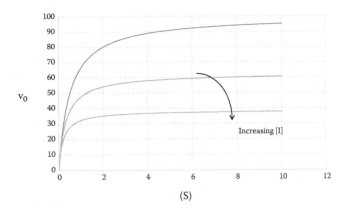

FIGURE 5.6 Anticompetitive inhibition decreases V. In the presence of inhibitor I, V decreases, but V/K is unchanged.

two forms: E and ES. As substrate concentration is increased, more of the enzyme is in the ES form, which is the only form that can be bound to the inhibitor. This explanation for why the anticompetitive inhibitor is more potent as substrate concentration is increased is the exact opposite of the why competitive inhibition is more potent as substrate concentration is decreased. It is this parallelism that explains the name (anticompetitive) and simplifies the interpretation. Notice that Figure 5.6 shows a convergence of the control curve with the curves with inhibitor present as the substrate concentration approaches zero. This means that there is no effect on the V/K, that is, at low substrate the inhibitor does not affect the first order rate constant governing the velocity.

If we instead attempt to interpret this form of inhibition on the basis of the Km, we have an immediate problem, because Km *decreases* in the case of anticompetitive inhibition. There is no escaping unusual interpretations if we assume Km is an equilibrium constant: substrate binding must improve! Consider the Li^+ inhibition of inositol monophosphatase [10], which is anticompetitive with respect to inositol monophosphate. The inhibition was said to have "remarkable consequences" and provides an interaction that is a "positive feedback loop" [3]. There, an analogy is made to a retractable seat belt, which responds more firmly with greater pressure. The analogy is partly true; the basis of the statement is that more inhibition is apparent as the substrate concentration is increased. However, this image does not address how this type of inhibition is fundamentally the opposite of competitive. It also obscures the fundamental reason for the inhibition, as attention is diverted to an equilibrium situation that does not exist. Focusing on the effects of uncompetitive inhibition on Km caused led one investigator to describe this form as being "unusual and counterintuitive," particularly since it reduces the Km for the substrate which "would normally cause an activation" [11].

From another vantage point, uncompetitive inhibition is considered rare, and various reasons explored for why that is the case [12]. It may also be the case that the inhibition is uncommon because it is not well understood and thus not widely recognized. In any event, there are clearly some examples, as noted above. As for this class of inhibitors being useful as drugs, the approach of designing inhibitors directed towards the enzyme-substrate complex has been successful [13–15].

5.2.3 MIXED INHIBITION (NONCOMPETITIVE)

There is a third type of inhibition, usually called noncompetitive. Distinguishing this linguistically from uncompetitive is difficult. As a popular term, it often carries a sinister connotation, as if competition were good, and anything else underhanded. Perhaps it has earned this, as the examples I have been able to invent to explain uncompetitive all involve some sort of shady or at least greedy behavior. Enzymes being without human qualities, we would have to conclude that the reason they are not as well explored is not that they are doing something underhanded, but because we are having difficulty coming to terms with them. Let us ease part of the linguistic crisis by renaming this *mixed inhibition*.

This third type is in fact really nothing new, but rather a combination of the other two. This is what happens when an inhibitor can bind both the free enzyme and the enzyme-substrate intermediate forms. The equation for initial velocity is then derived in the usual way, from the reactions:

$$E + S \rightleftharpoons ES \rightarrow E + P \tag{5.11}$$

$$E + I \rightleftharpoons EI \tag{5.12}$$

$$ES + I \rightleftharpoons ESI \tag{5.13}$$

where the inhibitor equilibria, taken in the reverse direction, become dissociation equations with the constants K_{is} for the equilibrium in Equation 5.12 and K_{ii} for the equilibrium in Equation 5.13.[iii] The equation for the initial velocity with the usual assumptions is then

$$v_0 = Vmax\,[S]/(Km * + [S]*) \tag{5.14}$$

with

$$[S]* = [S](1 + [I]/K_{is}) \tag{5.15}$$

$$Km* = K_m(1 + [I]/K_{ii}) \tag{5.16}$$

Let us first analyze a simple case in which both inhibitor constants are the same:

$$K_{is} = K_{ii} = K_i \tag{5.17}$$

In this special case of mixed inhibition (Figure 5.7), it is clear that both the V and the V/K are decreased over the entire range of substrate concentration. The inhibitor removes both forms of the enzymes equally from the reaction pathway and leaves a portion in a dead-end complex. This is because both enzyme forms are in equilibrium with the same inhibitor and have the same dissociation constant. Since

$$[E]_{tot} = [E] + [ES] \tag{5.18}$$

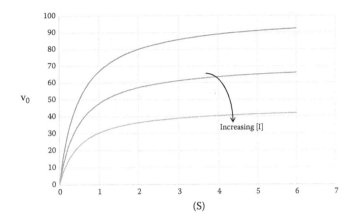

FIGURE 5.7 Mixed inhibition decreases V/K and V. The situation pictured is mixed inhibition in which the inhibitor I binds the E and ES form equally. As a result, both V and V/K decrease by the same amount.

without inhibitor, and the inhibitor removes the same fraction of each form, it is the same as just lowering the total amount of enzyme. This could also be accomplished by a decreased genetic expression of the enzyme, or by the presence of an irreversible inhibitor (see below). Kinetically, these situations are indistinguishable.

The Km is unchanged in this type of inhibitor. This drives home the point that this parameter is not a useful one in understanding inhibition. It is apparent that both forms of the enzyme are involved in the inhibition, so that any notion that we can ascribe Km to an affinity is not possible. All three forms of inhibition considered here are summarized in Table 5.1, along with the effects on V, V/K, and Km.

A remaining situation is a mixed inhibition, where the K_i values for E and ES are unequal. This has a simple explanation: whichever value of K_i is dominant (i.e., smaller) will cause the inhibition curve to be more like competitive or anticompetitive. Two cases are plotted in Figure 5.8. In Figure 5.8a, the K_{is} is lower (dominant), and thus the inhibition is greater at low substrate concentration (greater decrease in V/K), but still evident at high concentration (lesser decrease in V). The opposite situation is illustrated in Figure 5.8b, in which K_{ii} is lower (dominant), and thus

TABLE 5.1
Effect of Inhibition Type on Kinetic Constants

Inhibitor Type		Effect On:		
Name	Synonym	Vmax	Km	V/K
Competitive	–	–	↑	↓
Uncompetitive	Anticompetitive	↓	↓	–
Noncompetitive	Mixed	↓	–	↓

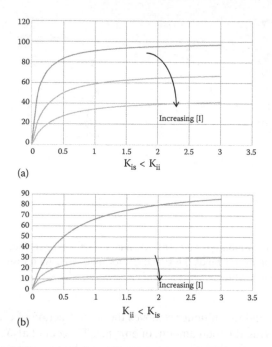

FIGURE 5.8 Mixed inhibition with different affinities. (a) Binding to E is stronger (K_{is} is lower), so the inhibitor effect on V/K is greater. (b) Binding to ES (K_{ii} is lower), so the inhibitor effect on V is greater.

inhibition is greater at high substrate concentration (greater decrease in V/K), but still present but less inhibited at low substrate concentration (lesser decrease in V/K).

Considering the broad implications of noncompetitive inhibition with varying K_{ii} and K_{is} values, it is evident that this form of inhibition encompasses the others. Competitive and anticompetitive are just special cases where one or the other inhibition constant is of no quantitative importance. The terminology of "mixed inhibition" is clearly more descriptive of the situation than *noncompetitive*. We can also conclude that there are just two conceptual types of inhibition, linked to binding the free enzyme E and the intermediate ES.

5.3 DOUBLE RECIPROCAL PLOT

The more usual presentation of enzyme inhibition is to plot the velocity–substrate profiles in linearized form. The most popular of these is the double reciprocal plot. However, introducing enzyme inhibition in terms of the double reciprocal plot is a barrier to understanding; the distinction between types just turns into pattern recognition [16]. Just as association constants are avoided for equilibria (and dissociation constants used instead), reciprocal plots require that we think in reciprocal space. It is even worse: we have to imagine how a reciprocal substrate concentration affects a reciprocal velocity. Prior to the ready availability of computers, the double-reciprocal plot was used to linearize the Michaelis-Menten plot and graphically extrapolate the values of V and K. At present these are routinely measured by computer calculated nonlinear regression. The problem with the double-reciprocal plot is that the worst data (lowest velocity and substrate concentration) is given the greatest weight in the analysis. This complication was recognized by the authors of the original paper describing the plot in 1934 [17].

The modern use of the double reciprocal form is two-fold. First, the patterns are recognizable and distinct. The competitive, anticompetitive and mixed forms in double-reciprocal form are straight lines in a $1/v_0$ versus $1/[S]$ plot that intersect on the $1/v_0$ axis, are parallel, or intersect to the left of the $1/v_0$ axis, respectively. The unique patterns are striking and easily distinguished. The second use of the double recipro-cal form is in the analysis of multi-substrate enzymes. Realistically, most enzymatic reactions use two substrates (and occasionally more). Use of the double reciprocal plot for these enzymes provides a visualization of changes in different substrates and products (slope and intercept effects) for the more complex analysis of multi-substrate enzymes. It should be noted that even with this ability to discriminate between simple visual patterns, there is a difficulty inherent in the comparison of parallel lines as opposed to those intersecting to the left of the $1/v_0$ axis. It is never certain that lines that *appear* parallel might actually be lines that eventually inter-sect. Further kinetic analysis is needed to resolve such issues.

5.4   APPLICATION OF ENZYME KINETICS: CALCINEURIN

Calcineurin is one of the protein phosphatases also known as PP2C, and is nota-bly the only phosphatase requiring Ca^{2+} for its activation [18]. One aspect of this enzyme is its involvement in the action of immunosuppressants such as the natural product cyclosporine or the subsequently developed synthetic drug FK506 (tacro-limus). These compounds act by forming a binary complex with prolyl isomerase. This drug/enzyme complex is an inhibitor of calcineurin. Results with either drug leads to essentially identical results, except that tacrolimus is about an order of mag-nitude more potent. In T cells of the immune system, calcineurin (once activated by Ca^{2+} as a result of an engaged T cell receptor) activates transcription of proteins such as interleukin-2, essential for clonal expansion of the T cell line. Thus these drugs act as immunosuppressors.

Calcineurin can also catalyze the dephosphorylation of small phosphorylated molecules such as p-nitrophenol-phosphate (pNP). Surprisingly, with this substrate, the cyclosporine:prolyl isomerase complex caused an *activation* of calcineurin activ-ity. Etzkorn et al. [19] found that the inhibition in the case of phosphorylated protein substrates was of the noncompetitive type, and suggested that this meant action at a "non-active-site." Left unclear was why this can account for the stimulation with the small non-protein substrate pNP.

Having just examined the meaning of enzyme inhibition, it is evident that this analysis does not show where the binding to the enzyme is, but rather indicates the enzyme form that engages the inhibitor. In an attempt to resolve this paradox, we used the FK505 complex (that is an equimolar mixture of FK506 with prolyl isom-erase) and replicated the finding of inhibition with protein substrate and stimulation with pNP [9]. Calcineurin, like many phosphatases, has a shift in the rate limiting step with the different substrates. We suggested that the action of the complex of cyclosporine (or FK506) ("the complex") with prolyl isomerase has one action: it increases the binding of the substrate to the enzyme. Certain small molecules like pNP bind poorly; the complex causes activation because it improves binding of enzyme to this substrate. The limiting step under these conditions is binding. However, with large-molecular-weight substrates such as proteins, product release (and not substrate binding) is rate limiting. Thus the complex by improving binding in this case impedes release. Thus, the increased binding of substrate to enzyme provides the explanation for both events: activation with pNP and inhibition with the phosphorylated protein.

5.5   IRREVERSIBLE INHIBITION

We have already encountered the special case of mixed inhibition with equal K_i values. As mentioned, the observed velocity-substrate profile is the same as if we just removed total enzyme because the action is to effectively lower $[E]_{total}$. Thus, it is no surprise that kinetically it is not possible to distinguish this form of mixed inhibition from irreversible inhibition. In the latter case, the action is similar to inactivation of a portion of the total enzyme. To separate these possibilities, it is necessary to do additional experiments, such as determining if inhibition can be reversed by dilution or filtration. With very tight-binding mixed inhibitors, this may not be a simple task. For example, many hours are required to dissociate ryanodine from its eponymous calcium release channel in the endoplasmic reticulum [20].

Classical cases of irreversible inhibition are the covalent modification of enzymes by the nerve gas isopropyl fluorophosphate [21] and the acetylation of prostaglandin synthase by acetylsalicylic acid (aspirin) [22]. What these modifications have in common is that the enzyme function is irretrievably lost; only new protein synthesis can restore the original cellular levels. Physiologically, protein degradation pathways such as those involving the lysosome or the proteasome, are similar to the external agents. Thus, all three cases: a poison, a drug, and protein degradation require new protein synthesis for replacement.

There is another type of irreversible inhibition, however, that is frequently encountered in cells, and different in its metabolic behavior. For example, phosphorylation of many proteins can be reversed in just one or a few steps. To distinguish this from the more drastic destruction of the protein just discussed, I introduce two new terms. The case exemplified by the phosphorylated enzyme which is returned to its dephosphorylated state in one or just a few steps, we will call **pathway-reversible**, introduced in Chapter 1. Further examples of this type of modification are the acetylation of proteins that is reversed by a deacetylase [23]. The other type of modification that involves degradation of the protein, requiring the much longer time scale protein synthesis to restore it, we will consider to be **pathway-irreversible**. The two reaction types are both irreversible inhibition, so they cannot be analyzed by the tools of enzyme kinetics we have applied to inhibitors binding at equilibrium. However, in the case of *pathway-reversible* steps, there is a small cycle, often just two enzymes, that form a modified enzyme and return it to its original form.

Note that each enzymatic reaction involved in a given process of a pathway-reversible cycle must itself be metabolically irreversible.

5.6 INHIBITORS AFFECTING METABOLIC PATHWAYS

In Chapter 2, we introduced that distinction between the reaction view and the pathway view. For enzymatic analysis, the reaction view is a common frame of reference. However, the pathway view is an important one if our focus is metabolism.

As an example of the distinction, in metabolic studies, the inhibition type is not always of importance. In the aforementioned example of titrating glucose formation with mercaptopicolinate [24], the inhibition mode did not factor in the analysis. Here, as often is the case in a pathway inhibitor, it is only the site of action that is of significance. A pathway titration was also used [25] to assess ATP exchange by mitochondria. In that study the experimenters used both atractyloside (a competitive inhibitor) as well as carboxyatractyloside (a noncompetitive inhibitor) with similar results.

5.7 NONSELECTIVE INHIBITION

Selectivity in inhibition complicates analysis and is a major concern of drug development, as off-target effects are common. We consider two types: inhibitors known to be non-selective, because they act as *class inhibitors* rather than specific enzyme blockers, and inhibitors known to be nonselective but in use nonetheless.

5.7.1 CLASS INHIBITORS

The compound aminooxyacetate (AOA, Figure 5.9) is commonly used as a transaminase inhibitor because of its ability to form a Schiff base with pyridoxal phosphate,

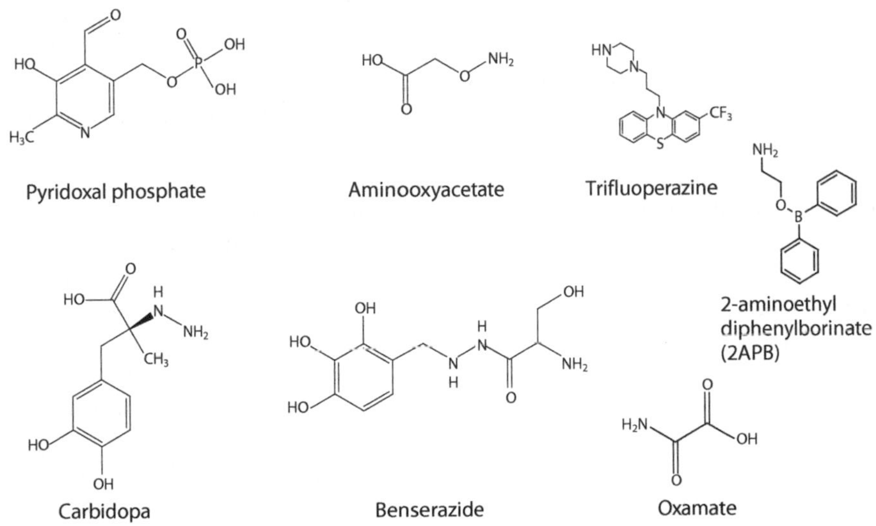

FIGURE 5.9 Enzyme inhibitor structures.

the bound cofactor in transaminase enzymes. As an example of one type of non-selectivity, AOA inhibits gluconeogenesis in the liver of chickens despite the fact that this pathway has no transamination steps [26]. This inhibition was traced to a second pathway: the malate/aspartate redox shuttle required to remove excess cytosolic NADH. This forms in excess under the experimental conditions used, as some lactate had to be oxidized for energy to support gluconeogenesis. Accordingly, in the additional presence of fatty acids to provide the energy (but not substrate) for gluconeogenesis, AOA inhibition was attenuated.

This type of nonselectivity is not one in which the inhibitor found a separate enzyme class, but rather where the same target is involved in a separate metabolic event. In fact, AOA has a much broader spectrum of action, inhibiting a wide number of transaminases, as well as those catalyzed by other enzymes that use pyridoxal phosphate: decarboxylases [27] and racemases [28].[iv]

As an example of class inhibitors that are medically important, two pyridoxal binding compounds, carbidopa and benserazide (Figure 5.9) are used to inhibit the decarboxylation of DOPA in the periphery. Thus administration of carbidopa with DOPA supplies brain DOPA while preventing peripheral decarboxylation [30]. While AOA would also serve as a DOPA decarboxylase inhibitor [31], its wider nonspecificity precludes its use. A similar selectivity profile is evident in the tryptophan pathway of catabolism through kynurenin. Here, too, AOA inhibits both a decarboxylase and a transamination step, whereas carbidopa and benserazide inhibit only the former [32].

5.7.2 More Nonselective Inhibitors

While the specificity of enzyme inhibitors is in some ways a subjective determination, there are many compounds that are broader in their action, acting across different classes of enzymes. One example is the fluoride ion. While this is used as an inhibitor of glycolysis at the enolase step [33], it appears instead to be an inhibitor of phosphatases [34–36]. While this appears not to affect tyrosine dephosphorylation [36], since there are a large number of alkyl phosphatases (ser/thr), it is a rather broad inhibitor. It is also used at very high concentrations, in the tens of millimolar. This may be because the true inhibitor is a complex with Al^{3+} in which case the concentration of inhibitor required is reduced several orders of magnitude [37,38].

An inhibitor that is widely used yet known to be non-selective is aminoethoxy-diphenyl borinate (2APB, Figure 5.9), used to block certain routes of Ca^{2+} entry into cells [39] (see also Chapter 7). The complexity of this inhibitor is well known: it blocks a large number of other processes that make its interpretation extremely difficult [40,41]. Nonetheless, it remains in wide use, largely because there are few alternatives.

A further example of an inhibitor that is nonselective and yet finds wide application is trifluoperazine (Figure 5.9). This compound was shown to be a blocker of calmodulin association enzymes [42]. To appreciate this action, we need only be aware that Ca^{2+} can stimulate some enzymatic reactions only in the presence of a Ca^{2+}-binding protein called calmodulin. The compound trifluoperazine does prevent the interaction of the Ca^{2+}-bound calmodulin with its target proteins *in vitro*, but it is not a useful inhibitor in intact cells, because the reason for the inhibition is a hydrophobic interaction. For example, the compound causes inhibition of hormone binding to receptors, clearly unrelated to calmodulin [43]. Recently, trifluoperazine was used as an agent to cause monomerization of calsequestrin, the endoplasmic reticulum Ca^{2+} buffer [44]. Once again, this is unrelated to calmodulin, but simply depends upon protein–protein interactions.

Recently, several studies have used oxamate (Figure 5.9) as a selective inhibitor of lactate DH in cells as a glycolytic inhibitor and anticancer agent. However, this inhibitor is also nonselective, inhibiting other targets in glycolysis as well as in other pathways, such as oxidative phosphorylation [45]. Thus, while the results of using an inhibitor may be consistent with the original hypothesis, unless the selectivity is clearly established, this may be coincidental rather than mechanistic.

5.8 TRANSPORTERS AS ENZYMES

Kinetically, transporters are the equivalent of enzymes. In fact, a transporter can be considered to be an enzyme that catalyzes a translocation of a molecule from one space to another, without chemical transformation.^v The kinetic properties of transporters are indistinguishable from enzymes. Transporters may be categorized as metabolically irreversible or near-equilibrium, and they can be inhibited. There is a unique feature of transport, however: some species, in particular monovalent ions, move through channels rather than exchanges. In that case, they are quite different from the transporters we are used to considering in metabolic sequences, and are treated as such. In what follows, we consider first a few features of transporters, and then a brief consideration of channels.

5.8.1 Most Intracellular Transporters Are Near Equilibrium

Like their enzymatic counterparts, most intracellular transporters exchange at near-equilibrium. This was the conclusion of a considerable number of investigations by JR Williamson's lab in the 1960s and 1970s [46–49], who initiated their studies to determine potential control elements in metabolism, but concluded that they were largely not regulatory. For example, many exchanges across the mitochondrial membrane—the major intracellular organelle in metabolic terms—are near-equilibrium, such as malate/Pi, α-ketoglutarate/malate, citrate/malate, and Pi/H+. Some are metabolically irreversible, at least under conditions where the mitochondria maintains a membrane potential, which would cover essentially all metabolic conditions. This includes the glutamate/aspartate exchanger and the ADP/ATP exchanger (Chapter 3). These achieve their mIRR status by exchanging charged forms that are imbalanced. Aspartate exits mitochondria as a –1 charged species, and glutamate enters uncharged, meaning that, upon exchange, the mitochondria has become more positive, utilizing a portion of the electrical gradient.

In some cases, transporters are part of an overall pathway, such as gluconeogenesis, fatty acid oxidation, or lipogenesis; these have reactions in both cytosol and mitochondria. While glycolysis is entirely cytosolic, complete oxidation of glucose requires a connection between the cytosol and mitochondria. Thus, the NEQ pyruvate transport step is part of the overall oxidation of glucose into CO_2.

With the combination of routes, balancing all cofactors requires the respiratory chain and ATP formation, as well as pathways that utilize the ATP. There is also the need to import the NADH; as previously mentioned in the context of gluconeogenesis above, this transport step involves several other enzymatic reactions and transport steps, so that the pathway assumes a much greater complexity. Thus, transporters are part of overall pathways and, as we have noted, potential points of inhibition.

5.8.2 TRANSPORT IN THE PLASMA MEMBRANE

Because of the steep gradient between intracellular and extracellular spaces, many transporters embedded in the outer cell membrane are metabolically irreversible. An exception is glucose utilizing GLUT2 in a few cells like liver, intestine, pancreas, and kidney [50,51]. Additionally, transport of the carboxylic acids lactate, pyruvate, β-hydroxybutyrate and acetoacetate are at near-equilibrium [52,53]. Uptake of many substrates such as amino acids are metabolically irreversible, such as N-transporters that utilize the Na^+ gradient established by the NaKATPase.

5.9 ION CHANNELS

One distinctive feature of ion channels is the far greater rates that can be achieved compared to an exchanger [54]. Ion channels are of most obvious interest in excitable cells, nerve and muscle. There are also cells that are not excitable, and yet use the membrane potential for signaling, such as the pancreatic endocrine cells. There are also ion channels used for net ion transport in cells that have no change in membrane potential, such as the CFTR found in pancreatic duct epithelia and in lung epithelia. For these cells, transport is limited by the requirement of charge balance for both the intracellular and extracellular solutions. In what follows, I consider a few cases of ion channels of importance to differentiate them from the transporters that we have considered so far. An important resource for this field is the monograph by Hille [55].

5.9.1 NA CHANNEL

The action of the Na channel in excitable cells signals membrane depolarization, as uncompensated charge from the entering positive ion rapidly renders the intracellular cytosolic space less negative.[vi] Voltage-gated Na channels open in response to a charge difference encountered by the protein due to the depolarization event. Unlike most metabolic events, the amount of Na^+ entering the cell that is needed to effect the depolarization is extremely small, and makes virtually no contribution to the total pool of intracellular Na^+. Thus, it takes many depolarization events before there is appreciable change in the intracellular $[Na^+]$, which is ultimately redressed by the NaKATPase pump. The connection between the depolarization events and the energy utilization through the NaKATPase is indirect (see also Chapter 8). Inhibition of the NaKATPase will eventually limit depolarization but it will be only after many thousands of depolarization events. Thus, in the normal course of considering depolarizations, ATP supply is not a critical factor, as it is say for a biosynthetic pathway that requires stoichiometric quantities of ATP.

Like most glucose transporters discussed above, the Na channel is always *downhill*, but metabolically irreversible. It is maintained in that state and thus is never a route for the export of Na^+ from inside the cell to the outside. Another route restores the Na, albeit at a far different rate than its uptake. For completeness, it should be pointed out that Na channels can also be activated by two other means: ligand binding, and mechanical displacement. These provide cells with the ability to respond to neurotransmitters and to mechanical force.

5.9.2 K Channels

Whereas Na channels are excitatory, K channels are inhibitory, or, generally speaking, serve the purpose of returning the membrane potential to baseline. Once again, the actual amount of K^+ ion that is exchanged is miniscule for each event, and the balance is ultimately redressed by the NaKATPase, just as with Na channels. What is of interest from a metabolic viewpoint is that the gradient of K^+ across the membrane at rest is very close to its equilibrium state. In this resting state, the channel is closed. It is only when the potential increases due to depolarization that the channel opens and K^+ flows down its gradient out of the cell. In the case of the K channel, the essential flow is still unidirectional, achieving an export of K^+, so, from our vantage point, it is like the Na channel: both downhill and metabolically irreversible.

There is a variety of K channels that can be considered to shape the action potential of excitable cells. Aside from the common voltage-activated K channels of nerve and muscle, responsible for restoration of the resting potential, there are others that are important in non-excitable cells, including Ca^{2+} activated K channels [56] and ATP-binding K channels [57,58] that are present in non-excitable cells. The latter, also known as KATP channels are linked to metabolic events, although the regulation is not likely due directly to ATP binding. Since ATP is essentially unchanged in concentration in cells as elaborated in prior chapters, it is a separate event that controls the opening of these channels. They are of particular interest in the endocrine pancreas cells, as their inhibition in the presence of increased metabolic flow (i.e., increased glucose entry) leads to cell depolarization and hormone secretion (Chapter 6).

5.9.3 Ca Channels

While most of the discussion of Ca^{2+} will be deferred to Chapter 7, Ca channels, like Na channels, can initiate some excitation events, usually at the end of a stimulus sequence in nerve and muscle. These channels share the qualities discussed for the Na and K channel, namely, downhill transport, metabolically irreversible status, and the fact that very small, inconsequential concentrations of Ca^{2+} are required to achieve the electrical alterations in the membrane. Large increases in intracellular Ca^{2+} concentration results from the flow of this ion from intracellular pools, largely the endoplasmic reticulum. These channels are not electrically stimulated; further elaboration of their activation is in Chapter 7. Plasma membranes also have many distinct Ca^{2+} channels, which are often less selective for Ca^{2+} itself and contribute to alteration of membrane potential for specialized purposes, such as taste receptors.

5.9.4   Cl Channels

These channels do not participate in as many physiological events as the previously considered ones. They are usually considered as charge-balancing channels, operating at near-equilibrium. Still, this ion can serve as a messenger of negative neurotransmitters, to hyperpolarize cells in the manner of K^+ except that the ion flow of Cl^- is in the opposite direction. One Cl channel mentioned at the onset of this section is the CFTR, named for its absence in cystic fibrosis [59]. This channel, localized to the luminal membrane, is important in ion balance in duct cells such as those in pancreas and lung. Defective channels compromise the subsequent exchange and release of bicarbonate and water to the luminal exterior of the cell, thus concentrating ions and proteins normally present in the ducts.

5.10   LIPID ENZYMOLOGY

To characterize enzymes acting on lipid substrates, a new feature must be considered that dominates its kinetics: the lipid–water interface. An alternative terminology for this is **interfacial catalysis**. Two enzyme examples are described here. Both interact with membranes and are related to hydrolysis of phospholipases (Figure 5.10). The first is the enzyme catalyzing hydrolysis in the 2-position of the phospholipid, Phospholipase A_2 (PLA). Due to the large number of studies on this enzyme, it is widely considered to be the model for interfacial catalysis itself. The second example is not a phospholipase, but is activated by the lipid product of phospholipase C. This is the enzyme protein kinase C, a membrane associated protein having much in common with PLA.

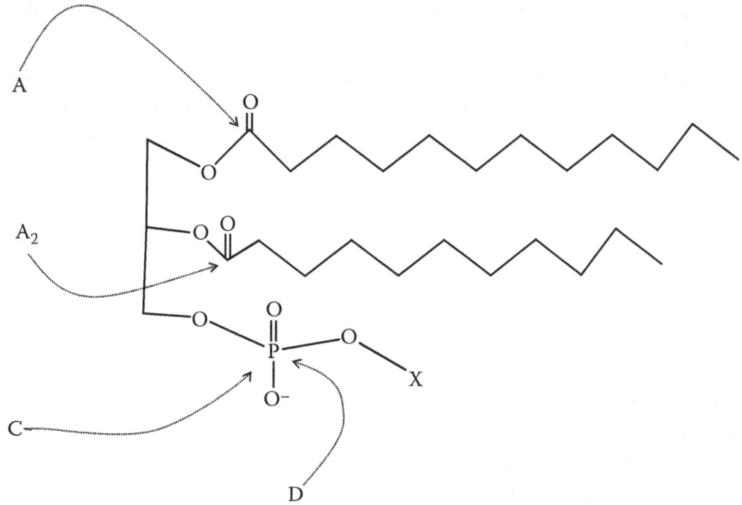

FIGURE 5.10   Sites of action for phospholipases.

5.10.1 PHOSPHOLIPASE A_2

Phospholipase A_2 (PLA) is found in both membrane-bound and soluble (secreted) forms; the latter is the most widely studied, and serves as a model of interfacial catalysis [60]. The enzyme is of interest as arachidonate is commonly found in the (sn)-2 position of membrane phospholipids, and this is released upon PLA action. As this fatty acid is the precursor to the eicosanoids, the enzyme can be considered the initiator for this entire broad category of signaling molecules. A current clinical interest in PLA is the observation that the enzyme associates with lipoproteins (particularly LDL), and this association correlates with increased risk of cardiovascular disease [61]. While inhibition of the PLA activity had some benefit in animal studies, clinical evaluation did not bear out this expectation [62]. Thus, the presence of PLA serves at present as a biomarker portending increased likelihood of broad cardiovascular damage.[vii]

What is known about PLA is that its catalysis is distinct from water–soluble enzymes. Perhaps the most striking difference is apparent in the substrate–velocity curve for PLA, illustrated in Figure 5.11. The unusual substrate dependence can be interpreted as two separate saturation curves [63]. The first region of very low activity represents activity of the monomer phospholipid. Once the concentration reaches critical micelle concentration (CMC) there is a burst of activity and then a second saturation. This can be interpreted as a new surface for the enzyme, that is, the membrane itself that allows superior anchoring of the enzyme and thus far greater catalytic activity. It is in this second phase that most of the enzyme activity is expressed, and it is also in this phase that interfacial catalysis is expressed. As examined by Jain and Berg, PLA interacts with the surface by sequential catalytic cycles as well as by release and rebinding (or, just "hopping and scooting") [64].

Because PLA depends heavily on the interface, the charge and physical form of the substrate can modify the observed activity. An early study suggested that a cellular protein could regulate PLA activity and account for subsequent changes in regulatory

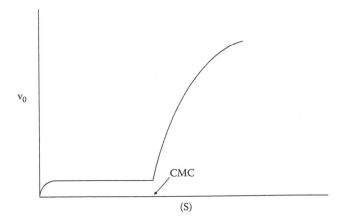

FIGURE 5.11 Burst kinetics for phospholipase A_2. Velocity is low until the substrate concentration—phospholipid—reaches CMC (critical micelle concentration). At that point, large rate increases ensue.

response; the protein was named lipocortin [65]. However, it happened that many proteins were able to affect the activity of PLA and their action depends on the very qualities that alter the interface: the charge of the interface, the type of PLA, and even the form of substrate. In fact, three types of effects of protein on PLA can be obtained: an inhibition (with very low concentration liposomes); an activation (with high substrate concentration liposomes); or a biphasic action (micelle substrate). Not only that, but virtually any globular protein has the same effect. Thus, the action of lipocortin is unlikely to reflect a physiological regulation [66]. As lipocortins are also known as annexins, structural proteins found in high concentrations, it is apparent that they are unlikely PLA regulators. Direct analysis of the interaction between lipocortins and PLA confirms little physical binding between the proteins [67]. As there was a great deal of interest at the time of its proposal, some current secondary literature (texts and reviews incidental to the topic) still refers to lipocortin and its relationship with PLA. Still, no further studies appear to have been performed on PLA in this context.

5.10.2   PROTEIN KINASE C

Like PLA, protein kinase C (PKC) exhibits membrane association. There is also a structural resemblance: both have a similar C2 domain [68]. The C2 domain is in the active site of PLA, but involved with lipid regulators of PKC. It might be supposed that, just as catalysis is dominated by the lipid:water interface in PLA, that regulation might be approached by considering those same interactions in the case of PKC. However, that is not generally the case: PKC regulation is usually interpreted directly in terms of ligand binding to the enzyme [69]. This notion has not been entirely ignored, however. For example, there have been studies that consider lipid-water interactions as a key to understanding the regulation of PKC [70]. In this last section of the chapter, I extend this perspective for PKC.

Experimentally, it has long been known that PKC appears to shift its cellular location upon activation between a soluble fraction in the unstimulated state and a pellet fraction in the stimulated state [71]. There are three activators: Ca^{2+}, diacylglycerol, and acidic phospholipids [72]. Isozymes of PKC have been found that require all three (classical forms), are Ca^{2+} independent (novel) or respond to none (atypical). The latter categorization presents difficulties as the identification of an enzyme as a PKC is structural rather than functional. Thus, some kinases that are structurally unlike other PKC isozymes can also respond to diacylglycerol (or at least the more potent analogs, phorbol esters), and these agents can bind other proteins as well [73,74].

If we consider the classical PKC isozymes, there remains the difficulty of how the various activators function and relate to the enzyme structure. An accepted idea is that one domain (C1) is responsible for binding the membrane, and a second (C2) the diacylglycerol. The latter may be Ca^{2+} independent in the novel forms, consistent with a lack of acidic amino acids in those proteins [68].

However, an unusual interaction exists between these activators: an increase in one enhances the potency of the others. Investigators interested in the physiological control have understandably focused on just two of the three activators: diacylglycerol and Ca^{2+}. However, there are reasons to conclude that Ca^{2+} is not an activator.

The amount of Ca^{2+} needed for activation is well above the maximal concentration achieved during signaling (about 1 µM). Moreover, despite the assumption that Ca^{2+} and diacylglycerol are produced together at levels that activate PKC, careful studies measuring the amount, time course, and source of the diacylglycerol refute that supposition: in fact, the true source of most diacylglycerol is phosphatidylcholine [75,76]. It may be supposed that protein-membrane association generally requires Ca^{2+} but this may not be the case, as demonstrated for membrane fusion [77]. A separate reason for believing that Ca^{2+} is an important regulator may be the name itself: the "C" in PKC does in fact refer to Ca^{2+}. However, the observation that led to that name was that a Ca^{2+} dependent protease led to activation [78]. Nishizuka and coworkers found that PKC could also be stimulated by high concentrations of Ca^{2+} without proteolysis, albeit in the presence of phospholipids [79]. Some studies have shown that the interactions of Ca^{2+} and phospholipids are distinct from those of DAG, but still suggest a possible role for Ca^{2+} regulation. Other proteins that mimicked the interaction with Ca^{2+} and phospholipids were the lipocortins [80] (also known as annexins), the same proteins thought to be PLA regulators as discussed above. Even if we accept the notion that the principal regulator of PKC is DAG, the fact remains that no hypothesis consistently explains why the apparent translocation occurs between soluble and pellet fractions, why the three *in vitro* regulators are synergistic, and most importantly, exactly how DAG activates the enzyme.

5.10.2.1   Physical States of Diacylglycerol

Cell membranes form bilayers because the majority of their lipids are bilayer formers, such as phosphatidylcholine. This property can be observed in simple experimental systems: pure phosphatidylcholine and some mixtures form a bilayer structure non-enzymatically in aqueous solutions. Lysolecithins instead form micelles. The term "critical micelle concentration" actually applies to all lipid species, bilayer former and micelle former alike, and is a term we have encountered before in describing the behavior of phospholipase A_2 as an enzyme.[viii]

The means by which amphipathic lipids interact with water [81] are illustrated in Figure 5.12. The bilayer form indicated for phosphatidylcholine is shown as Figure 5.12a. This separates two water phases: for cells we take these as intracellular and extracellular fluids. Other lipids such as lysophospholipids assume a micellar form, in which there is a hydrophobic core and one water phase (Figure 5.12b). This can be explained on the basis of stacking interactions: the bilayer-formers stack like bricks; the micelles form like three dimensional pizza slices. Neither of these forms are found for DAG [82]. Diacylglycerols instead form *inverse micelles* in which the hydrophilic portion is directed to the interior and the hydrophobic to the exterior (Figure 5.12c). This requires that the medium in which the DAG is found is itself hydrophobic. There is such a medium: the interior of a phospholipid bilayer. Figure 5.13 shows how this structure can form. It has been suggested that this type of intermediate is a requirement for membrane fusion [82]. It is important to notice that the presence of DAG within the membrane as presented in Figure 5.13 has an extra, internal water phase, contributed by the hydroxyl groups of DAG. For the present purposes, it can provide a model to explain how PKC can be activated and also explain the unusual kinetic interaction of the three effectors.

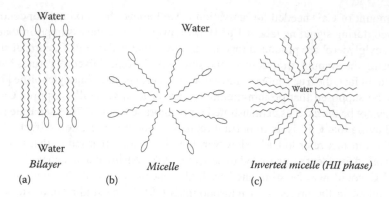

FIGURE 5.12 Lipid–water interactions. (a) Bilayer lipids such as phosphatidylcholine create a hydrophobic interior separating two water layers. Ovals represent water soluble head groups, such as phosphocholine. (b) Micelles such as lysophosphatidylcholine create just two phases: an internal hydrophobic phase consisting of fatty acid tails and the exterior containing the water-soluble head group. (c) Inverted micelles form with certain lipids such as diacylglycerol, in which an interior water layer including the glycerol moiety is formed and the exterior is hydrophobic.

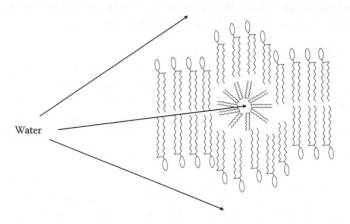

FIGURE 5.13 Diacylglycerol-inverted micelle in a membrane. The bilayer is interrupted by the formation of the diacylglycerol-inverted micelle. In this structure, there are three water phases as indicated: two on the opposing sides of the bilayer, and a third interior to the inverted micelle.

5.10.2.2 How Diacylglycerol Activates Protein Kinase C

The structure of PKC is a clamshell in which the two lobes of the protein are connected by a hinge region. It is this hinge that is the site of proteolytic cleavage *in vitro* to yield a constituently active kinase called protein kinase M. These regions are illustrated in Figure 5.14. One lobe of protein contains a pseudosubstrate that can bind the kinase active site in the other lobe. In the unstimulated state in cells, the enzyme exists with the lobes attached to each other as the active site binds the pseudosubstrate site and the protein attaches to the membrane as a surface interaction protein as illustrated in Figure 5.14b. Upon stimulation of PLC, a small amount of membrane

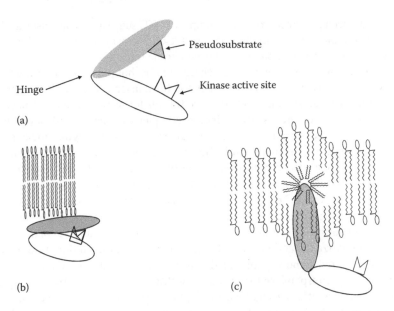

FIGURE 5.14   Protein kinase C interactions with lipids. (a) Enzyme is a bi-lobed molecule, with a pseudosubstrate in the regulatory portion and the kinase active site in the activity portion. These are connected by a flexible hinge region that is susceptible to *in vitro* proteolysis. This separates the two in isolated preparations, leading to a constitutively active kinase portion ("protein kinase M"). (b) Proposed inactive enzyme associates with the plasma membrane as a surface binding protein. (c) After the conversion of some membrane phospholipid to DAG, the latter forms an inverted micelle, enabling the solubilization of the regulatory subunit into the membrane. This frees the catalytic region to act on cytosolic substrates. The action is transitory as the constant movement of DAG to the cytosolic membrane provides a substrate to enzymes that can remove it, such as the DAG kinase and DAG phosphatase, which removes the inverse micelle phase, causes the protein to resume its surface membrane orientation as in b, and terminates the signaling system.

is hydrolyzed to produce DAG, which remains associated with the membrane. Since a known form of interaction between these is an inverted micelle as illustrated in Figure 5.13, the additional presence of a protein may produce a situation drawn as Figure 5.14c. In this situation, a portion of the enzyme does exist *within the membrane phase*. This is possible because of the new aqueous phase created within the membrane that can satisfy the hydrogen bonding requirements for a peptide turn.

   The ability of the protein to exists as a partially membrane embedded entity is entirely due to the ability to solubilize a water phase inside the membrane. In prior discussions of trimeric G proteins [83], it was pointed out that these are strictly surface-binding proteins. Moreover, that review pointed out that no known protein can be partially inserted into membranes, as this would require a turn of the protein backbone, exposing amide and carbonyl groups to a hydrophobic interior. The authority for that conclusion was Singer [84], who published a review of the subject. It is ironic that, ten years later, Singer published a landmark paper detailing biological membranes, which contained a sketch of one protein that was in fact inserted part way through the membrane [85]. For this reason, many such diagrams followed, but no actual examples of proteins fitting this diagram have been presented. Yet, the example of PKC suggests that at least transiently, there is an example of a protein that can insert itself part way into the membrane from just one side due to the creation of an intermembrane water space.

We can now bring in the negatively charged phospholipids and explain why they can serve as *in vitro* activators of PKC: they too form inverse micelles [81]. In fact it is really just the neutral phospholipids that are bilayer formers. Because the negatively charged phospholipids are usually minor participants in the overall population of membrane lipids, the overall structure is a bilayer. Yet, with enough of the negatively charged phospholipid, an inverse micelle can once again form, and this can serve to activate PKC. Ca^{2+} too can form a region of inverse micelle [81] and thus force activation of PKC. The ability of any one of the three agents to create the conditions for internalization of PKC explains the otherwise baffling result from kinetic studies that any one of the three activators reduces the concentration of the others that is needed for activation. This unusual form of synergism is difficult to explain if we are trying to rationalize the actions of PKC activation by specific binding interactions.

We have offered some support for the notion that a lipid interaction rather than selective binding is responsible for activation of PKC by showing a differential quenching of fluorescence of a phospholipid fluorophore upon activation of the enzyme by diacylglycerol or phorbol esters [86]. Moreover, the model also explains why diacylglycerols and phorbol esters are distinctive in their inhibition potency. Diacylglycerols can be rapidly removed by hydrolysis or phosphorylation, but phorbol esters are resistant to such metabolic transformation. This also explains the *in vitro* observations of the apparent translocation between cytosol and membrane, but places them in a different context than the usual interpretation. The resting-state soluble fraction results from a release of the protein from its normal surface attached location. The activated-state pellet fraction results from the insertion of the protein into the membrane. This is easier to obtain by the phorbol esters than by the diacylglycerol, as the latter is more transient.

Diacylglycerol is distinctive in another way from the most membrane lipids: it can "flip" within the membrane, although this is somewhat modified in biological systems [87]. Thus, while phospholipids are effectively anchored on one face, diacylglycerols can rapidly translocate between membrane faces.

REFERENCES

1. Cook, P. F., and Cleland, W. W. (2007) *Enzyme Kinetics and Mechanism*, Garland Science, London, New York.
2. Cornish-Bowden, A. (2004) *Fundamentals of Enzyme Kinetics*, 3rd ed., Portland Press, London.
3. Berridge, M. J., Downes, C. P., and Hanley, M. R. (1989) Neural and developmental actions of lithium: A unifying hypothesis. *Cell* 59, 411–419.
4. Merkler, D. J., and Schramm, V. L. (1993) Catalytic mechanism of yeast adenosine 5'-monophosphate deaminase. Zinc content, substrate specificity, pH studies, and solvent isotope effects. *Biochemistry (Mosc)*. 32, 5792–5799.
5. Soling, H. D., Bernhard, G., Kuhn, A., and Luck, H. J. (1977) Inhibition of phosphofructokinase by fructose 1 6-diphosphatase in mammalian systems: Protein–protein interaction or fructose 1 6-diphosphate trapping? *Arch. Biochem. Biophys.* 182, 563–572.

6. Hu, H., Juvekar, A., Lyssiotis, C. A., Lien, E. C., Albeck, J. G., Oh, D., Varma, G., Hung, Y. P., Ullas, S., Lauring, J., Seth, P., Lundquist, M. R., Tolan, D. R., Grant, A. K., Needleman, D. J., Asara, J. M., Cantley, L. C., and Wulf, G. M. (2016) Phosphoinositide 3-kinase regulates glycolysis through mobilization of aldolase from the actin cytoskeleton. *Cell* 164, 433–446.

7. Kim, J. H., Kim, E. L., Lee, Y. K., Park, C. B., Kim, B. W., Wang, H. J., Yoon, C. H., Lee, S. J., and Yoon, G. (2011) Decreased lactate dehydrogenase B expression enhances claudin 1-mediated hepatoma cell invasiveness via mitochondrial defects. *Exp. Cell Res.* 317, 1108–1118.

8. Tang, H., and Goldberg, E. (2009) Homo sapiens lactate dehydrogenase c (Ldhc) gene expression in cancer cells is regulated by transcription factor Sp1, CREB, and CpG island methylation. *J. Androl.* 30, 157–167.

9. Yin, M., and Ochs, R. S. (2003) Mechanism for the paradoxical inhibition and stimulation of calcineurin by the immunosuppresive drug tacrolimus (FK506). *Arch. Biochem. Biophys.* 419, 207–213.

10. Phiel, C. J., and Klein, P. S. (2001) Molecular targets of lithium action. *Annu. Rev. Pharmacol. Toxicol.* 41, 789–813.

11. Fell, D. (1997) *Understanding the Control of Metabolism*, Portland Press, London.

12. Cornish-Bowden, A. (1986) Why is uncompetitive inhibition so rare? A possible explanation, with implications for the design of drugs and pesticides. *FEBS Lett.* 203, 3–6.

13. Schramm, V. L. (2015) Transition States and transition state analogue interactions with enzymes. *Acc. Chem. Res.* 48, 1032–1039.

14. Thomas, K., Cameron, S. A., Almo, S. C., Burgos, E. S., Gulab, S. A., and Schramm, V. L. (2015) Active site and remote contributions to catalysis in methylthioadenosine nucleosidases. *Biochemistry (Mosc).* 54, 2520–2529.

15. Zoi, I., Motley, M. W., Antoniou, D., Schramm, V. L., and Schwartz, S. D. (2015) Enzyme homologues have distinct reaction paths through their transition states. *J. Phys. Chem. B* 119, 3662–3668.

16. Ochs, R. S. (2010) The problem with double reciprocal plots. *Curr. Enzyme Inhibit.* 6, 164–169.

17. Lineweaver, H., and Burk, D. (1934) The determination of enzyme dissociation constants. *J. Am. Chem. Soc.* 56, 658–666.

18. Rusnak, F., and Mertz, P. (2000) Calcineurin: Form and function. *Physiol. Rev.* 80, 1483–1521.

19. Etzkorn, F. A., Chang, Z. Y., Stolz, L. A., and Walsh, C. T. (1994) Cyclophilin residues that affect noncompetitive inhibition of the protein serine phosphatase activity of calcineurin by the cyclophilin.cyclosporin A complex. *Biochemistry (Mosc).* 33, 2380–2388.

20. Pessah, I. N., and Zimanyi, I. (1991) Characterization of multiple [3H]ryanodine binding sites on the Ca^{2+} release channel of sarcoplasmic reticulum from skeletal and cardiac muscle: Evidence for a sequential mechanism in ryanodine action. *Mol. Pharmacol.* 39, 679–689.

21. Kettlun, C., Gonzalez, A., Rios, E., and Fill, M. (2003) Unitary Ca^{2+} current through mammalian cardiac and amphibian skeletal muscle ryanodine receptor Channels under near-physiological ionic conditions. *J. Gen. Physiol.* 122, 407–417.

22. Kalgutkar, A. S., Kozak, K. R., Crews, B. C., Hochgesang, G. P., Jr., and Marnett, L. J. (1998) Covalent modification of cyclooxygenase-2 (COX-2) by 2-acetoxyphenyl alkyl sulfides, a new class of selective COX-2 inactivators. *J. Med. Chem.* 41, 4800–4818.

23. Walsh, C. (2006) *Posttranslational Modification of Proteins: Expanding Nature's Inventory*, Roberts and Co. Publishers, Englewood, CO.

24. Rognstad, R. (1979) Rate-limiting steps in metabolic pathways. *J. Biol. Chem.* 254, 1875–1878.

25. Stubbs, M., Vignais, P. V., and Krebs, H. A. (1978) Is the adenine nucleotide translocator rate-limiting for oxidative phosphorylation? *Biochem. J.* 172, 333–342.

26. Ochs, R. S., and Harris, R. A. (1980) Aminooxyacetate inhibits gluconeogenesis by isolated chicken hepatocytes. *Biochim. Biophys. Acta* 632, 260–269.

27. Snell, E. E. (1990) Vitamin B6 and decarboxylation of histidine. *Ann. N. Y. Acad. Sci.* 585, 1–12.

28. Okazaki, S., Suzuki, A., Mizushima, T., Kawano, T., Komeda, H., Asano, Y., and Yamane, T. (2009) The novel structure of a pyridoxal 5'-phosphate-dependent fold-type I racemase, α-amino-ε-caprolactam racemase from *Achromobacter obae*. *Biochemistry (Mosc.)* 48, 941–950.

29. Helmreich, E. J. (1992) How pyridoxal 5'-phosphate could function in glycogen phosphorylase catalysis. *Biofactors* 3, 159–172.

30. Montioli, R., Voltattorni, C. B., and Bertoldi, M. (2016) Parkinson's disease: Recent updates in the identification of human dopa decarboxylase inhibitors. *Curr. Drug Metab.* 17, 513–518.

31. John, R. A., and Charteris, A. (1978) The reaction of amino-oxyacetate with pyridoxal phosphate-dependent enzymes. *Biochem. J.* 171, 771–779.

32. Badawy, A. A., and Bano, S. (2016) Tryptophan metabolism in rat liver after administration of tryptophan, kynurenine metabolites, and kynureninase inhibitors. *Int. J. Tryptophan Res.* 9, 51–65.

33. Szewczyk, A., and Piku–a, S. a. (1998) Adenosine 5'-triphosphate: An intracellular metabolic messenger. *Biochim. Biophys. Acta—Bioenerg.* 1365, 333–353.

34. Mahmmoud, Y. A. (2008) Capsaicin stimulates uncoupled ATP hydrolysis by the sarcoplasmic reticulum calcium pump. *J. Biol. Chem.* 283, 21418–21426.

35. Strnad, C. F., Parente, J. E., and Wong, K. (1986) Use of fluoride ion as a probe for the guanine nucleotide-binding protein involved in the phosphoinositide-dependent neutrophil transduction pathway. *FEBS Lett.* 206, 20–24.

36. Chan, C. P., Gallis, B., Blumenthal, D. K., Pallen, C. J., Wang, J. H., and Krebs, E. G. (1986) Characterization of the phosphotyrosyl protein phosphatase activity of calmodulin-dependent protein phosphatase. *J. Biol. Chem.* 261, 9890–9895.

37. Lange, A. J., Arion, W. J., Burchell, A., and Burchell, B. (1986) Aluminum ions are required for stabilization and inhibition of hepatic microsomal glucose-6-phosphatase by sodium fluoride. *J. Biol. Chem.* 261, 101–107.

38. Blackmore, P. F., Bocckino, S. B., Waynick, L. E., and Exton, J. H. (1985) Role of a guanine nucleotide-binding regulatory protein in the hydrolysis of hepatocyte phosphatidylinositol 4 5-bisphosphate by calcium-mobilizing hormones and the control of cell calcium. Studies utilizing aluminum fluoride. *J. Biol. Chem.* 260, 14477–14483.

39. Dobrydneva, Y., and Blackmore, P. (2001) 2-Aminoethoxydiphenyl borate directly inhibits store-operated calcium entry channels in human platelets. *Mol.Pharmacol.* 60, 541–552.

40. Missiaen, L., Callewaert, G., De Smedt, H., and Parys, J. B. (2001) 2-Aminoethoxydiphenyl borate affects the inositol 1,4,5-trisphosphate receptor, the intracellular Ca^{2+} pump and the non-specific Ca^{2+} leak from the non-mitochondrial Ca^{2+} stores in permeabilized A7r5 cells. *Cell Calcium* 29, 111–116.

41. Maruyama, T., Kanaji, T., Nakade, S., Kanno, T., and Mikoshiba, K. (1997) 2APB, 2-aminoethoxydiphenyl borate, a membrane-penetrable modulator of Ins(1,4,5)P-3-induced Ca^{2+} release. *J. Biochem. Tokyo.* 122, 498–505.

42. Vaca, L., and Sampieri, A. (2002) Calmodulin modulates the delay period between release of calcium from internal stores and activation of calcium influx via endogenous TRP1 channels. *J. Biol. Chem.* 277, 42178–42187.

43. Blackmore, P. F., El-Refai, M. F., Dehaye, J. P., Strickland, W. G., Hughes, B. P., and Exton, J. H. (1981) Blockade of hepatic alpha-adrenergic receptors and responses by chlorpromazine and trifluoperazine. *FEBS Lett.* 123, 245–248.

44. Wang, L., Zhang, L., Li, S., Zheng, Y., Yan, X., Chen, M., Wang, H., Putney, J. W., and Luo, D. (2015) Retrograde regulation of STIM1-Orai1 interaction and store-operated Ca^{2+} entry by calsequestrin. *Sci. Rep.* 5, 11349.

45. Moreno-Sanchez, R., Marin-Hernandez, A., Del Mazo-Monsalvo, I., Saavedra, E., and Rodriguez-Enriquez, S. (2017) Assessment of the low inhibitory specificity of oxamate, aminooxyacetate and dichloroacetate on cancer energy metabolism. *Biochim. Biophys. Acta* 1861, 3221–3236.

46. LaNoue, K. F., and Williamson, J. R. (1971) Interrelationships between malate-aspartate shuttle and citric acid cycle in rat heart mitochondria. *Metabolism.* 20, 119–140.

47. Murphy, E., Coll, K. E., Viale, R. O., Tischler, M. E., and Williamson, J. R. (1979) Kinetics and regulation of the glutamate-aspartate translocator in rat liver mitochondria. *J. Biol. Chem.* 254, 8369–8376.

48. Hoek, J. B., Coll, K. E., and Williamson, J. R. (1983) Kinetics of glutamate efflux in rat liver mitochondria. *J. Biol. Chem.* 258, 54–58.

49. Williamson, J. R., and Cooper, R. H. (1980) Regulation of the citric acid cycle in mammalian systems. *FEBS Lett.* 117 Suppl, K73–K85.

50. Olson, A. L., and Pessin, J. E. (1996) Structure, function, and regulation of the mammalian facilitative glucose transporter gene family. *Annu. Rev. Nutr.* 16, 235–256.

51. Kellett, G. L., Brot-Laroche, E., Mace, O. J., and Leturque, A. (2008) Sugar absorption in the intestine: The role of GLUT2. *Annu. Rev. Nutr.* 28, 35–54.

52. Rogatzki, M. J., Ferguson, B. S., Goodwin, M. L., and Gladden, L. B. (2015) Lactate is always the end product of glycolysis. *Front. Neurosci.* 9, 22.

53. Dennis, S. C., Kohn, M. C., Anderson, G. J., and Garfinkel, D. (1985) Kinetic analysis of monocarboxylate uptake into perfused rat hearts. *J. Mol. Cell. Cardiol.* 17, 987–995.

54. Nicholls, D. G., and Ferguson, S. J. (2013) Bioenergetics. 4th ed. Ed.

55. Hille, B. (2001) *Ion Channels of Excitable Membrane*, Sinauer Associates Inc., Sunderland, MA.

56. Goforth, P. B., Bertram, R., Khan, F. A., Zhang, M., Sherman, A., and Satin, L. S. (2002) Calcium-activated K^+ channels of mouse beta-cells are controlled by both store and cytoplasmic Ca^{2+}: Experimental and theoretical studies. *J. Gen. Physiol.* 120, 307–322.

57. Flagg, T. P., Enkvetchakul, D., Koster, J. C., and Nichols, C. G. (2010) Muscle KATP channels: Recent insights to energy sensing and myoprotection. *Physiol. Rev.* 90, 799–829.

58. Rorsman, P., Ramracheya, R., Rorsman, N. J., and Zhang, Q. (2014) ATP-regulated potassium channels and voltage-gated calcium channels in pancreatic alpha and beta cells: Similar functions but reciprocal effects on secretion. *Diabetologia* 57, 1749–1761.

59. Cotton, C. U., Stutts, M. J., Knowles, M. R., Gatzy, J. T., and Boucher, R. C. (1987) Abnormal apical cell membrane in cystic fibrosis respiratory epithelium. *J. Clin. Invest.* 79, 80–85.

60. Berg, O. G., Gelb, M. H., Tsai, M.-D., and Jain, M. K. (2001) Interfacial enzymology: The secreted phospholipase A_2-paradigm. *Chem. Rev.* 101, 2613–2654.

61. Cai, A., Zheng, D., Qiu, R., Mai, W., and Zhou, Y. (2013) Lipoprotein-associated phospholipase A_2 (Lp-PLA(2)): A novel and promising biomarker for cardiovascular risks assessment. *Dis. Markers* 34, 323–331.

62. Kokotou, M. G., Limnios, D., Nikolaou, A., Psarra, A., and Kokotos, G. (2017) Inhibitors of phospholipase A_2 and their therapeutic potential: An update on patents (2012–2016). *Exp. Opin. Therap. Pat.* 27, 217–225.

63. Verger, R., and de Haas, G. H. (1976) Interfacial enzyme kinetics of lipolysis. *Ann. Rev. Biophys. Bioeng.* 5, 77–117.

64. Jain, M. K., and Berg, O. G. (1989) The kinetics of interfacial catalysis by phospholi-pase A_2 and regulation of interfacial activation: Hopping versus scooting. *Biochim. Biophys. Acta* 1002, 127–156.

65. Hirata, F., Schiffmann, E., Venkatasubramanian, K., Salomon, D., and Axelrod, J. (1980) A phospholipase A_2 inhibitory protein in rabbit neutrophils induced by gluco-corticoids. *Proc. Natl. Acad. Sci.* 77, 2533–2536.

66. Conricode, K. M., and Ochs, R. S. (1989) Mechanism for the inhibitory and stimulatory actions of proteins on the activity of phospholipase A_2. *Biochim. Biophys. Acta* 1003, 36–43.

67. Ahn, N. G., Teller, D. C., Bienkowski, M. J., McMullen, B. A., Lipkin, E. W., and de Haen, C. (1988) Sedimentation equilibrium analysis of five lipocortin-related phospho-lipase A_2 inhibitors from human placenta. Evidence against a mechanistically relevant association between enzyme and inhibitor. *J. Biol. Chem.* 263, 18657–18663.

68. Pappa, H., Murray-Rust, J., Dekker, L. V., Parker, P. J., and McDonald, N. Q. (1998) Crystal structure of the C2 domain from protein kinase C-δ. *Structure* 6, 885–894.

69. Bullen, J. W., Jr., Bluher, S., Kelesidis, T., and Mantzoros, C. S. (2007) Regulation of adiponectin and its receptors in response to development of diet-induced obesity in mice. *Am. J. Physiol. Endocrinol. Metab.* 292, E1079–E1086.

70. Bazzi, M., and Nelsestuen, G. L. (1987) Role of substrate in determining the phospho-lipid specificity of protein kinase C activation. *Biochem* 26, 5002–5008.

71. Darbon, J. M., Oury, F., Clamens, S., and Bayard, F. (1987) TPA induces subcellular translocation and subsequent down-regulation of both probol ester binding and protein kinase C activities in MCF-7 cells. *Biochem. Biophys. Res. Commun.* 146, 537–546.

72. Hannun, Y. A., Loomis, C. R., and Bell, R. M. (1986) Protein kinase C activation in mixed micelles. Mechanistic implications of phospholipid, diacylglycerol, and calcium interdependencies. *J. Biol. Chem.* 261, 7184–7190.

73. Kazanietz, M. G. (2005) Targeting protein kinase C and "non-kinase" phorbol ester receptors: Emerging concepts and therapeutic implications. *Bba. Proteins Proteom.* 1754, 296–304.

74. Kazanietz, M. G. (2000) Eyes wide shut: Protein kinase C isozymes are not the only receptors for the phorbol ester tumor promoters. *Molec. Carcinog.* 28, 5–11.

75. Exton, J. H. (1994) Phosphatidylcholine breakdown and signal transduction. *Biochim. Biophys. Acta Lipids Lipid Metab.* 1212, 26–42.

76. Exton, J. H. (1990) Signaling through phosphatidylcholine breakdown. *J. Biol. Chem.* 265, 1–4.

77. Lerner, I., Trus, M., Cohen, R., Yizhar, O., Nussinovitch, I., and Atlas, D. (2006) Ion interaction at the pore of Lc-type Ca^{2+} channel is sufficient to mediate depolarization-induced exocytosis. *J. Neurochem.* 97, 116–127.

78. Inoue, M., Kishimoto, A., Takai, Y., and Nishizuka, Y. (1977) Studies on a cyclic nucleotide-independent protein kinase and its proenzyme in mammalian tissues. *J. Biol. Chem.* 252, 7610–7616.

79. Takai, Y., Kishimoto, A., Iwasa, Y., Kawahara, Y., Mori, T., and Nishizuka, Y. (1979) Calcium-dependent activation of a multifunctional protein kinase by membrane phos-pholipids. *J. Biol. Chem.* 254, 3692–3695.

80. Bazzi, M. D., and Nelsestuen, G. L. (1991) Proteins that bind calcium in a phospholipid-dependent manner. *Biochemistry (Mosc.)* 30, 971–979.

81. Jain, M. K. (1988) *Introduction to Biological Membranes*, 2nd ed., Wiley, New York.

82. Basanez, G., Nieva, J. L., Rivas, E., Alonso, A., and Goni, F. M. (1996) Diacylglycerol and the promotion of lamellar-hexagonal and lamellar-istropic phase transitions in lip-ids: Implications for membrane fusion. *Biophys. J.* 70, 2299–2306.

83. Chabre, M. (1987) The G protein connection: Is it in the membrane or the cytoplasm? *TIBS* 12, 213–215.

84. Singer, S. J. (1962) The properties of proteins in nonaqueous solvents. *Adv. Protein Chem.* 17, 1–68.
85. Singer, S. J., and Nicolson, G. L. (1972) The fluid mosaic model of the structure of cell membranes. *Science* 175, 720–731.
86. Yin, M., and Ochs, R. S. (2001) A mechanism for the partial insertion of protein kinase C into membranes. *Biochem. Biophys. Res. Commun.* 281, 1277–1282.
87. Ueda, Y., Ishitsuka, R., Hullin-Matsuda, F., and Kobayashi, T. (2014) Regulation of the transbilayer movement of diacylglycerol in the plasma membrane. *Biochimie* 107 Pt A, 43–50.

ENDNOTES

i Years ago, a speaker commented on the wide use of this term, and pointed out that it is usually advanced by those without much actual housekeeping experience. For example, she noted that housekeeping itself is variable; if company is expected, the effort must be ramped up.

ii In more realistic enzyme mechanisms, more forms are present. For example, we might assume that ES leads to an EP complex before it dissociates into E plus P. In some cases, a covalent intermediate is formed, such as a phosphorylated enzyme. The fact remains that the simple mechanism usually represents the situation well enough, simply because only kinetically significant intermediates are important in a mechanism. Moreover, inhibitor binding to ES is similar to inhibitor binding to any intermediate enzyme form; any of these produce anti-competitive inhibition.

iii The naming of the two inhibition constants refers to their behavior in the double-reciprocal plot of $1/v_0$ versus $1/[S]$. The resulting patterns of straight lines that result can show either a change in the slope of the lines once inhibitor is present, which is the K_{is}, or a change in the intersection at the $1/v_0$ axis, K_{ii}.

iv It is a curious feature of pyridoxal phosphate that by far the majority of this cofactor is found attached (again by Schiff base formation) to muscle glycogen phosphorylase. However, in this case, AOA is not an inhibitor, as the Schiff base is not reactive in the course of catalysis: it is the phosphate itself that is used in the phosphorylase catalytic cycle [29].

v An exception is the glucose transporter of bacteria which both catalyze translocation as well as phosphorylation so that its enzymatic properties are evident.

vi For a thorough discussion of electrical changes in cells, virtually any text on physiology will provide insight. Deeper understanding can be achieved by considering the aforementioned treatise of Hille [55]. However, the origin of the membrane potential is a relatively simple matter: the combination of the NaKATPase, which extrudes more Na^+ than it takes in K^+, and the greater permeability of the membrane for K^+ than Na^+ (due to membrane channels, not the endogenous lipid), the inside is about 75 mV more negative than the extracellular space.

vii A separate issue of ongoing clinical interest is unrelated to the enzymatic activity of PLA *per se*: PLA is an immunoreactive component of venoms, such as those from bees and snakes. These secretions can be considered analogous to pancreatic PLA secretion; all of these enzymes are of the soluble type.

viii The use of the term CMC, which strictly should only apply to micelle formation is also broadly used for lipid aggregate formation, which is the meaning here. It is also known that CMC is a somewhat variable measure, as it depends strongly on conditions, in particular salt concentration and temperature.

6 Tissue-Specific Pathways

The pathways presented so far are present in virtually all cells. Actual pathways in cells or within comparable cells of different species are distinct, reflecting the cell's biological role. In this chapter, I focus on the differences between cell types, along with a consideration of species specificity.

The specialized metabolism of cells provides an awareness of what makes a cell unique. It is clearly not just a single enzymatic reaction or even a few in sequence that determines cell physiology; otherwise it would be possible to just force the expression of an enzyme by an exogenous vector and transform one known cell type into another.[i]

The examples chosen for this chapter represent well-studied cases. They are also selected to illustrate different uses of similar metabolic pathways.

6.1 LIVER

Despite its bland appearance, the liver has extensive metabolic complexity. In our discussion of liver metabolism, we are considering only the hepatocytes (parenchymal cells), rather than others, such as the second-most populous cells, the Kupffer cells, the fixed macrophage population for the organ.

From a physiological perspective, after ingestion of food, all of the metabolites absorbed by the digestive tract apart from the lipid-soluble ones are delivered first to the liver by the specialized portal blood circuit. Every major energy source involves the liver: glucose, lipid, and nitrogen metabolism have critical pathways in the liver that are required for the overall metabolic economy of the body. For carbohydrate metabolism, the liver is the first line in supplying the blood with more glucose when that energy substrate is in decline, through breakdown of glycogen and through gluconeogenesis. The liver is also the single source of urea formation in disposing of nitrogen and the catabolism of amino acids, as well as for the synthesis of most of the blood proteins. The liver can oxidize fatty acids and is the only source of the ketone bodies for the rest of the body, and is also a major site of triglyceride biosynthesis.

Liver is the source of bile formation and first-pass drug metabolism, although the later pathways are generally outside the purview of intermediary metabolism. Each of the principle tasks of the liver will be represented in the specialized pathways discussed.

As a guiding principle for understanding the liver, it is useful to consider the distinction between the fed and the fasted state. Typically, pathways and their control enzymes are more active in one state than the other. Thus, we consider "fed pathways" as glycogen synthesis, glycolysis, fatty acid biosynthesis, and cholesterol biosynthesis, and "fasted pathways" as glycogenolysis, gluconeogenesis, fatty acid oxidation, and urea synthesis.

6.1.1 GLYCOGEN METABOLISM

Using the feeding/fasting paradigm for glycogen metabolism is a straightforward matter. In the fed state, glycogen synthesis removes glucose from the circulation to provide a storage pool in the liver. In the fasted state, at the first stages, glycogenolysis produces glucose to increase the blood concentration to supply this sugar to peripheral cells. As fasting proceeds into several hours and beyond, gluconeogenesis assumes a greater importance, as we will consider in the next section.

The pathways of both glycogen synthesis (Figure 6.1) and breakdown (Figure 6.2) are relatively simple ones. This is made conceptually even simpler when we consider that it is just the key enzymes of each process—glycogen synthase for synthesis and glycogen phosphorylase for degradation—that are principally responsible for the regulation of glycogen metabolism.

6.1.2 GLYCOGEN SYNTHESIS

In the pathway shown in Figure 6.1, the formation of glycogen appears as a branch from the glycolytic pathway. It may be considered that glycogen arises from

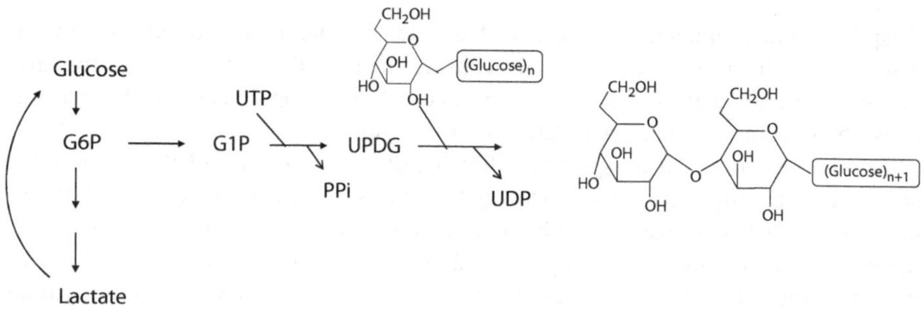

FIGURE 6.1 Glycogen synthesis. The pathway is shown to connect to glycolysis and its reverse, gluconeogenesis, from the intermediate G6P (glucose-6-P).

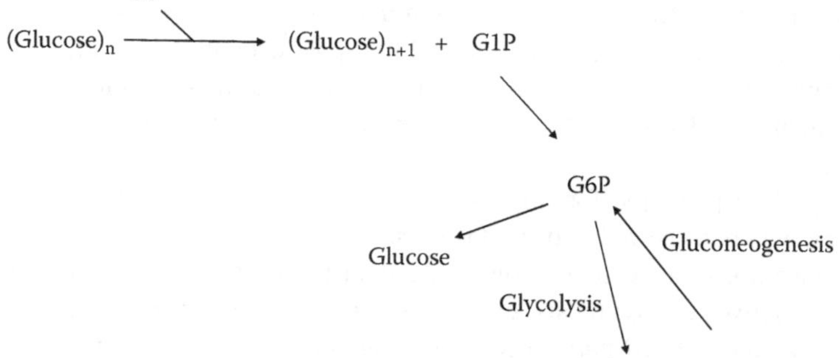

FIGURE 6.2 Glycogenolysis. Breakdown of glycogen produces G1P (glucose-1-P), which is in turn converted to G6P, connected to glucose output, glycolysis, and gluconeogenesis.

extracellular glucose, or from other precursors that form G6P. There are two further reactions in the pathway. First, there is the transfer of a portion of the outer several glucose units to the interior, a branching enzyme. Second, glycogen starts with a protein known as glycogenin, to which the glucose residues are attached at a tyrosine residue [4]. There is no evidence that either of these are events are subject to metabolic regulation; however, that does not rule out the possibility that there are conditions in which these ancillary steps play an important role.[ii] Still, the focus is largely on the pathway of Figure 6.1, and in particular on the control of glycogen synthase.

The synthase enzyme is well studied, subject to covalent modification as well as to allosteric control, an activation by glucose-6-P [7]. The covalent modification includes multiple sites of phosphorylation, including the classical as well as the more recently discovered GSK3 (glycogen synthase kinase 3), which has a wide range of cellular targets [8]. The covalent modulation of glycogen phosphorylase is simpler, as it is subject to phosphorylation on just one reside and has only one kinase acting on it [9–11].

In liver, glycogen synthesis—and correspondingly, glycogen synthase—is active under fed conditions, and enables the conversion of blood glucose into glycogen stores. This becomes the source of blood glucose supply in the post-prandial state, and is also partially responsible for the lowering of blood glucose after a carbohydrate-containing meal; the uptake into muscle (see below) is responsible for most of the total uptake.

As a conceptual matter, some may wonder why the cell needs glycogen storage; it would appear to be an elaborate construct to store glucose residues; perhaps just free glucose could accomplish the same thing. This, however, is a problem in physical chemistry: the osmolarity of large numbers of individual glucose molecules would be overwhelming. Instead, the massive glycogen molecule which can contain hundreds of thousands of glucose residues counts for purposes of osmotic pressure, as just one molecule.

6.1.3 GLYCOGENOLYSIS

The breakdown of glycogen to glucose phosphates (Figure 6.2) is also a well-regulated pathway in the liver [11]. Like the opposing synthesis pathway, glycogenolysis consists of only a few steps, and just one is the focal point, in this case, glycogen phosphorylase. The breakdown of the $\alpha1\rightarrow6$ glycosidic bonds (branch points) requires separate enzymatic machinery: a debranching enzyme to remove a stretch of glucosyl residues, and then a glucosidase to remove the last, single glucosyl residue. Thus, direct formation of glucose arises just from the branch point residues, but due to the presences of glucose phosphatase in liver, the major product of glycogenolysis is typically glucose released to the blood.

Glycogenolysis largely reflects the activity of glycogen phosphorylase, which is regulated by phosphorylation catalyzed by glycogen phosphorylase kinase [12]. The latter enzyme is regulated by phosphorylation by PKA resulting from an increase in cAMP, and is allostericaly activated by a rise in cytosolic Ca^{2+} concentration. Once activated, the phosphorylase kinase can activate glycogen phosphorylase.

Ca^{2+} activation results from a direct binding to calmodulin, which exists as a subunit of phosphorylase kinase.

We will consider the broad action of hormones and signaling in Chapter 7. Here it is important to note that the ability of glucagon to activate the breakdown of glycogen is through an increase in cAMP, then PKA, phosphorylase kinase, and phosphorylase activation. In fasting, a decline in blood glucose signals the rise in glucagon and the concomitant fall in insulin release from the endocrine pancreas, conditions which lead to glycogen breakdown in liver. Most of the glucosyl units are released to the blood stream under fasting conditions, so that the route taken between glucose production and glycolysis depicted in Figure 6.2 is determined by prevailing hormonal conditions. In the fasted state, G6P represents a point of convergence of glycogenolysis with gluconeogenesis, the hepatic pathway we consider next.

6.1.4   Gluconeogenesis

This pathway is commonly defined as the formation of glucose from non-carbohydrate precursors. The definition is a bit of a problem as, for example, fructose metabolism can produce glucose in the liver using a portion of the gluconeogenic pathway. Dihydroxyacetone is certainly a sugar, and while not present in significant quantities physiologically, is also a gluconeogenic substrate. Usually, we think of lactate or an amino acid like alanine as the pathway substrate and they fit this definition. Broadly, a better definition would include both non-carbohydrate precursors and monosaccharides other than glucose.

The pathway from lactate is presented in Figure 6.3. Here, lactate is converted in the cytosol to pyruvate, which then enters the mitochondria for conversion to oxaloacetate. As there is no mitochondrial oxaloacetate transporter, conversion to a molecule that has a transporter is necessary. Three possibilities are illustrated in

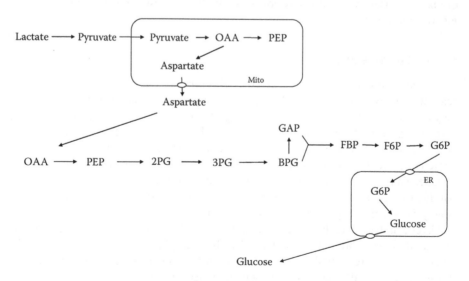

FIGURE 6.3   Gluconeogenesis from lactate.

Figure 6.4. First, a mitochondrial form of PEPCK can catalyze formation of PEP, which has a transporter to move it into the cytosol. However, there is in addition a cytosolic form of PEPCK, for which cytosolic oxaloacetate is required. In order to transport oxaloacetate carbon out of the mitochondria, the other two reactions of Figure 6.4 are possibilities. These are malate formation (via malate dehydrogenase) and aspartate formation (via aspartate aminotransferase). There are transporters for both malate and aspartate, and corresponding cytosolic enzymes to regenerate oxaloacetate. The malate route constitutes a redox shuttle, bringing reducing equivalents in the form of NADH from mitochondria to cytosol. Notice how this is the reverse of the malate/aspartate shuttle involved in moving reducing equivalents in the form of NADH from cytosol to mitochondria [13,14]. The aspartate route does not export reducing equivalents. In the overall pathway of gluconeogenesis from lactate (Figure 6.3), it is evident that it is mostly the reverse of glycolysis, except for bypasses at the metabolically irreversible steps: glucokinase, phosphofructokinase, and pyruvate kinase. Just as in glycolysis, glyceraldehyde phosphate DH involves the NAD/NADH cofactor pair. In gluconeogenesis, this reaction runs in the direction of oxidation, so that NADH is required for each three-carbon molecule flowing towards glucose formation. NADH is derived from the lactate DH step when lactate is the substrate, so there is a balance to the reducing equivalents. Cofactor balance is incomplete, however, because we can also see that ATP is required for the formation of glucose, and all of that ATP must be obtained from the mitochondria.[iii]

The shared enzymes of glycolysis and gluconeogenesis are near-equilibrium steps, but it is a mistake to imagine that they can actually go in both directions at the same time. There is a key distinction between "near-equilibrium" and "equilibrium." What near-equilibrium status means is that the reaction is poised to go in either direction. When gluconeogenesis is operating, flow through glyceraldehyde phosphate DH is in the direction of GAP formation, and there cannot be simultaneous flow in the opposing direction. That is also true of the ATP-generating reactions.

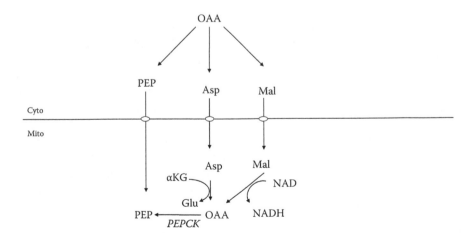

FIGURE 6.4 Fates of mitochondrial oxaloacetate. Three possibilities are: decarboxylation to PEP through PEPCK, transamination to aspartate (asp), and reduction to malate (mal).

Thus, all ATP must be generated by mitochondria. To do so, there must be an oxidizable substrate to supply electrons to the respiratory chain. If the only substrate available is lactate, then pyruvate will be oxidized by the mitochondria rather than being converted to oxaloacetate, and this means that an extra NADH will accumulate in the cytosol.

As noted in the prior chapter in the context of class inhibitors, aminooxyacetate, a transaminase inhibitor, blocks lactate gluconeogenesis in chicken hepatocytes [15,16]. The full significance of this can be appreciated now that gluconeogenesis is developed. Chickens have exclusively a mitochondrial PEPCK. Thus, gluconeogenesis involves stoichiometric PEP exit, after which the remainder of the pathway is the same in all species. While direct PEP exit obviates any need for oxaloacetate transfer, there remains an observed inhibition by the transaminase inhibitor, but only because lactate presented as the sole substrate requires export of its reducing equivalents to the mitochondria. The reason for this need is that some of the lactate carbon must enter the mitochondria as pyruvate and be oxidized to provide ATP for gluconeogenesis. The further addition of a fatty acid to supply that energy obviates this requirement, and relieves aminooxyacetate inhibition. The redox shuttle, not the gluconeogenic pathway, requires aspartate aminotransferase.

In rat liver, PEPCK is entirely cytosolic, the other extreme situation. Here, PEP export is not an option, but malate or aspartate are. For lactate gluconeogenesis in the rat liver, aspartate exit is the required pathway, due to the redox balance for this substrate.

Guinea pig and human liver have PEPCK in both the mitochondria and cytosol. Only the cytosolic form can be hormonally induced and thereby regulate gluconeogenesis. There are no known allosteric or covalent short-term modifiers of PEPCK. Rather, there is extensive control of gene expression. Several hormones, including glucagon and steroids, increase the mRNA and protein level of the enzyme [17–19].

From the point of view of pathways, the increase in PEPCK will cause an increase in the formation of PEP and thus drive the near-equilibrium reactions in the gluconeogenic direction.

The remaining steps of gluconeogenesis, while also subject to regulation, are less critical to the control of the pathway. The other regulated gluconeogenic steps are fructose bisphosphatase and glucose-6-phosphatase.

The first of these enzymes catalyzes the reverse flow of pathway intermediates of glycolysis, and represents in our terminology a *pathway-reversible* step from FBP back to F6P. Alternatively, the combination of PFK and FBPase can be considered one of the *substrate cycles* [20] of the glycolytic/gluconeogenic routes, because F6P and FBP constitute a very short cycle. The other name for this activity was more demonstrative: *futile cycle*, because adding up the full reactions in both directions amounts to a net hydrolysis of ATP to ADP and Pi, which, of course, seems futile. It has been argued that what is accomplished by this is a sensitive means of regulation of both pathways by simply altering one or both of the enzyme activities [21].

6.1.4.1 Pyruvate Gluconeogenesis and Its Significance

Having examined lactate gluconeogenesis up to the point of PEP formation, the route from pyruvate diverges only slightly. Clearly, providing very high concentrations to pyruvate (usually 10 mM compared to a blood concentration of about 0.1 mM) is not a physiological situation, but understanding this artificial pathway and its consequences is illuminating in surprising ways.

The route is illustrated in Figure 6.5. Pyruvate at high concentrations supports a substantial rate of gluconeogenesis, but produces an extremely oxidized state of NADH/NAD in both the cytosol and mitochondrial matrix. These are determined using the lactate/pyruvate ratio and the β-hydroxybutyrate/acetoacetate ratios, respectively. These ratios reflect the redox states of NADH/NAD in those spaces due to the established near-equilibrium nature of the enzymes lactate DH and β-hydroxybutyrate DH in their respective compartments. With lactate as substrate, the cytosolic ratio is about 10; with pyruvate as substrate, it is 1 or less. This is because presenting pyruvate as a substrate forces the lactate DH in the direction of lactate formation (as illustrated for the LDH reaction of Figure 6.5), which in turn oxidizes cytosolic NADH. Pyruvate entering mitochondria becomes carboxylated to OAA, after which reduction to malate is the only pathway available when cytosolic PEPCK is used. It is, of course, also possible to use mitochondrial PEPCK, but that enzyme is not inducible, so it can only provide PEP for gluconeogenesis under conditions when the pathway is least active. With the exit of malate, NADH is provided to the cytosol, and the remaining steps are the same as for lactate as substrate.

The route of malate exit with pyruvate as substrate and aspartate exit with lactate as substrate are the two extremes that are also of importance for other pathways, in particular urea synthesis, as discussed in the next section. However, the response of pyruvate gluconeogenesis to the presence of cyclic AMP (generated in the hepatocyte upon the binding of the hormone glucagon and the β-adrenergic action of epinephrine) provides some justification for studying an apparently non-physiological situation [22].

Glucagon is well known to stimulate gluconeogenesis as part of its overall action in increasing blood glucose. Through its ability to increase cellular cAMP, and subsequently PKA (protein kinase A), this action is known to be exerted in the short term (that is, without protein synthesis involvement) at pyruvate kinase and PFK.

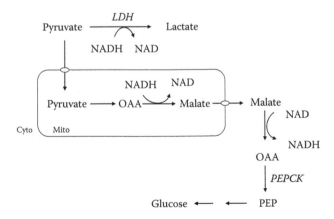

FIGURE 6.5   Gluconeogenesis from pyruvate.

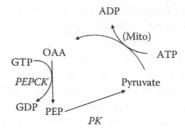

FIGURE 6.6 Cycle between pyruvate and PEP. For gluconeogenesis, PEP formation is catalyzed by PEPCK; for glycolysis, PEP is converted to pyruvate by PK (pyruvate kinase). However, in both pathways, all of the activities shown are active, constituting a substrate cycle.

The first to be established was the inhibition of pyruvate kinase, which is known to be directly phosphorylated by PKA and inactivated. However, *in vitro* demonstration of this action at the enzyme level alone is not sufficient evidence for its action in an intact cell. While a crossover plot showed evidence of inhibition by cyclic AMP between PEP and pyruvate, this could involve any of the steps illustrated in Figure 6.6.

Experiments were conducted with rat hepatocytes, which have only the cytosolic form of PEPCK, so that only malate or aspartate exit is possible for the provision of cytosolic PEP (Figures 6.4 and 6.5). Pyruvate gluconeogenesis and lactate production by rat hepatocytes is *inhibited* by cyclic AMP [22]. In lactate gluconeogenesis, diminishing substrate cycling through pyruvate kinase leads to stimulation of lactate gluconeogenesis, because less carbon is diverted from glucose formation. The same substrate cycling causes inhibition of pyruvate gluconeogenesis, because the cycling itself provides additional carbon (that is beyond gluconeogenic flux) to bring out more malate to the cytosol. This is because the situation with high pyruvate is depleting the cytosolic NADH through the lactate DH reaction, and more than one malate molecule is needed to provide enough NADH for GAPDH. From this interpretation, we would expect that another effect of cyclic AMP on the hepatocyte would be to cause a further oxidation of cytosolic NADH, which was also observed.

6.1.4.2 Lag in Lactate Gluconeogenesis

A similar problem in redox supply exists for lactate gluconeogenesis, which shows up as a lag in the early phase of the incubation. It was observed [23] that in the initial 15 min or so of incubation of lactate with rat hepatocytes that there is a very low rate of gluconeogenesis, which rises to a steady state during the subsequent incubation. A clue to the situation was that there was a relatively high NADH/NAD redox state that also leveled off corresponding to the rate of gluconeogenesis. This could be obviated by providing a small amount of pyruvate as well as lactate; with a 10:1 ratio of lactate to pyruvate, the lag was obviated. The conclusion was that the intermediates needed to provide the transamination steps for lactate so that aspartate exit could ensure were at low levels and that they are being generated during that lag

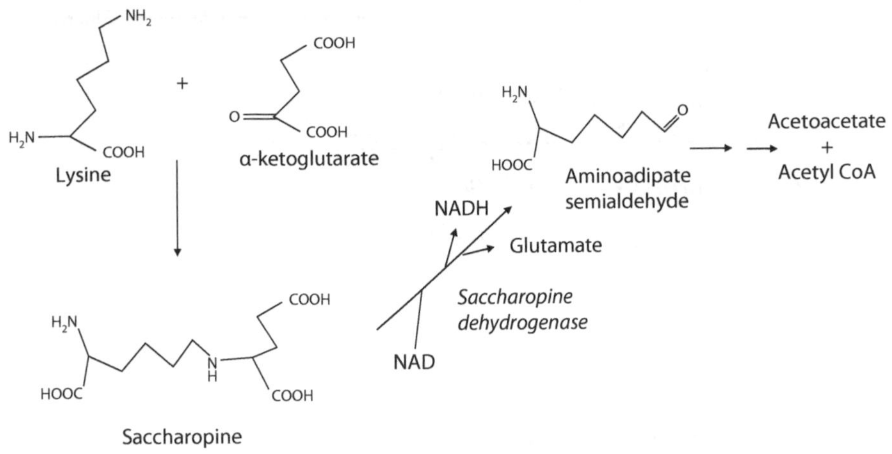

FIGURE 6.7 Lysine degradation through the saccharopine pathway.

phase. One way to provide them was to add lysine to the incubation [23], which itself can be catabolized to provide a small amount of glutamate through the saccharopine pathway (Figure 6.7). In a similar way to the inclusion of a small amount of pyruvate with lactate as substrate, lysine addition eliminated the lag phase. Presumably, during the lag phase in the presence of only lactate as substrate, proteolysis eventually provided amino acids that were able to supply the intermediates of redox shuttle. Note that only small amounts of these intermediates are necessary as they are regenerated in the pathway. They play a role similar to cofactors or metabolic cycles.

6.1.4.3   Gluconeogenesis from Dihydroxyacetone and Glycerol

Dihydroxyacetone and glycerol differ only by a two-electron reduction. The first substrate is used experimentally; the second is physiologically important as it arises from triglyceride hydrolysis that occurs largely in adipocytes.

Glycerol is converted to glycerol phosphate by the triose kinase, and subsequently to DHAP by glycerol phosphate dehydrogenase. That enzyme is also a part of one of the reducing equivalent shuttles, moving cytosolic NADH electrons into QH_2 in the mitochondrial membrane (Figure 3.9). Supplying the liver cells with glycerol produces a very high concentration of cytosolic NADH, attenuating glycolytic flow through GAPDH. The flow is thus upwards towards glucose (Figure 6.8). NADH in this case is not needed for gluconeogenesis, but can be transferred to mitochondria, by either the glycerol-P or the malate/aspartate shuttle. Glycerol cannot be metabolized in the adipocyte because this tissue lacks glycerol kinase; in fact, glycerol concentration in the blood is used as an index of lipolysis [24]. While this ignores the fact that glycerol can be consumed by liver in the formation of glucose, the assumption is that the rate of lipolysis by the adipocyte exceeds the rate of hepatic gluconeogenesis from glycerol [25].

As evidence that glycerol can utilize the GOL shuttle for the transfer of reducing equivalents, stimulation of glycerol gluconeogenesis in rat hepatocytes can be demonstrated by hormones that increase cytosolic Ca^{2+}, such as vasopressin and the

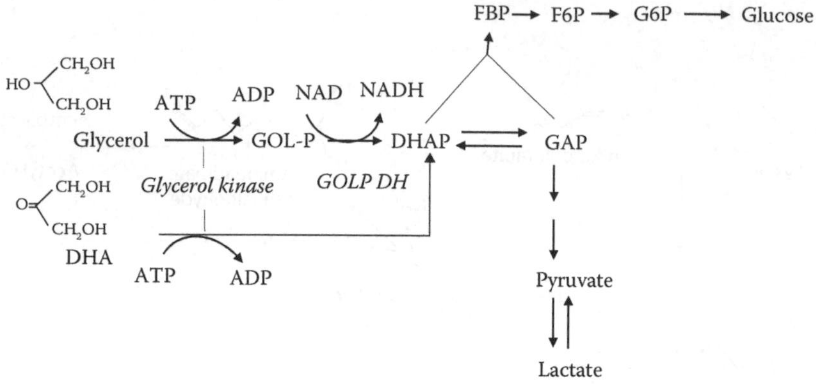

FIGURE 6.8 Glycerol and DHA metabolism.

α_1-linked catecholamines [26]. This is likely the result of stimulation of the mitochondrial portion of this shuttle, the glycerol phosphate oxidase, which catalyzes the metabolically irreversible conversion of glycerol phosphate to DHAP on the outer face of the mitochondrial membrane, while Q is reduced to QH_2 within the membrane [27,28]. In the case of glycerol gluconeogenesis [29], not only does an increased NADH prevent flow in the glycolytic direction through glyceraldehyde phosphate DH, but the first step of the pathway itself is limited by the removal of NADH. Thus, this nucleotide regulates flow in both directions from glycerol-P.

With DHA (dihydroxyacetone) a substrate, significant amounts of lactate (glycolytic flow) and glucose (gluconeogenic flow) form (Figure 6.8). Lactate (and pyruvate) formation from DHA is slight underestimate of glycolysis, as some pyruvate carbon is oxidized. Measured in livers from fasted animals, with glycogen essentially depleted, interpretation of glucose formation can be entirely attributed to gluconeogenesis. Since ATP is generated through the glycolytic limb, less ATP is needed with DHA than with lactate as a substrate for gluconeogenesis. In fact, for DHA gluconeogenesis, the two ATPs needed for the triose phosphates can be supplied by glycolytic portion. Some NADH must be exported from the mitochondria, as with pyruvate gluconeogenesis, but there is a difference. In the case of DHA metabolism, there is no dead end to accumulate: in place of lactate formation, DHAP is in near equilibrium only with phosphorylated intermediates.

As an example of the use of DHA gluconeogenesis, consider the experiment in which the fatty acid oleate is added to determine its action on these hepatic pathways [30]. Glucose formation increases, lactate formation decreases, and citrate accumulates. These actions can be replicated by adding the ketone β-hydroxybutyrate, so that fatty acid oxidation *per se* and not the fatty acid molecule is responsible (the ketone body can form only mitochondria NADH in liver as there is no further metabolism in this tissue). Inhibition of the glycolytic portion results from a smaller demand on cellular ATP by that pathway. An increase in gluconeogenesis is driven by the diversion of the carbon in that direction, and a stimulation by citrate of FBPase.

A similar experiment with more sophisticated analysis was performed more recently [31]. In this case, 2-^{13}C-labelled DHA was injected into mice and perfused livers analyzed in an nmr tube. The technique of hyperpolarized nmr was used, which both increases sensitivity by about four orders of magnitude, and requires very rapid measurement as the hyperpolarized state decays.[iv] The result was that glycolysis from DHA was greater in livers from fed animals, and gluconeogenesis greater in those from fasted animals. There was also label in the gluconeogenic direction in feeding and in glycolysis in fasting. While the investigators suggested that this might indicate two compartments for each process, it could instead simply a labelling phenomenon of intermediates. For example, given the existence of substrate cycling, it is not surprising to find label in reaction steps that oppose the steady-state flux direction, as discussed above. It is difficult to determine with the technical limitation of the experimental setup (analysis must be completed in seconds) if steady state is actually achieved. The future of this type of measurement—which already greatly improves sensitivity of this nondestructive method—may resolve such issues.

6.1.4.4 Gluconeogenesis from Fructose

Fructose metabolism in liver resembles DHA in that both enter the pathway at the triose level and produce glucose and lactate [33]. Fructose enters cells via the GLUT4 transporter and uses two tissue specific enzymes, fructokinase and aldolase B (Figure 6.9). As a substrate of fructokinase (and a poor substrate of glucokinase), fructose is phosphorylated at the 1-position rather than the 6. The resulting fructose-1-P is a substrate for aldolase B, which requires anchoring at only one phosphoryl group, forming the products DHAP and glyceraldehyde. The glyceraldehyde is phosphorylated by triose kinase[v] and we now have the two triose phosphates that are at near-equilibrium in the triose phosphate isomerase reaction. While fructose and

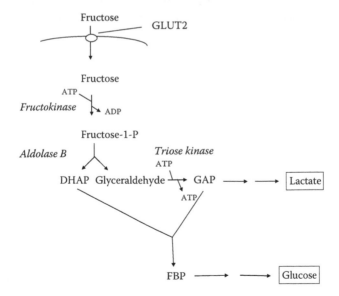

FIGURE 6.9 Fructose utilization.

DHA share the same entry points, fructose is unique as a substrate as it can cause ATP depletion, an action illustrated by an untoward use of this sugar as a nutrient for alcoholics.

Because fructose entry into the glycolytic pathway bypasses the PFK step, it was believed that it would provide readily available nutrition for alcoholics [34]. It is common that alcoholics skip meals, and a large dose of fructose seemed like a way of preventing a hypoglycemia that is characteristic of this particular state. Ethanol inhibits gluconeogenesis from lactate because it depresses the oxaloacetate concentration through near-equilibrium displacement at malate dehydrogenase [35,36] (Figure 6.10). The decreased OAA concentration limits gluconeogenesis at PEPCK. While fructose therapy would work in principle, in these patients, it caused liver failure and death. The problem stems from fructose metabolism. In bypassing the PFK reaction, fructose is limited instead by fructokinase, which is more active than the subsequent reaction, aldolase B. The fructokinase product fructose-1-P accumulates and ATP levels are lowered correspondingly. It can be demonstrated that liver ATP depletion occurs within minutes [37]. As discussed in the prior chapter, ATP is normally at a constant concentration in cells (despite early and some recent studies which suggest ATP as an energy barometer) and its depletion can be used to track cell viability. It is possible that for very short periods of time a transient drop in ATP can be restored, at least in isolated cell experiments, and there is also evidence that this might be possible under certain conditions in heavily exercised muscle. In the case of fructose metabolism, isolated cells have a lag phase after which the fructose-1-P has dropped sufficiently for the cells to perform gluconeogenesis by the same pathway as for DHA. The distinction is, the two routes are separated in time: glycolysis for the first several minutes and then gluconeogenesis. Fructose-1-P can also activate pyruvate kinase during the first phase. Normally, pyruvate kinase is activated by FBP in glycolysis, a feed-forward activation in the pathway from glucose. In fructose metabolism, F1P can also activate pyruvate kinase due to its high concentration.

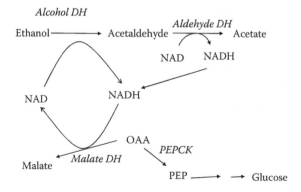

FIGURE 6.10 Ethanol lowers OAA concentration and inhibits gluconeogenesis. NADH production in the oxidation of ethanol reduces OAA to malate, lowering the substrate for PEPCK and inhibiting gluconeogenesis.

It should be noted that fructose metabolism is more generally a concern because of its conversion to lipids in the liver, for similar metabolic reasons to its actions in gluconeogenesis: bypass of the PFK reaction and depletion of cellular ATP [38].

6.1.4.5   Galactose Gluconeogenesis

Galactose is an unusual substrate for gluconeogenesis. It might be considered instead as a glycolytic precursor since it enters the pathway at G6P; as a gluconeogenic precursor, the pathway requires just one more step. However, as pointed out above, a substrate can be either gluconeogenic or glycolytic. With our more inclusive definition of gluconeogenesis, we consider substrates that lead to an elevated blood glucose from the liver through steps leading to G6P and its subsequent hydrolysis. The formation of G6P from galactose was shown in Chapter 3 (Figure 3.18). Note that it has a similarity to fructose metabolism in that the 1-position is phosphorylated by a dedicated hepatic enzyme, in this case, galactokinase. The formation of a UDP sugar and subsequent conversion to UDP-glucose use distinct hepatic enzymes. Unlike fructokinase, galactokinase does not proceed as rapidly, and ATP is not depleted. As most galactose arises from milk products as lactose, it is unlikely that a problem similar to fructose arises. More to the point, flow through the glycolytic route requires PFK, which means that its entry into the pathway is modulated in a similar way to glucose itself.

6.1.5   Urea Synthesis and Nitrogen Metabolism

As a disposal molecule for excess nitrogen, urea stands in contrast to ammonia excretion (fish), or urate excretion (birds). Mammalian excretion of both of these molecules also occurs, but on a far smaller scale. The pathway presented in Figure 6.11 shows a

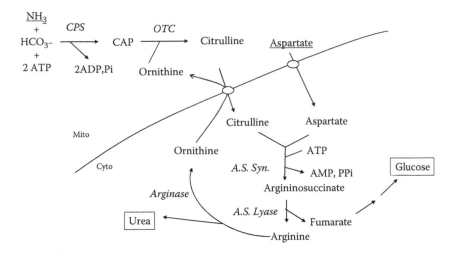

FIGURE 6.11   Urea cycle. The precursors NH_3 and aspartate, underlined in the diagram, are each precursors to half of the nitrogen of urea. The two are required in equal amounts at the arginosuccinate synthetase reaction (A.S. Syn.). A.S. Lyase, argininosuccinate lyase.

cyclic flow that requires two cytosolic inputs: ammonia and aspartate. The unique enzymes of ureogenesis are indicated in italics in the figure. In terms of adaptation to the pathway, all five are induced under conditions of increased pathway flow, such as glucagon elevation due to fasting [39]. The increased muscle protein breakdown during fasting produces an increased amino acid substrate to the liver. The hormones glucagon and cortisol are known to cause an increase in the abundance of all ureogenesis enzymes as well as the rate of the overall pathway [40]. There are also known short-term controls on the pathway, with their relative importance presumably dependent upon the circumstance. For example, the carbamoyl phosphate synthetase (CPS) enzyme is activated by the allosteric regulator N-acetylglutamate, which may in turn respond to the levels of either of the components required for the synthesis of the regulator, acetyl CoA and glutamate [41].

There are two common misconceptions about the urea cycle, both of which relate to how it interacts with other pathways. The first of these is the disposition of the fumarate produced by the argininosuccinate lyase (A.S. Lyase) step. The second is the means by which the stoichiometric requirement of equal amounts of aspartate and citrulline is provided for the argininosuccinate synthetase (A.S. Syn.) step. For the first, answer is simple: fumarate is converted by cytosolic fumarase to malate, which is subsequently converted to glucose by the cytosolic portion of the gluconeogenic pathway. The reason for the misconception is the "paper chemistry" treatment in many textbooks, in which fumarate is recognized as a Krebs cycle intermediate and the pathway appears to end there. The problem is that fumarate is strictly an *intermediate* of the Krebs cycle, and cannot be oxidized; only acetyl CoA is a substrate for this pathway. Understanding the second issue of equal amounts of citrulline and aspartate production for their condensation requires a more detailed explanation of how glucose and urea synthesis are related.

6.1.6 RELATIONSHIP OF UREA SYNTHESIS TO GLUCONEOGENESIS

Hans Krebs' first cycle was the urea cycle (in 1932), and his group elucidated the connections between urea synthesis and gluconeogenesis much later [42]. Much of the analysis was based on the stoichiometry between glucose and urea formed from different amino acids. To ensure that substrates were not also utilized for energy, the incubations of hepatocytes included a fatty acid (oleate) to provide a source of mitochondrial energy, sparing the carbon of amino acids from oxidation.

When we consider gluconeogenesis as an independent pathway, with no connections to nitrogen metabolism, the major pathway variations were based on the redox state (Figures 6.3 through 6.5). One extreme is pure lactate, which provides NADH in the cytosol before contributing its carbon to the mitochondria, and thus can re-enter the cytosol as aspartate. The other extreme is pure pyruvate, which oxidizes cytosolic NADH, and so its entry into the mitochondria is followed by malate formation. Even here, it should be pointed out that there is a connection to nitrogen metabolism: lactate gluconeogenesis requires transamination reactions in mitochondria and cytosol.

The other preliminary we must consider is that the two "carriers" of OAA— aspartate and malate—are actually at the *same* redox level. The only reason that aspartate exit is considered redox-neutral in terms of contribution of NADH to the cytosol is because it is converted back to OAA. Thus, each half-reaction in the trans-amination reaction is a redox reaction.[vi] Thus, if aspartate is not converted back to OAA in the cytosol, it carries reducing equivalents. This is evident if we trace the metabo-lism of the aspartate in the urea cycle (Figure 6.12). Because there is no corresponding transamination of aspartate, it is converted to malate, at the same redox level.

Now let us consider the general case: the liver is presented with a variable mixture of amino acids and ammonia, and these must form equal amounts of mitochon-drial aspartate and mitochondrial citrulline (the latter representing ammonia) so that both can exit the mitochondria for condensation at cytosolic argininosuccinate syn-thase. This feat is accomplished by the near-equilibrium reactions glutamate DH and aspartate aminotransferase (Figure 6.13). Note how the reaction of the glutamate

Aspartate

Mito

Cyto   Aspartate

Citrulline

Argininosuccinate

Fumarate   Arginine

Malate

FIGURE 6.12   Malate formation in the urea cycle. The portion from mitochondrial aspartate to cytosolic malate is illustrated.

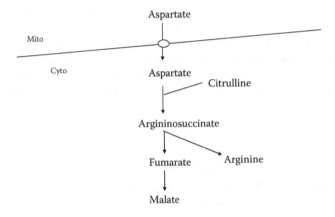

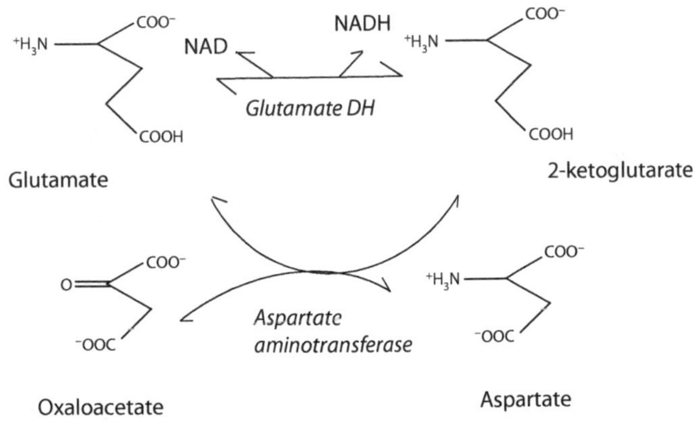

FIGURE 6.13   Two near-equilibrium enzymes interconvert glutamate and α-ketoglutarate. Glutamate DH and aspartate aminotransferase share the pathway intermediates glutamate and α-ketoglutarate, and proceed in different directions depending on the nature of the nitro-gen containing substrate.

DH explicitly involves redox transfer with NADH, showing the distinction in redox levels between glutamate and α-ketoglutarate, which is not evident in the aspartate aminotransferase reaction.

The fact that these reactions are at near-equilibrium is essential to their operation and to the understanding of how the pathway can be flexible in accommodating any amino acid load. The problem can be reduced to just two extremes: an excess of aspartate, or an excess of ammonia. In each case, the mitochondria must produce equal amounts of these to provide the condensation partners aspartate and citrulline (the ammonia contribution) at argininosuccinate synthetase. The two situations are illustrated in Figure 6.14, following the format of the two near-equilibrium enzymes drawn in Figure 6.13. In Figure 6.14a (excess aspartate), the aspartate is directly exported to the cytosol, and also drives the transamination reaction in the direction of oxaloacetate and glutamate formation. The glutamate in turn is deaminated through glutamate DH in the oxidative direction, producing NH_3 for conversion to citrulline. If there is an excess of NH_3 (Figure 6.14b) GLDH is driven in the direction of glutamate formation, and aspartate formed by transamination of that glutamate with OAA. The other distinction between these pathways is the extent of glucose formation tied to the process of urea synthesis. For the case of aspartate excess, the

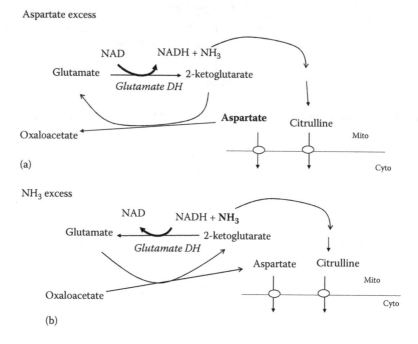

FIGURE 6.14 Extremes of nitrogen supply for the urea cycle. The two nitrogen atoms of urea arise from NH_3 and aspartate in equal amounts. To achieve this with various substrates, two general cases are shown. (a) Aspartate excess: in the figure aspartate is indicated in bold; its nitrogen is used to form the NH_3 by running the oxidative direction of glutamate DH. (b) NH_3 excess: this drives glutamate DH in the reductive direction, providing the nitrogen for aspartate by transamination of glutamate with oxaloacetate.

fumarate generated in the cytosol at argininosuccinate synthase forms malate and ultimately glucose. Not only that, but the OAA formed in the mitochondria through the illustrated aminotransferase reaction must also form glucose, through conversion first to malate and then entry into the cytosol. However, the situation of NH_3 excess is very different. The same formation of fumarate ensues from argininosuccinate synthase, but it can enter the mitochondria, be oxidized to OAA and balance the aminotransferase.

6.1.7 GLUTAMINE AND ASPARAGINE

As an example of how this view of metabolic flexibility can be put to use, consider the pair of substrates glutamine and asparagine. They are both amides of acidic amino acids, and yet have distinct pathways in liver catabolism. Both are first deaminated and produce similar products: an acidic amino acid and ammonia. However, asparagine deaminase is cytosolic, while glutaminase is mitochondrial. Thus, asparagine deaminase (Figure 6.15a) directly produces the needed aspartate for the key condensation step in ureogenesis. The ammonia is produced in the cytosol, but no metabolic reactions exist in the cytosol for ammonia; it passes into the mitochondria for conversion into the ammonia donor citrulline, which is then transported back to the cytosol. Unique among the amino acids, aspartate directly produces the equal amounts of ammonia and aspartate needed for condensation. The carbon backbone

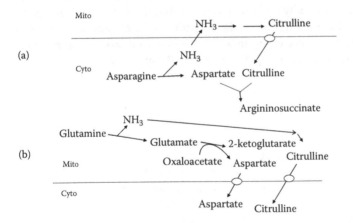

(c) α-ketoglutarate ⟶ Succinyl CoA ⟶ Succinate ⟶ Fumarate ⟶ Malate ⟶ Oxaloacetate

FIGURE 6.15 Asparagine and glutamine catabolism. The substrates both form NH_3 as the first step in their metabolism. (a) Asparagine is converted to aspartate and NH_3 by cytosolic asparaginase. Subsequently, aspartate is used along with the NH_3 (after conversion to citrulline) in the condensation step to form arginosuccinate. (b) Glutamine is converted to glutamate and NH_3 by mitochondrial glutaminase. The nitrogen from glutamate is converted by transamination to mitochondrial aspartate, which is then transported into the cytosol. (c) Balance of the glutamine pathway requires traversal of a portion of the Krebs cycle between α-ketoglutarate and oxaloacetate.

of the citrulline is cycled back to the mitochondria (as ornithine in the operation of the urea cycle, Figure 6.11), so that there is little mitochondrial metabolism of asparagine.

Glutamine is deaminated by mitochondrial glutaminase, so that glutamate and ammonia are produced (Figure 6.15b). The aspartate needed for ureogenesis is directly produced by transamination against oxaloacetate, so that the two substrates for argininosuccinate synthetase are, like asparagine deaminase, produced in equal amounts. However, the resupply of oxaloacetate requires further metabolism of the α-ketoglutarate through a portion of the Krebs cycle (Figure 6.15c), so that mitochondrial metabolism, and reoxidation of the produced NADH through those steps, is required for the metabolism of glutamine.

Making use of this distinction, I showed some time ago that Ca^{2+} linked hormones, that is, those eliciting a rise in cytosolic Ca^{2+} and subsequently mitochondrial Ca^{2+} (see Chapter 7) cause a stimulation of glutamine but not asparagine gluconeogenesis in hepatocytes [43]. Since the amino acids share the same pathway in the cytosol as outlined above, the Ca^{2+} sensitive step(s) must be localized to the mitochondria. There was a crossover at the α-ketoglutarate dehydrogenase reaction, indicated by a decrease in α-ketoglutarate and an increase in succinyl CoA. As Ca^{2+} is known to stimulate α-ketoglutarate DH, this enzyme was suggested as the site of Ca^{2+} stimulation for gluconeogenesis.

Glutamine has a metabolic importance in the whole body as it is a major nitrogen carrier, having the highest concentration of amino acids in the blood. Moreover, it is used as a substrate for rapidly dividing tissues, including both normal and pathological. Cancer cells have not just an oxidative pathway for glutamine metabolism, but also a reductive one in some cases, in which the flow from α-ketoglutarate in the direction of citrate overlaps that of lipid metabolism (Chapter 10).

6.1.8 FATTY ACID OXIDATION

Most fatty acids present in blood are long-chain, about 16 to 18 C. Entry into hepatocytes may require a protein (such as CD36) but this point is unclear.[vii] Entry into the cell is not a regulatory step, as the disposition of the fatty acid may be in either of two disparate directions synthesis or oxidation. For either of those pathways, the first step is conversion to a CoA ester ("activation"), after which the fate of this intermediate diverges (Figure 6.16). The first step in the oxidation of fatty acids is the conversion of the fatty acyl CoA ester into a fatty acyl carnitine ester, catalyzed by CAT I (carnitine acyltransferase I). This enzyme is inhibited allostericaly by malonyl CoA, produced by the lipogenic enzyme acetyl CoA carboxylase (discussed in the following section). The notion that an intermediate of *de novo* fatty acid synthesis suppresses fatty acid oxidation [48] is a pleasing regulatory concept.

The mitochondrial steps for the oxidation of saturated fatty acids are illustrated in Figure 6.17. The reactions illustrate the formation of a carbon chain shortened by two carbons (n-2), and the two-carbon fragment itself, acetyl-CoA. The latter is the substrate of the Krebs cycle, as well as of ketone body formation (exclusively in liver), as detailed in a latter section.

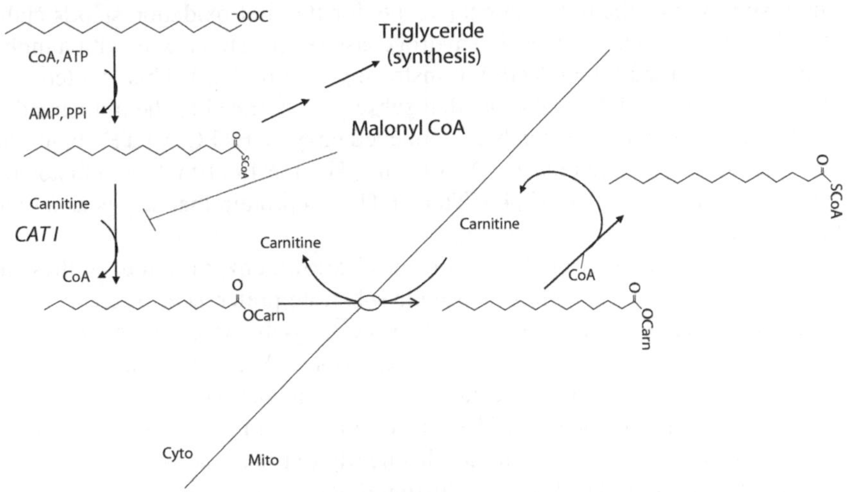

FIGURE 6.16 Early steps of fatty acid oxidation. Also shown is the inhibition of CAT I by malonyl CoA and the connection to triglyceride synthesis.

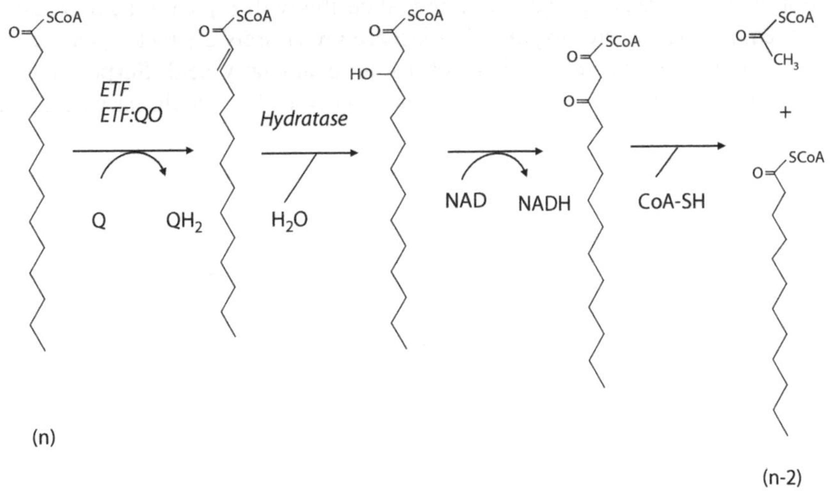

FIGURE 6.17 Mitochondrial (later) steps of fatty acid oxidation. The first oxidative step transfers its electrons to the mobile protein carrier ETF (electron transferring flavoprotein), which moves to the mitochondria and transfer electrons to the membrane bound complex ETF:QO (ETF-Q oxidase), and then to Q. One further oxidation, with electrons transferred to NAD occurs in the pathway. One of the stages of fatty acid to a shorter carbon chain (n-2) and acetyl-CoA is shown.

The disposition of the reducing equivalents for the first oxidation step is elaborated in Figure 6.18. The fatty acyl dehydrogenase of this step reacts with a mobile protein cofactor called ETF (electron transferring flavoprotein). This is a tetramer, containing four bound FAD cofactors that subsequently travel to the mitochondrial membrane and react with the membrane-attached enzyme ETFQO, ETF-ubiquinone oxidoreductase. The latter reduces Q to form QH_2, the lipid mobile cofactor that carries electrons to Complex III [49]. Thus, ETF is a protein that serves as a redox mobile cofactor.

ETF itself is an acceptor of electrons from 11 known enzymes; four of these are acyl-CoA dehydrogenases, differing in their substrate preference for chain length. The enzyme preferring the longest (called very long-chain acyl CoA dehydrogenase) is membrane attached; the others are cytosolic. Apart from these enzymes, other donors of electrons to ETF are part of amino acid and one-carbon metabolism [50,51]. The subsequent reactions of fatty acid oxidation in Figure 6.17—the hydratase and alcohol oxidation—are similar chemically to the fumarase and malate DH steps of the Krebs cycle. The last step illustrated, the thiolase, generates the product acetyl CoA and the two-carbon shortened acyl CoA.

Repeating the steps shown is not quite a cycle, as the new substrate is shorter than the first; some have suggested the term *spiral*. However, there is another distinction: the four separate dehydrogenases that generate the double bond, and at least two separate hydratase isozymes.

There are other issues of fatty acid oxidation that we will consider in turn. First, there is the possibility that the chain is odd-numbered. Second, the fatty acid in question may be already unsaturated, and since this will typically be a *cis* double bond it requires a separate enzyme. Third, there are entirely distinctive sets of reactions required when the substrate is not a long-chain fatty acid. Surprisingly, the extreme cases—short- to medium-chain acids and very long-chain fatty acids—are

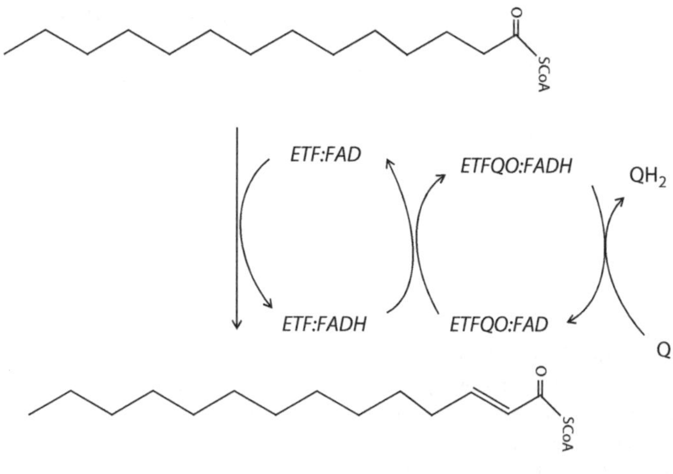

FIGURE 6.18   Steps of electron transfer to Q using ETF and ETFQO.

both independent of regulation at CAT I; these fatty acids are considered "carnitine independent."

6.1.8.1 Odd-Chain Fatty Acids

While unusual in mammalian metabolism, odd-chain fatty acids can arise in bacterial and plant metabolism. The same β-oxidation pathway of Figure 6.17 proceeds until the last three-carbon chain is released, as propionyl CoA. As illustrated in Figure 6.19, this is carboxylated to S-methylmalonyl CoA and then converted to the R-form through an epimerase (also known as a racemase). The final reaction in the sequence is an unusual one, requiring the cofactor B12 to produce the straight-chain molecule succinyl CoA. Since this is a Krebs cycle intermediate, it cannot itself be oxidized and thus those fatty acids that are odd-chained must form glucose through the gluconeogenic pathway. This is also the case of other metabolites that converge on this pathway by the formation of propionate, such as valine, isoleucine, threonine, and methionine. Thus all of these sources may be considered anapleurotic, or filling reactions to the Krebs cycle [52]. As a general rule, fat cannot be converted to sugar, since virtually all fatty acids are even chains and thus converted to acetyl CoA, which is oxidized rather than converted to glucose. While odd-chain fatty acids are rare, ruminants have extensive bacterial metabolism and may contribute a larger quantity through milk.[viii]

6.1.8.2 Oxidation of Unsaturated Fatty Acids

Fatty acids in nature have *cis* double bonds; those that appear as intermediates in fatty acid oxidation are *trans*. The position of the double bond may be at an even or an odd number position from the C1 of the fatty acid. In order to oxidize these double bonds, two reactions are needed in addition to those already considered in β-oxidation. The first of these is required for unsaturated fatty acids in which the double bond appears at an odd number position, such as oleate ($18\Delta^9$). Following three rounds of beta-oxidation, the intermediate as shown in Figure 6.20 occurs. This

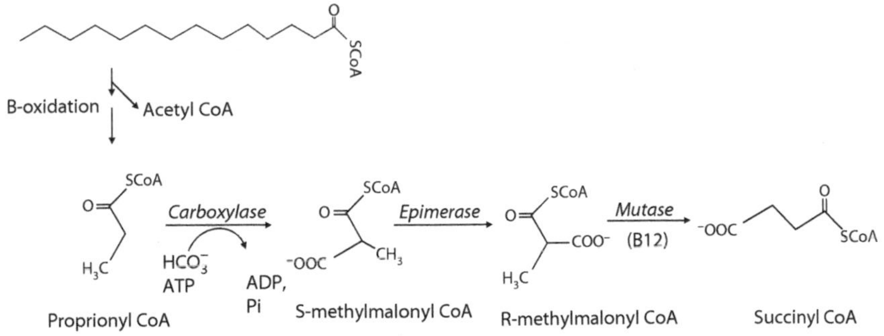

FIGURE 6.19 Odd-chain fatty acid oxidation. The unique step is in the last round of oxidation, in which the three-carbon propionyl CoA is formed instead of acetyl CoA. This is carboxylated to methylmalonyl CoA, an epimerase reverses its stereochemistry, and a B12-dependent mutase converts the branched molecule to succinyl CoA.

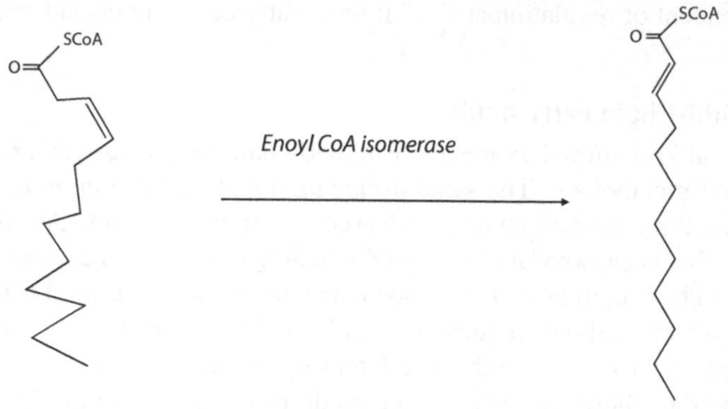

FIGURE 6.20 Enoyl CoA isomerase. Double bonds arising from polyunsaturated fatty acids require a movement of the double bond (isomerization) to a form that is a substrate of the hydratase step.

is an intermediate which has a *cis* double bond at the 3-position. The intermediate is a substrate of the enoyl CoA isomerase [54]. The second situation, in which a double bond occurs at an even numbered carbon, is always the case with polysaturated fatty acids, as these have the additional unsaturation three carbons from the first, such as the 9,12-desaturation in linoleic acid ($18\Delta^{9,12}$). The double bond at position 9 is metabolized in the same way as oleate, but the product in this case is the 4-cis double bond species illustrated in Figure 6.21. This is a substrate for an acyl CoA DH of the standard β-oxidation pathway, producing the cis-trans diene product. This species

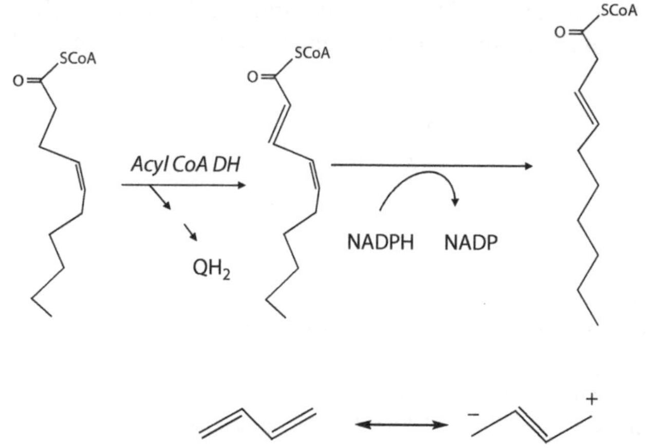

FIGURE 6.21 Fatty acyl diene reduction. With fatty acids such as oleate ($18\Delta^{9}$), the standard reactions of fatty acid oxidation ending with the acyl CoA DH as shown leads to the conjugated diene intermediate shown. This is reduced by a reductase using NADPH as the hydride source. Note the resonance forms below that explain the product formation (the hydride of NADPH adds to the carbocation).

is converted to the 3-*trans*-double bond, catalyzed by the NADPH dependent diene reductase. The resonance form of the butadiene portion is illustrated to explain how the hydride addition produces this product. The last intermediate, the 3-*trans* enoyl CoA is also a substrate of the enoyl CoA isomerase (Figure 6.20).

6.1.8.3 Shorter- and Longer-Chain Fatty Acids

Fatty acids much shorter or longer than the usual 16- or 18-carbon chain have distinctive pathways. The short-chain (about 12 or fewer carbons) fatty acid metabolism is illustrated for octanoate [55] in Figure 6.22. Octanoate is not converted to the CoA ester in the cytosol as it is a poor substrate for the activation enzyme. Moreover, the physical quality of the fatty acid is such that it can be transported as the fatty acid itself into the mitochondria. This is the case for both medium-chain (about C8) and short-chain (C4 or less) fatty acids, although each class has a separate activation reaction to form the CoA ester in the mitochondria. Once these esters are formed, they enter the β-oxidation pathway described in the previous section. Because these fatty acids are not converted to carnitine esters, they do not involve CAT I, are not regulated by malonyl CoA, and are independent of the activity state of acetyl CoA carboxylase.

The other class is more unusual: the very long-chain fatty acids. An example of this class is the 22 C erucic acid, which has a single double bond at C13 [56]. This is derived from rapeseed oil, a European plant, and is associated with mitochondrial damage.[ix] Erucic acid is a substrate for cytosolic activation, and is converted to the CoA ester. However, the latter is not a substrate for CAT I, but rather is transported into the peroxisome [58]. This pathway can also oxidize long-chain fatty acids, and can be induced independently of the mitochondrial pathway by the addition of the peroxisome proliferators, such as clofibrate [59]. In fact, clofibrate is a lipid lowering compound, now known to affect synthesis in liver through a receptor known as the PPARα.

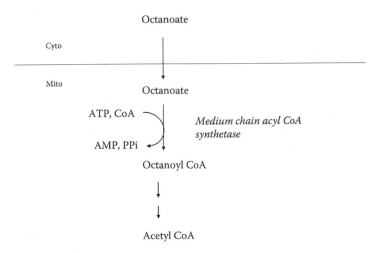

FIGURE 6.22 Short-chain fatty acid oxidation: octanoate. Direct diffusion of short-chain fatty acids like octanoate into the mitochondria is followed by their activation by a mitochondrial synthetase.

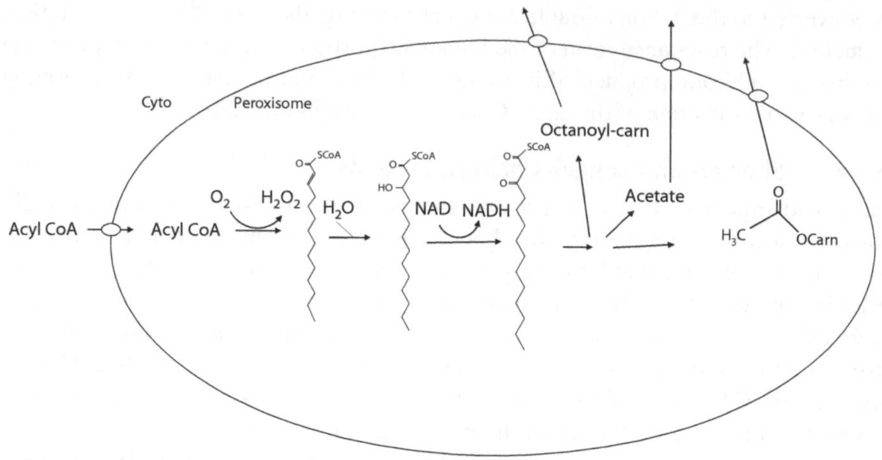

FIGURE 6.23 Peroxisomal fatty acid oxidation. A reduced-energy capture is evident as the first oxidation step of the acyl CoA transfers electrons from O_2 to H_2O_2. This is necessary as the peroxisome has no respiratory chain. The products are also distinct from mitochondrial oxidation. In the peroxisome, they are octanoyl-carnitine, acetate, and acetyl-carnitine.

Long-chain acyl-CoA molecules can be directly imported into the peroxisome by transporters analogous to ABC exchangers, after which a pathway for fatty acid oxidation exists that is similar to that in the mitochondria [60]. The first major distinction is that the formation of the enoyl CoA in the first oxidation step is linked to H_2O_2 production rather than the formation of QH_2 (Figure 6.23). Like its mitochondrial counterpart, this oxidation also involves a bound FAD that extracts the electrons from the fatty acid prior to their transfer to the mobile cofactor, which in this case is O_2. The peroxide formed is converted to water by the action of catalase (the enzyme marker of the peroxisome), so that the mobile cofactors involved in the first oxidation step are oxygen and water. In the second oxidation step, reducing equivalents as NADH are believed to be removed to the cytosol by a redox shuttle, possibly involving a peroxisomal form of lactate DH. The final products of peroxisomal oxidation are acetate [61], acetyl-carnitine, and medium chain fatty acids, such as octanoyl carnitine. All of these can be further oxidized by mitochondria through CAT I independent steps.

Thus, oxidation in the peroxisome only partially contributes to energy formation by its ability to contribute NADH and the shorter-chain fatty acids that are subsequently oxidized by mitochondria. In general, the peroxisomes have detoxification reactions in higher organisms, although in some organisms it plays a key role in fatty acid oxidation. For example, in yeast, it is the exclusive site of fatty acid oxidation [60].

6.1.8.4 Ketone Body Formation

In liver mitochondria, acetyl CoA has an alternative fate to oxidation by the Krebs cycle: it may be used for the synthesis of ketone bodies. These that can exit the liver for oxidation in other tissues [59]. This can be viewed colloquially as an expression of the altruistic nature of liver, providing substrate for other tissues, in particular muscle, heart, and brain. The pathway is particularly short (Figure 6.24) and

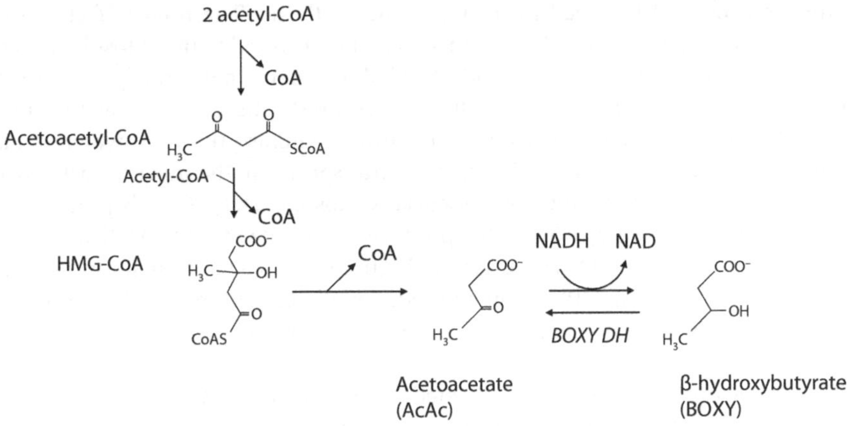

FIGURE 6.24 Ketone body biosynthesis.

culminates in a dehydrogenase that is analogous to the cytosolic lactate DH. The β-hydroxybutyrate DH of the ketogenesis pathway also achieves near-equilibrium and also provides an estimate of mitochondrial free NAD/NADH status [62]. Both ketone body products are metabolic dead ends in the liver cell, and exit both mitochondria and the cell for extra hepatic oxidation.

In terms of its regulation, the pathways for both fatty acid oxidation and ketone body synthesis are primarily regulated by malonyl CoA [63]. Beyond that, there are no known regulators of the enzymes for formation of AcAc, and the reduction step is near-equilibrium. This is why the pathway is generally thought of as a "spillover" route of fatty acid oxidation in the liver in which the excess acetyl CoA produced by fatty acid oxidation ends up in ketone bodies.

The ketone bodies leaving the liver, BOXY (β-hydroxybutyrate) and AcAc (acetoacetate) have a particular ratio that reflects the near-equilibrium nature of BOXY DH [64]. It has been suggested that the ratio itself is a means of carrying a specific redox state into the tissue utilizing these molecules for energy, a form of shuttle for redox communication between cells [65]. A similar case can be made for the pair of molecules produced by glycolysis and exported to the extracellular space: lactate and pyruvate. These are metabolites of the lactate dehydrogenase reaction in the cytosol, which equilibrates with cytosolic NAD and NADH. The ratio of lactate to pyruvate thus presented to peripheral tissues can be considered to be a reflection of the NAD/NADH of the tissue of origin (i.e., liver).

6.1.9 FATTY ACID SYNTHESIS

Most fatty acid synthesis (lipogenesis) takes place in the liver, although adipocytes can also synthesize fatty acids [66]. As the liver also oxidizes fatty acids (unlike adipocytes), a key site of regulation occurs between these pathways.

Carbon for the biosynthesis of fatty acids originates largely with carbohydrates, although amino acids can make a substantial contribution as protein consumption is often in excess of the need for new protein synthesis. The routes of utilization merge in mitochondrial acetyl CoA, which must be exported to the cytosol for incorporation into the long-chain fatty acid. As CoA esters cannot directly traverse the mitochondrial membrane, a shuttle system is required. This is known as the citrate shuttle (Figure 6.25). The first step is the citrate synthase reaction, which is also the first step of the Krebs cycle. Citrate has a transporter in the inner mitochondrial membrane, and so can exit the mitochondria. Cytosolic acetyl CoA is produced by the action of citrate lyase, which also produces oxaloacetate (OAA). The ensuing steps in the disposition of OAA are seen in Figure 6.25 as a pathway for transhydrogenation of NADH to NADPH, using the sequential action of malate dehydrogenase and malic enzyme. The NADPH formed can be used in later reductive steps of fatty acid synthesis.

The balance of exchangers is indicated in Figure 6.25b. Citrate can exchange with malate, malate with P_i, and pyruvate cotransports with protons. The issue of whether these exchanges were merely "paper chemistry" or actual routes was raised by Conover [67] However, the early evidence that is the basis for the route shown and the utilization of NADPH from malic enzyme was discussed in letters responding to that challenge [68,69].

The committed step for fatty acid biosynthesis is the cytosolic conversion of acetyl CoA to malonyl CoA, catalyzed acetyl CoA carboxylase (Figure 6.26). Early studies suggested regulation may be exerted by citrate stimulation and long-chain acyl CoA inhibition [70]. However, the correlation of citrate with the activity state of this enzyme has not been consistent [71], suggesting this "feed-forward" activation by citrate that appears in the pathway diagram may not apply in cells. Not only that, it is inconsistent with the fact that citrate inhibits phosphofructokinase [72], which is a necessary step for glucose conversion to fatty acids.

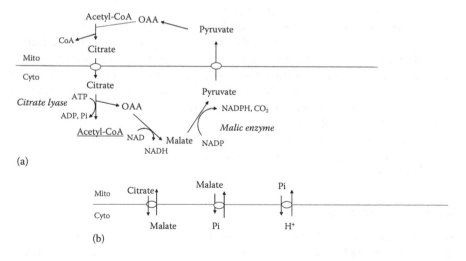

FIGURE 6.25 Citrate export cycle for cytosolic acetyl CoA. (a) Pathway produces cytosolic acetyl CoA and converts cytosolic NADH into cytosolic NADPH that is used for fatty acid synthesis. (b) Exchangers needed in the cycle for balance.

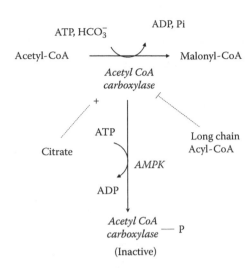

FIGURE 6.26 Acetyl CoA carboxylase. The key enzyme in fatty acid biosynthesis, regulated by phosphorylation (AMPK) and allosterically inhibited by long-chain acyl CoA. Also shown is the classical activation by citrate, but it is uncertain if it plays this role in cells.

The other regulator indicated in Figure 6.26 is long-chain acyl CoA, which has support in more recent studies as an inhibitor of the enzyme [73]. In addition, a long-chain fatty acid analog TOFA (tetradecyloxy furoic acid) is a potent inhibitor of the enzyme and of fatty acid synthesis [74]. A product of TOFA, olumacostat glasaretil, is an even more potent inhibitor of acetyl CoA carboxylase and may be of significant clinical use [75].

It has been known that phosphorylation of the enzyme leads to its inactivation [76], and that glucagon leads to a phosphorylation mode in liver that would be consistent with cyclic AMP-dependent phosphorylation and inactivation of this enzyme and thus fatty acid synthesis overall. At least some studies have suggested that cyclic AMP-dependent protein kinase is responsible for the phosphorylation of acetyl CoA carboxylase [77]. However, more recent studies have shown that the residue known to be phosphorylated that exerts inhibition of the carboxylase is instead the target of AMP kinase [78]. In one study, it was shown that AMPK inhibition *in vivo* was diminished in cells that synthesize more enzyme; this effect could be overcome by TOFA inhibition [79].

For subsequent steps of lipogenesis, both acetyl CoA and malonyl CoA are converted to esters of the small protein ACP (acyl carrier protein), as indicated in Figure 6.27. The actual steps of synthesis and reduction are all part of a single enzyme with seven catalytic sites: palmitate synthase. The acetyl-S-ACP reacts with a cysteine side chain of the synthase to form a thioester. This reacts in the condensation reaction with a malonyl-S-ACP to form acetoacetyl-ACP and CO_2 (released during the condensation). NADPH reduction of the carbonyl, dehydration of the alcohol, and reduction of the double bound by another NADPH produces butyryl-ACP. This species can substitute for acetyl CoA-S-ACP, in that it can react with the cysteine

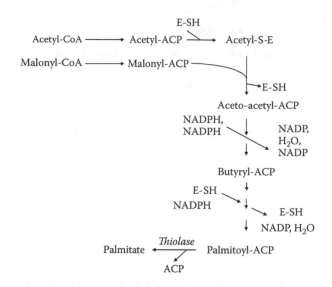

FIGURE 6.27 Cytosolic steps of fatty acid biosynthesis. Also called palmitate synthase since the C16 is the product, the pathway shows two rounds of this pathway. The first uses acetyl CoA and malonyl CoA. In the second round the C4 butyryl CoA substitutes for the acetyl CoA and a second malonyl CoA is used to provide the two-carbon addition, and release of CO_2. All subsequent steps use malonyl CoA, so that one acetyl CoA and seven malonyl CoA molecules are used for each molecule of palmitate produced.

side chain, and a second molecule of malonyl CoA. The cycle repeats until the chain reaches 16 carbons, at which point it is a substrate for thiolytic cleavage, and palmitate is produced (Figure 6.27).

It is of chemical interest, but not tremendous metabolic interest, that the seven synthetic steps (but not the ancillary loading reactions) are catalyzed by the protein fatty acid synthase, with all enzymes joined into one protein strand, which is a dimer having two of each active site [80]. It does serve as an illustration of why enzymes exist in a complex. In this case, we have a different expression of a complex than the case of the pyruvate dehydrogenase complex (where the enzymes are associated but not part of the same strand), but functionally they are the same. When the intermediates are passed from one enzymatic site to another directly, there are no metabolic branch points. In fatty acid synthase, this is achieved by reactions taking place along a series of enzymatic sites that exist covalently bound to one another. In enzyme complexes such as the pyruvate dehydrogenase complex, it is achieved instead by passing the products of one reaction to the next enzyme that is not itself covalently bound but nonetheless strongly attached.

For comparison, consider fatty acid synthase of bacteria. In *E. coli*, for example, the steps of fatty acid synthesis exist as individual enzyme reactions, and there is no passing of one to the next [81]. This allows for the possibility of intermediate reactions. It also argues against an alternative explanation—that having clustered or bound enzymes passing one reaction to the next is more metabolically efficient. It is difficult to imagine any mammalian cell being more efficient than bacteria, which are evolutionarily far more advanced than mammals. The entire life span of bacterial cells can be shorter than half-an-hour. From the fact that mitochondria are evolved from bacteria, it is perhaps not surprising to find that mitochondria are known to have the enzymes of fatty acid synthesis as well, and each is a separate entity. There is some evidence that the end product is octanoyl-S-ACP [49].

6.1.10 Neutral Lipid Formation

The conversion of fatty acids to triacylglycerols and phosphatidyl choline is known as the neutral lipid pathway; this is outlined in Figure 6.28. The long-chain acyl CoA molecules are formed in the activation step, the same reaction used in the utilization of exogenous fatty acids for synthesis from exogenous fatty acids, or for oxidation. While most of the regulatory focus for lipid biosynthesis (i.e., feeding/fasting transition) is exerted on the formation of fatty acids, there is evidence that subsequent steps have regulatory significance, such as the phosphatidic acid phosphatase enzyme [82]. One intriguing hypothesis is that there is a connection between the neutral lipid pathway and protein kinase C (discussed in Chapter 7), since increased glucose can activate the enzyme and is also a precursor to DAG synthesis by the biosynthetic route [83]. However, this pool of DAG is unlikely to be a regulator. First, it is an intermediate in the process and not a rapidly formed signal. Second, the pool of DAG that is formed occurs typically at the plasma membrane, not the ER membrane where

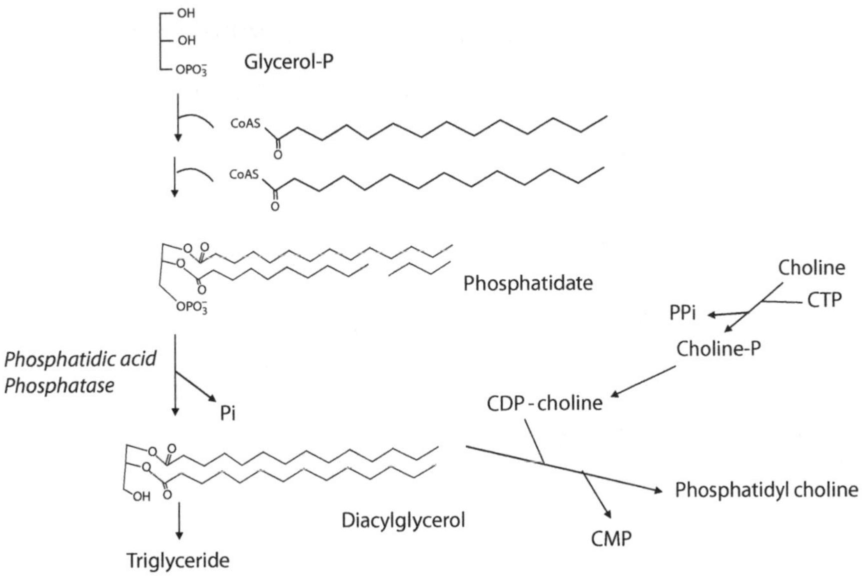

FIGURE 6.28 Neutral lipid biosynthesis. The common precursor of triglycerides and neutral phospholipids is diacylglycerol, the product of phosphatidic acid phosphatase. The most common phospholipid is phosphatidyl choline; the formation of CDP-choline from choline is illustrated.

triglyceride formation occurs. Thus, this likely represents a metabolically separate pool of DAG.

6.1.11   CHOLINE METABOLISM

Choline metabolism plays a prominent role in liver, partly because of the requirement for VLDL export, and shares some functions with kidney [49,51,84]. Choline is derived from the diet, or from the phospholipase D cleavage of membrane phosphatidylcholine. The pathway of choline catabolism is shown as Figure 6.29. Choline enters the mitochondria, where it reacts on the matrix face of the inner mitochondrial membrane at the choline dehydrogenase reaction. The details of this step are uncertain [51], but the indication of electron transfer to ubiquinone (Q) as shown in Figure 6.29 is consistent with the findings that the reaction must occur on the matrix surface, and the recent observation that choline dehydrogenase has a bound pyrroloquinoline quinone [85] (Figure 6.30). The finding that electron transfer (measured as reactive oxygen species) to complex II as well as complex III is suppressed by their respective inhibitors [86] means electron transfer *in vivo* is likely to Q and subsequently Complex III. This makes choline dehydrogenase analogous to glycerol phosphate oxidase, although the latter reacts with its substrate on the cytoplasmic side of the membrane.

Subsequent steps of choline metabolism are well established, but unusual. The product of choline dehydrogenase, betaine aldehyde, is oxidized to betaine (sometimes called glycine betaine) by the matrix enzyme betaine aldehyde dehydrogenase.

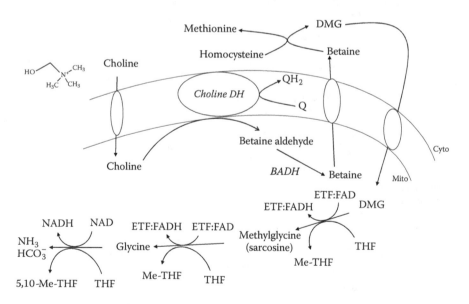

FIGURE 6.29   Choline oxidation. The pathway of oxidation of choline is shown, involving three exchanges across the mitochondrial membrane, a connection to 1-carbon metabolism (methionine formation and Me-THF formation) and several oxidative steps. Like fatty acid oxidation, intermediate electrons are passed to the mobile cofactor ETF.

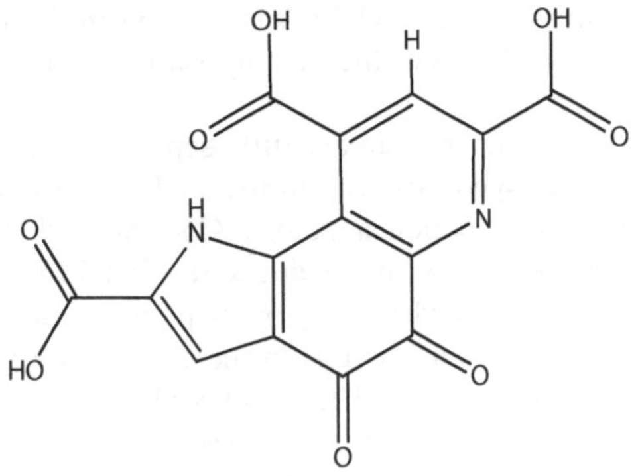

FIGURE 6.30   Pyrroquinoline quinone. This is the bound cofactor in choline dehydrogenase.

The redox cofactor is the protein redox carrier ETF discussed in the prior section in the context of fatty acid oxidation. The reduced form of ETF reduces the mitochondrial membrane protein ETFQO (ETF Q oxidase, as in fatty acid oxidation), so electrons enter the respiratory chain at QH_2. Next, betaine is transported across the mitochondrial membrane and reacts with homocysteine to produce dimethyl glycine plus methionine. This methyltransferase is an alternative means of converting homocysteine to methionine; the methyltransferase using methyl-THF (tetrahydrofolate) is the more commonly used one. As methionine is converted to SAM (S-adenosylmethionine) in a subsequent synthetase, it means the methyl group from betaine is used as a one-carbon methyl donor in many reactions, including those which convert phosphatidylethanolamine to phosphatidylcholine [49].

Dimethylglycine is not further metabolized in the cytosol, but is instead transported back into the mitochondria. Within the matrix, two sequential reactions catalyze the removal of a methyl group to THF, and electrons to ETF. These reactions produce first sarcosine (monomethyl glycine) and then glycine. Finally, glycine is converted to 5,10-Me-THF plus ammonia by the action of the glycine cleavage system, which also converts NAD to NADH. Thus, the catabolism of choline intersects the metabolism of the amino acids methionine, homocysteine, and glycine, and thus one-carbon metabolism.

An alternative fate to the betaine production in the cytosol is its transport across the plasma membrane and thus out of the cell. Betaine accumulation itself can increase extracellular osmolarity. While this effect may contribute to the osmotic gradient in kidney, it has also been argued to play a role in the viability of sperm cells [84,86].

6.1.12   RELATIONSHIP OF FATTY ACID OXIDATION TO FATTY ACID SYNTHESIS

Having stated above that the central control of fatty acid oxidation in the liver is at the transport step (malonyl CoA inhibition), and that the key regulatory enzyme of fatty acid synthesis produces malonyl CoA, the regulatory connection for these pathways is clear. Historically, it was the search for a connection between these pathways that inspired the discovery of malonyl CoA regulation in the first place [63]. The opposing lipid pathways differ from the glycolysis/gluconeogenesis pair in several ways.

First, the routes of lipid metabolism are fully separate. The opposing lipid pathways are largely cytosolic (synthesis) and mitochondrial (oxidation), converging in the cytosol for their mutual regulation at malonyl CoA. Second, there are no substrate cycles in the lipid pathways. This is largely due to the first difference, in occurring in separate compartments, but even where they overlap there is no similar utilization of the same pathway intermediates as there is in the sugar pathways. Third, the redox cofactors of the lipid pair are distinct. The NAD/NADH and NADP/NADPH redox states are extremely divergent [87], so that synthesis and oxidation of fatty acids are effectively unconnected in terms of their reduction cofactors.

It is true that the pair of lipid pathways respond similarly to pair of carbohydrate pathways in response to the feeding/fasting transitions. The onset of fasting depresses lipogenic enzymes through the fall in insulin and rise in glucagon, and causes an increased fatty acid oxidation. Reciprocal regulation of the two lipid pathways is also evident in situations leading to the activation of AMP Kinase (AMPK). As a regulation system at a fundamental level, activation of this enzyme, for example through a rise in AMP concentration, leads to phosphorylation of acetyl CoA carboxylase, inhibition of fatty acid synthesis, a fall in malonyl CoA level, and, subsequently, increased fatty acid transport and oxidation by mitochondria.

6.1.13 FATTY ACID METABOLISM AND GLUCONEOGENESIS

Following the transition from feeding to fasting, the predominant hepatic pathways are fatty acid oxidation and gluconeogenesis, so that the outputs of liver under the fasting condition are glucose, CO_2, and ketone bodies. It is of interest that in adult-onset diabetes (see also Chapter 10), there is a paradoxical situation in that gluconeogenesis and fatty acid oxidation are elevated, and yet the liver outputs large amounts of triglycerides as VLDL [88]. The explanation is that hepatic insulin insensitivity of these pathways is uneven. The divergence between glucose and triglyceride output reflects the activation of the transcriptional control protein for gluconeogenesis (FOXO1) without a suppression of the transcriptional activator for triglyceride output, SREBP-1c [89]. Thus, under these conditions, the pathways of gluconeogenesis, fatty acid oxidation, and triglyceride synthesis are operating simultaneously. This is supported by the fact that complete genetic ablation of the hepatic insulin receptor does attenuate triglyceride output [90].

However, there is still one metabolic aspect that should be considered. While the authors point out that SREBP-1c can stimulation acetyl CoA carboxylase and acetyl CoA carboxylase, which are enzymes of the fatty acid biosynthesis pathway [89], neither of those enzymes can be responsible for the increased VLDL output of triglycerides under this situation. The reason for this is that there is no carbon source for acetyl CoA to supply *de novo* fatty acid synthesis if gluconeogenesis is operating. For the usual carbohydrate supply of fatty acid biosynthesis, the reverse route of glycolysis is needed. Thus, the actual supply of triglycerides has to be the greatly increased fatty acids released from adipocytes in Type II diabetes. It is likely, therefore, that the increased formation of other genes beyond the formation of fatty acids by SREB-1c is responsible, such as acyltransferases or stearoyl-CoA desaturase [91].

6.1.14 OTHER HEPATIC LIPID ROUTES

The major outputs of lipid metabolism by the liver are the lipoprotein VLDL in the fed state and ketone bodies (acetoacetate and β-hydroxybutyrate) in the fasted state. The VLDL also contains substantial quantities of cholesterol, an extensive pathway that may be considered a branch point of cytosolic acetyl CoA (Figure 6.31). Also shown branching from the cholesterol pathway are intermediates like geranyl-PP and farnesyl-PP, branched chain lipids that themselves form covalent adducts with proteins, often modifying their cell localization behavior. The amount of flow of acetyl CoA to fatty acids is far greater than that through cholesterol (e.g., refs. [92,93]), which is in turn far greater than that of the branch points utilizing the intermediates shown. Nonetheless, virtually all of the cholesterol in the body is derived from the liver and taken up by the periphery in lipoprotein metabolism, suppressing extra hepatic cholesterol biosynthesis. While it is common that cases of hypercholesterolemia are treated by inhibiting the reaction utilizing the intermediate hydroxymethylglutaryl CoA (HMG CoA) shown (HMG CoA reductase), some of the side effects of these "statin" drugs are not obvious consequences of a low cholesterol (e.g., muscle rabhdomyelosis). These may rather be due to side pathways distal to the reductase step, such as those leading to dolichol or ubiquinone [94]. It should also be pointed out that AMPK, which catalysis phosphorylation and inhibition of acetyl

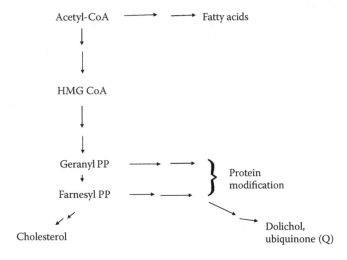

FIGURE 6.31 Cholesterol and branched chain lipid formation. Both cholesterol and branched chain lipids are derived from acetyl CoA. The formation of HMG CoA is similar to the ketone body pathway, except that the pathway shown here takes place in the cytosol, so the enzymes are distinct. Intermediates such as geranyl PP can be used to covalently modify proteins and are also transformed to other cellular lipids: dolichol (used in the transfer of sugars to proteins in the ER) and Q.

CoA carboxylase also inhibits HMG CoA reductase [95], so that both of these pathways are diminished under conditions of energy depletion.

In fact, despite the fact that acetyl CoA carboxylase is the key regulatory enzyme for lipogenesis, and this is largely a liver pathway, the enzyme is nonetheless found in a variety of other tissues. In those tissues, a distinct isoform exists that is more sensitive to lower concentrations of malonyl CoA and it is likely to exist simply to regulate fatty acid oxidation as there is little fatty acid biosynthesis. One tissue in which fatty acid oxidation plays a prominent role in energy formation is the muscle.

6.2   SKELETAL MUSCLE

It is difficult to resist an anthropomorphic characterization of muscle metabolism. If we think of the liver as the "altruistic" organ, providing substrates for the greater outside community of cells, the muscle is the "selfish" organ, utilizing exogenous and stored substrates for its own purposes. Those purposes are transparent: energy utilization is virtually entirely for two processes: contraction and one major ion pump that returns Ca^{2+} to the sarcoplasmic reticular (ER^x).

Not only is the utilization side of muscle metabolism extremely simple, the supply side is as well, at least compared to the liver. Amino acid metabolism is more limited and distinct, involving a few specialized routes. Similarly, there are fewer pathways in muscle for carbohydrates and lipids.

There are some specializations, established as skeletal muscle subtypes. As summarized in Table 6.1, muscle types are metabolically divided between more glycolytic (fast, Type I) and more oxidative (slow, Type IIb), and an intermediate type (mixed, Type IIa).[xi] Experimentally, the types are characterized by the specific myosin isoform that is present, but our concern here is metabolic. These two correspond directly to exercise regimens, the "anaerobic" (e.g. weight lifting) and "aerobic" (running).

6.2.1   ENERGETIC CONCERNS OF FAST MUSCLE

Fast muscle contains a store of creatine phosphate that should be viewed as a buffer to prevent substantial attenuation of ATP, rather than a separate energy source. The creatine phosphate pool is somewhat greater than the concentration of this metabolite in slow muscle, but both serve to largely maintain the concentration of ATP following exercise [96]. More extensive studies of the creatine phosphate system in the context of exercise are presented in a specialty monograph [97].

TABLE 6.1
Skeletal Muscle Types

Type	Metabolic	Contractile	Appearance
I	Glycolytic	Slow	Red
IIa	Mixed glycolytic and oxidative pathways	Intermediate	Pink
IIb	Oxidative pathways	Fast	White

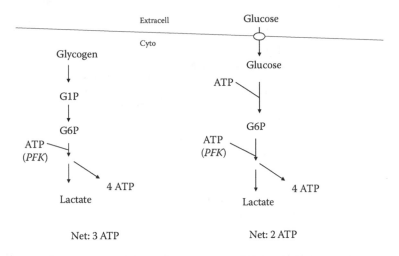

FIGURE 6.32 Glycolysis from glucose and glycogen. Glycolysis from glycogen forms more ATP than glycolysis from glucose. However, the former requires glycogen formation in the resting state (see also Figure 4.4).

Two substrates for the glycolytic pathway in muscle are extracellular glucose and intracellular glycogen [98]. While glucose utilization may appear to be the more common and favorable pathway, in fact glycogen utilization is a more efficient energy pathway to support contraction. Thus, to reach the common intermediate of G6P with glucose as substrate, there is the requirement of one ATP at hexokinase. The accounting for glycogen would appear to be greater, as to form the glucosyl unit in the first place requires three ATPs (one for hexokinase and the equivalent of two to form a UDP-glucose). Yet, considering that the glycogen formation can be "front-loaded" during resting conditions, the true comparison is between glucose → lactate and glycogen → lactate. From this standpoint, glycolysis-from-glycogen produces three ATP/glucosyl units, while glycolysis-from-glucose produces just two ATP/glucosyl unit (Figure 6.32). This was outlined from an energetic perspective in Chapter 4.

It is true that the glucose transport step is stimulated during muscle contraction, which implies a direct utilization of the substrate in this state. However, it is well established that glycogen, once depleted during exercise, imposes limitations on the exercise itself [99–101]. The use of glycolysis to supply muscle power is evident in the extreme case of alligator jaw muscle, which develops considerable power, albeit over a very short time period. Maximum power is obtained over a period of just a few minutes [102]. Following this effort, the lactate produced needs several hours to be reconverted to glucose by liver gluconeogenesis.[xii]

6.2.2 Effect of Insulin and Contraction on Glucose Utilization

Both the presence of insulin and contraction increase glucose metabolism of muscle. Four pathways—glucose uptake, glycogen synthesis, glycogenolysis, and

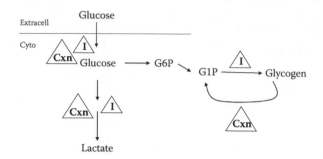

FIGURE 6.33 Contraction and insulin stimulate glycolysis and glycogenolysis. In muscle, insulin (I) stimulates glucose transport (GLUT4), glycogen synthase, and glycolysis. Contraction (Cxn) stimulates glucose transport (GLUT4), glycogenolysis, and glycolysis.

glycolysis—are presented in Figure 6.33, accompanied by the points of regulation by either insulin (represented by **I**) or contraction (represented by **Cxn**). In both cases, the upward-pointing triangles indicate stimulation. As illustrated, both conditions stimulate glucose transport and glycolysis, but insulin specifically activates glycogen synthesis and contraction selectively stimulates glycogen breakdown.

The actual situation is more involved, since glycolysis-from-glycogen is likely an important pathway for contraction, particularly in fast muscle. Thus, while it is true that glucose transport is stimulated by contraction, it is probably the case that it is not the major route of glycolysis in these cells. Additionally, while no effect of contraction on glycogen synthesis is indicated, the depletion of glycogen is itself a stimulus to enhanced glycogen synthesis. The opposing directions of the glycogen pathways are separated in time for muscle.

The signaling mechanism behind these events is discussed in the following chapter. However, it should be pointed out that it is well established that the mechanism for glucose transport stimulation by insulin is distinct from contraction [103]. This is the reason that insulin resistance of type II diabetes can be overcome to a large extent by exercise [104].

6.2.3 Can Near-Equilibrium Enzymes Be Regulatory?

The power of dividing enzymes between NEQ and mIRR classes is the ability to discriminate between mechanisms likely to be physiological from those which are not. As an example, lactate DH (LDH) is established as an NEQ enzyme, but this is not widely appreciated. Thus, some investigators have been drawn to examining its potential role in metabolic regulation.

Liang et al. [105] recently observed that exercise in human subjects can induce a switch in lactate DH isoforms from LDHa to LDHb. Moreover, transgenic mice in which LDHb is overexpressed in muscle produces an oxidative phenotype. Drawing on work from Kaplan and colleagues in the 1960s [106–108], the investigators suggest that the LDHb form (predominant in the heart) favors pyruvate formation, which the LDHa form favors lactate formation. When this idea was first proposed, it was a conjecture that fit with the oxidative nature of the heart and the glycolytic nature of (type II) muscle. However, it is no longer a tenable notion. The actual basis was a difference in pyruvate inhibition of the enzyme during assay. However, since lactate DH achieves near-equilibrium, its status in assay systems is irrelevant. Even the fact that it is induced is not proof that it is a regulatory enzyme. In the same study, the overexpression of LDHb also led to multiple other changes, including several nuclear coded mitochondrial enzymes [105]. Thus, the model itself is not one in which a single enzyme has changed, but several, even among those measured. The problem is in finding an alteration and assuming that it has regulatory significance. It is possible that it has some other meaning.

One example of this is the role of LDH and other enzymes such as enolase and alcohol dehydrogenase serving as a crystallin, proteins found in very high concentrations to achieve lens refraction [109]. These do not occur in humans but do in various other species, in which they are called "taxon-specific crystallins." Underscoring the fact that the enzyme activity itself is dispensable for this function, some of these proteins have evolved to have similar structure but without catalytic activity. Thus, it is possible that an increased expression of a protein might have a function other than altering metabolic flow. As a simple guiding principle, we can say that once an enzyme in a pathway has been demonstrated to achieve near-equilibrium status, we need go no further in examining its regulatory properties.

As a separate indication of a more complex role for LDH, a recent review [110] stated that "after years of controversy the existence of a mitochondrial LDH has finally been proved." The study cited was from 1999 [111] and examined LDH in isolated mitochondrial fractions from heart, skeletal muscle, and liver. The investigators found that respiration was observed when lactate was added to these preparations in the absence of added NAD, suggesting that lactate entered mitochondria, was oxidized inside by a mitochondrial LDH, and subsequently pyruvate was oxidized. The study, however, had other complications that led the authors to be less certain of their own conclusions. For example, no mitochondrial leader sequence was found for the LDH transcript, which they were unable to explain. In addition, when mitochondrial fractions were further treated with digitonin, which removes the outer membrane, the LDH content was largely eliminated. Thus, it would appear that LDH was adventitiously adhering to the mitochondrial fractions during isolation. A separate problem is the expectation that lactate could enter the mitochondria. While the authors suggested it would travel across the pyruvate transporter, it has been shown that lactate is not a substrate for this transport protein [112]. It was also shown that LDH is associated with the external membrane of mitochondria but not the matrix [113], so that the functional study of that original paper is also in doubt. It seems reasonable to conclude that LDH is a cytosolic protein, as classically demonstrated, and is near-equilibrium, and thus not a means of controlling glycolysis or the NAD/NADH redox state.

6.2.4 Glycolysis

The two established metabolically irreversible enzymes of glycolysis are PFK and PK. The muscle PFK is known to be regulated by fructose 2,6 P_2 [114]. This metabolite, first discovered in liver [115], is produced by the phosphorylation of

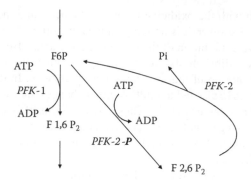

FIGURE 6.34 Formation of the PFK activator fructose-2,6-P$_2$. A separate enzyme, PFK2, catalyzes the formation of fructose-2,6-P$_2$ in the phosphorylated form (PFK2-P), and its conversion to fructose-6-P in the dephosphorylated form (PFK2).

F6P, catalyzed by PFK-2 (Figure 6.34). This enzyme catalyzes this conversion in its phosphorylated form; in its dephosphorylated form, it catalyzes the hydrolysis of the 2-phosphate of fructose 2,6 P$_2$, forming F6P. It is established that the rate of glycolysis corresponds to the concentration of fructose 2,6 P$_2$ under a variety of conditions [116], including contraction itself, which leads to phosphorylation of PFK2 [114]. A further control of control of glycolysis at PFK is the induction of distinct PFK2 isoforms in response to the hypoxia induced protein HIF-1α [117]. These forms of the enzyme (designated as PFKFB 1 to 4) favor glycolysis in a wide number of tissues, including muscle.

Experimentally, inhibition of PFK-2 is a means of experimentally attenuating glycolysis [118]. While fluoride inhibition of glycolysis due to its inhibition of enolase was established in early literature [119], it is not widely used currently, as it is nonspecific and a weak cellular inhibitor. A commonly used inhibitor of glucose entry into glycolysis—2-deoxyglucose—is also problematic as this will trap cellular ATP and lower its concentration [120].

While it is theoretically possible for other enzymes of glycolysis to alter the flow of the pathway, under normal conditions, they play essentially no role. That is because, under those conditions, these enzymes maintain near-equilibrium (see the previous discussion of LDH). It is equally unlikely that drug action, either direct or indirect, can influence the glycolytic pathway at steps other than the metabolically irreversible ones. This is because most of the glycolytic enzymes are NEQ, a condition which the cell achieves by producing large amounts of these enzymes. The correspondingly high concentration of drug needed to effectively inhibit them makes off-target effects more likely.

6.2.5 ENERGETIC CONCERNS OF SLOW MUSCLE

Contraction of slow muscle requires force generation over a longer time frame. This sustained-contraction requirement is met by oxidative metabolism, which can provide more energy, at a slower rate. As an extreme, migratory birds oxidizing lipid can power flight muscle for weeks to enable trips of over a thousand miles without rest [121,122].

While carbohydrate metabolism could provide energy for slow muscle, it is quantitatively less important fuel than fatty acids. The ketogenic amino acids do not provide significant fuel to muscle as their fate is largely transamination, with the ketoacid delivered to the liver for compete oxidation.

Lipid utilized for energy must be converted to fatty acids for their oxidation in mitochondria. As described previously for the liver, this fatty acid is principally long-chain fatty acid and thus subject to control of entry into the mitochondria. The major control point is once again the conversion of the long-chain acyl CoA into long-chain acyl carnitine by CAT I. The muscle has a distinctive form of CAT I from the liver, which is inhibited by far lower concentrations of malonyl CoA [123]. The formation of the malonyl CoA, as in liver, depends on acetyl CoA carboxylase. However, there is little fatty acid synthesis in muscle; the enzyme appears to be utilized strictly to provide malonyl CoA as a regulatory molecule. Then the controls can be exerted at the acetyl CoA carboxylase step. For example, it is established that AMPK directed phosphorylation targets the carboxylase, and in this way lifts inhibition of CAT I, permitting fatty acid oxidation by muscle. It is less clear how the lack of insulin achieves this same end, although the metabolic effect on the fatty acid is the same.

Within the mitochondria, it is less clear what controls are possible for this process. However, contraction is accompanied by a rise in cytosolic Ca^{2+} which will secondarily increase mitochondrial Ca^{2+}. This has the effect of activing mitochondrial dehydrogenases, such as the α-ketoglutarate DH, which enhances Krebs cycle flow and would thus respond to the degree of contractile force, which is determined by cytosolic Ca^{2+}. Further discussion of Ca^{2+} changes in cells are presented in the following chapter.

6.3   HEART METABOLISM

Like slow muscle, heart metabolism is principally oxidative, consuming both glucose and fatty acids [124,125]. Heart can also avidly consume the ketone bodies, acetoacetate and β-hydroxybutyrate [126]. As these are readily available, insulin-insensitive fuels, they may be useful as a therapy for pathological situations [127,128]. The pathway for ketone body utilization is illustrated in Figure 6.35. Oxidation of the ketone bodies requires the formation of succinyl CoA, used to convert acetoacetate to acetoacetyl-CoA. The two consequences of this use of succinyl CoA[xiii] are (a) the loss of high-energy phosphate formation at succinyl CoA synthetase (Figure 6.36), and (b) the need to drive α-ketoglutarate DH at a rate commensurate with the rate of ketone body formation. It has been observed long ago by Carafoli [131] that Ca^{2+} can stimulate ketone body oxidation by isolated rat heart mitochondria. He attributed this to a Ca^{2+} stimulation of β-hydroxybutyrate DH. However, Ca^{2+} also stimulates oxidation of acetoacetate, which bypasses the β-hydroxybutyrate DH step (unpublished data[xiv]). While this study obviously needs further investigation, it is clear that this situation parallels that of slow skeletal muscle in the prior section, in which Ca^{2+} activation of the α-ketoglutarate DH drives mitochondrial oxidation, which is also established for heart [132].

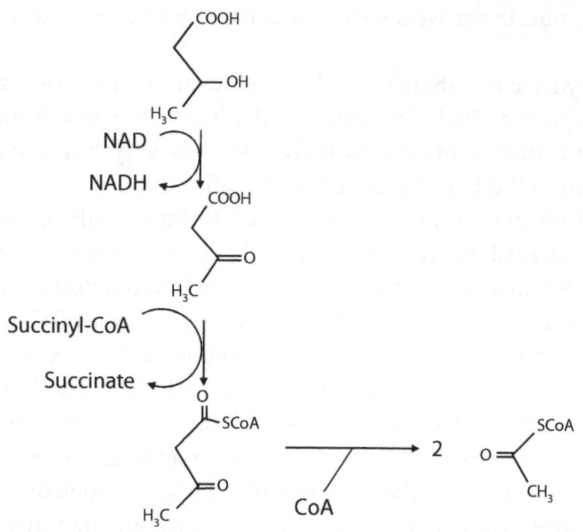

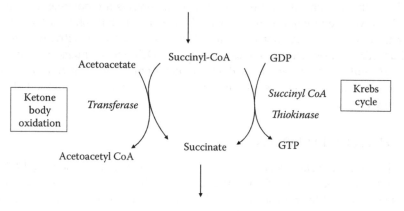

FIGURE 6.35   Ketone body oxidation.

FIGURE 6.36   Two reactions connecting succinyl CoA to succinate. The conversion of acetoacetate to the CoA ester uses succinyl CoA from the Krebs cycle. Thus, to the extent that
ketone bodies are oxidized, the energy capture of succinyl CoA synthetase is bypassed using
the reaction shown instead.

Heart can also oxidize glucose, so that insulin stimulation of glucose uptake
to heart leads to its complete oxidation. Curiously, there is a glycogen store in the
heart, as well as the attendant enzymes for its synthesis and degradation. There have
been suggestions that this glycogen may serve some role in development [133], but it
also is established that regulation of glycogen metabolism occurs, linking glycogen
metabolism to glucose uptake [134]. While heart is the quintessential slow muscle,
there is nonetheless both glycogen and the creatine phosphokinase system [97] to
supply energy for rapid contractile episodes.

6.4 ADIPOSE

The specialization of adipocytes is evident from the nondescript micrographs of this cell: the appearance of a large spherical oil drop with a thin outer layer of cytosol. Fat synthesis performed by adipocytes varies with diet and disease state [135–137]. Beyond their ability to both synthesize, take up, and store lipid, adipocytes would seem to have few activities. For example, they do not oxidize fatty acids, nor can they reuse the glycerol that is produced by triglyceride hydrolysis. However, these cells are now established as a major endocrine organ. In addition, the white and brown adipose (and intermediate forms) have distinctive metabolic features.

6.4.1 TRIGLYCERIDE METABOLISM

The principal metabolic pathways for fat cells are the formation and breakdown of triglycerides. For triglyceride synthesis, fatty acids are derived from one of three sources: dietary (chylomicrons), hepatic biosynthesis (VLDL), or endogenous de novo synthesis. As discussed above, the relative contributions are variable. For triglyceride breakdown, the products, fatty acids and glycerol, are mostly released to the blood and metabolized by other tissues. Fatty acids, bound to plasma proteins (largely albumin) are oxidized by other tissues; glycerol is metabolized by liver.

The synthesis of the glycerol backbone of triglycerides can be derived from glucose by the first portion of the glycolytic pathway. Glucose entry into the fat cell, like that into muscle, utilizes the GLUT4, an insulin-sensitive transporter. In fact, the notion of a transporter having a residence on an internal vesicle as well as the plasma membrane, and being shuttled in the presence of stimulus (usually, insulin) originated with the studies in fat cells [138]. A separate pathway for glycerol formation similar to hepatic gluconeogenesis was proposed based in part on the finding that adipocytes contain PEPCK [139]. This remains controversial as studies using C^{13} labelling suggest that this is not the major route of supply for this backbone [140].

What is undisputed is that the route of synthesis and breakdown of lipids dominates adipocyte metabolism. Thus, under fed conditions, insulin leads to suppression of the triglyceride lipolysis, and synthesis is the dominant pathway [141]. This provides storage of the fatty acids as lipid (which includes both triglycerides and cholesterol esters). Under fasting conditions, insulin levels decline, and epinephrine causes increased lipolysis by cAMP directed activation of TG lipase, and fatty acids and glycerol are produced, and both products are exported from the cell. Fatty acids have many organ targets, such as muscle, heart, and liver. However, glycerol is exclusively metabolized by liver, which is usually considered to have the most glycerol kinase activity, although the muscle also contains this enzyme [142]. Due to this simplification and the relatively constant rate of that utilization, adipocyte lipolysis can be estimated by the concentration of blood glycerol [24].

Three lipases along with accessory proteins are involved in the hydrolysis of the triglyceride. First is the production of diacylglycerol by ATAG (adipose triacylglycerol lipase). Next, diacylglycerol is converted to the monoester by HSL (hormone sensitive lipase), formerly believed to be the major activity for the hydrolysis of the first ester as well. Finally, a selective monoacylglycerol lipase removes the last fatty acid. HSL also catalyzes hydrolysis of the fatty ester to cholesterol. Thus, hydrolysis of cholesterol ester is also stimulated under the same conditions that lead to hydrolysis of triglycerides.

Much detail has been added to the knowledge of fat cell metabolism as it is now well established that it plays a major role in the obesity epidemic (see Chapter 10). There are known to be several proteins that can serve to transport the fatty acids across the membrane, including CD36, FABP4, and FATP1. It should be noted that the ability of a protein to bind fatty acid is not a selective process, as illustrated by the fact that albumin can be missing from the blood (analbuminemia) but compensated by an increase in other blood borne proteins [143].

Aside from the uptake of fatty acids, glycerol must be transported across membranes; as this is a water-soluble compound, it requires a transport protein. That is known to involve a specific form of aquaporin, AQP-7, which is able to transport glycerol [144,145]. Aquaporins are well known for their ability to transport water, but water can be transported into cells in some cases without a transporter, and does so across artificial membranes. The fatty acids can also be transported without carriers, so that the presence of carriers may be for the purpose of accelerating a process that can exist without the transporter but perhaps not rapidly enough in the case of a tissue specialized for fatty acid delivery to the blood.

6.4.2 ENDOCRINE FUNCTION OF THE ADIPOCYTE

The first indication that the adipocyte also had an endocrine role was the 1994 discovery of leptin, the hormone produced by adipocytes and acting on cells of the hypothalamus [146]. This communication between adipocyte and brain was unexpected, and immediately suggested the possibility that obesity could be cured.[xv] However, it happened that while adipocytes do produce more leptin as they become more laden with fat, the receptors in the hypothalamus are down-regulated, just like the insulin receptors (and most others). Thus, leptin receptor insensitivity is the hallmark of obesity, just like insulin receptor insensitivity, and this is not easily remedied by attempting to satiate patients with exogenous leptin [147].

Another major hormone synthesized by fat tissue is adiponectin, which is unusual in having short-term actions in muscle tissue, activating AMPK through upstream kinases involving as-yet-undetermined signaling systems [148,149]. Beyond these are several other endocrine products, including TGF-α, visfatin, and other putative secreted factors [150].

6.4.3 PPAR-GAMMA AND DRUG RESPONSE

Two isoforms of the PPAR nuclear receptor family are: the α (for which peroxisomal proliferation was named), predominately is found in liver; and the γ which is in many tissues but believed to be most prominent in the adipocytes. The latter bind tightly to the "glitazone" class of diabetes drugs, named for the distinctive ring structure present in these compounds (Figure 6.37).

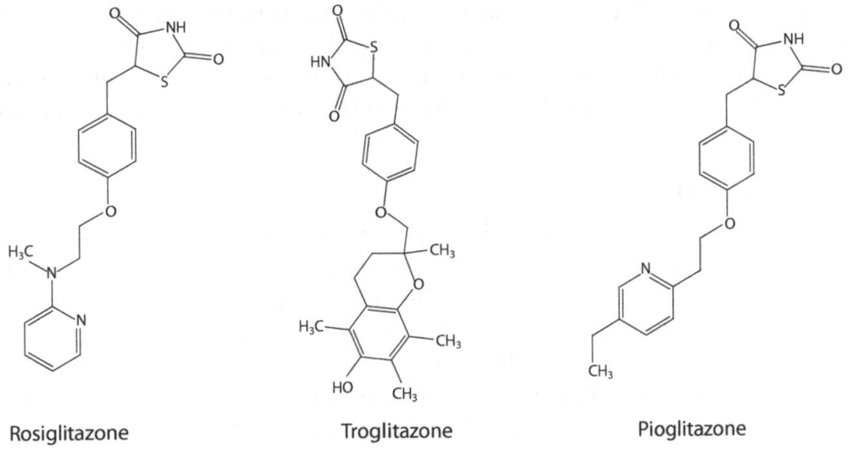

Rosiglitazone   Troglitazone   Pioglitazone

FIGURE 6.37 Structures of the glitazone drugs used for type II diabetes.

All of the glitazones are effective diabetes drugs, with the principle action of enhancing triglyceride synthesis in the adipocyte [151]. As a result, plasma fatty acid concentrations are diminished, and insulin sensitivity improves. The side effects for each are different: troglitazone, the earliest-used drug, has liver toxicity [152]; rosiglitazone has cardiovascular side effects [153], and pioglitazone is associated with elevated cancer risk.

6.4.4 Distinct Fat Cell Types

Fat cells display a variety of types within the body. Most fat cells are known as white adipose tissue (WAT) which can be broadly divided between subcutaneous and visceral fat, although this itself represents a simplification [154]. A further division of types is between WAT and brown adipose fat (BAT); the latter has recently become a more intriguing topic as it was discovered that this cell type is also present in adult humans [155].

6.4.4.1 Subcutaneous versus Visceral Fat

While the definition of the two fat types clearly refers to their locations, the metabolic distinctions have long been evident. Subcutaneous is the "good fat" and visceral is the "bad fat." This insight has been derived from observational studies but may also be based on the fact that the secretions mentioned above are distinct between the two populations. Visceral fat [156], due to its prominence in the abdominal cavity, preferentially drains into the portal vein, while subcutaneous fat enters the systemic venous system. Moreover, the visceral adipocytes are the major contributors to the characteristic low-level chronic inflammatory state characteristic of obesity and diabetes. Macrophages are attracted by secretion of a chemoattractant protein, and the macrophages in turn produce the inflammatory cytokines TNF-α and IL-6 [157]. Aside from increased fatty acid secretion, the formation of "crown" structures results, a mass of dead cells surrounded by inflammation producing peripheral cells. The other metabolic distinction of the visceral cells over the subcutaneous ones is a relative insensitivity to insulin and a greater uptake of glucose. Perhaps the only positive aspect of visceral fat depots is the fact that they are preferentially reduced with diet.

6.4.4.2   Brown Fat

It has long been known that a distinctive form of fat cells from the ones so far described has a characteristic brownish appearance, due to the presence of large numbers of mitochondria. The brown adipose tissue (BAT) has long been known to contribute to non-shivering thermogenesis, arousal from hibernation, and heat production in the newborn, including humans [158,159]. Epinephrine, through the β3-adrenergic receptor, leads to increased synthesis of an uncoupling protein (UNC). The cells then produce heat in place of ATP in a natural form of uncoupling.

Brown fat was recently shown to occur not only in newborn, but in adult humans [160]. A more controversial report is that a hormone is produced by muscle named irisin that converts white to brown adipocytes [161,162]. It also appears that a tissue of intermediate state may be involved, the so-called "beige" fat [163]. The conversion, appropriately enough, is called "browning." The interest in the uncoupling proteins that appear to play important roles in these and perhaps other cells, has also expanded with the discovery of several isoforms of the UNC proteins. For example, fat switching is implicated in mTOR signaling in adipocytes. In experiments in which the mTOR target S6K1 is depleted [164], fat cells change from lipogenesis to fatty acid oxidation. This is not a normal metabolic pathway for white fat cells, but is a major activity for brown fat cells.

6.5   INTESTINAL CELLS

The absorptive cells of the small intestine (enterocytes) meet at least part of their energy needs by using glutamine as substrate. This was demonstrated in a series of studies by Windmueller and Spaeth [165–167]. Subsequent work has shown that the amount of net glutamine utilization by enterocytes can vary with species [168], although a net intestinal utilization of glutamine by humans has been demonstrated [169].

While glutamine can contribute to at least a portion of the cell's energy provision, its metabolism can also provide nitrogen for the synthesis of nucleotides, directly provide mitochondrial NADH, as well as cytosolic NADPH via the reductive glutamine pathway (Figure 6.38). Also shown in this figure (modified from ref. [165]) is the formation of citrulline, using enzymes that form a portion of the urea cycle. It should be noted that the urea cycle itself is not operative in tissues apart from the liver, and that the Krebs cycle per se is not a pathway of glutamine metabolism. In each case, only a portion of these pathways is utilized.

The enterocyte also produces the lipoprotein particle known as the chylomicron, of similar density to the hepatic VLDL. In the case of the enterocyte, production of this lipoprotein represents dietary fat. Chylomicrons production requires, like VLDL production, the formation of phospholipid and apolipoprotein by the cell, and

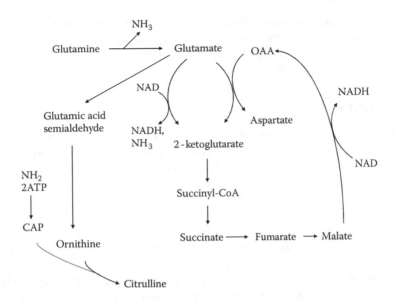

FIGURE 6.38   Oxidation of glutamine in the intestinal epithelia.

the ability to export it the extracellular space. Unlike liver, the extracellular space of the enterocyte on the basolateral side cannot communicate with the capillaries, which are not sinusoidal. Instead, the chylomicrons are delivered into the capillaries of the lymphatic system, which are the entry point of this circulation. Thus, the lipid materials alone bypass the portal circuit upon entry from the diet.

6.5.1   Production and Response to Hormones

While endocrine cells are present through much of the gastrointestinal tract, those of the duodenum are the most closely linked to the chemical composition of foods. This is because of the lengthy and variable residence time of food in the stomach (hours); the small quantities released through the pylorus into the duodenum are then sampled for composition, and appropriate hormonal signals released into the blood stream. For example, a high concentration of carbohydrates causes release of the GIP (gastric inhibitory peptide, or glucose insulinotrophic peptide) which triggers insulin release from the pancreatic β-cell [170,171]. This was named the "incretin effect" as it promotes more insulin release than that produced by the increase in glucose concentration alone (hence, an incremental increase). Drugs designed to mimic this action (e.g., ref. [172]) or preventing its degradation (selective peptidase inhibitors) as *incretins*.

Hormonal orchestration of the digestive process itself, while somewhat removed from the realm of metabolism, is nonetheless essential for bringing in nutrients to the body. Scattered throughout the digestive tract are several endocrine cells that produce these hormones to effect such activities as increased contraction of the stomach itself (gastrin), secretion of bile and of the enzymatic (duct) release from the pancreas to the duodenum (cholecystokinin), and the more recently discovered ghrelin, released from the stomach and acting at the hypothalamus to control feeding behavior. Ghrelin is at least in part responsible for the decrease in appetite following gastric bypass surgery [173].

6.6 KIDNEY

As the kidney plays the role of selective blood filter, returning specific compounds to the bloodstream (such as glucose), entirely removing others (such as xenobiotics), and partially retaining yet others depending on the overall body physiology (such as Na^+), its metabolic pathways are specialized and distributed between distinct portions of the kidney tubule. From a broad metabolic perspective, we may separate cell types into just two: the convoluted tubules (proximal and distal) and the medullary segments. The former have more mitochondria, and some specialized routes such as gluconeogenesis, and are aerobic cells. The latter have fewer specialized cells, encounter lower oxygen tensions, and are more glycolytic.

The reuptake of glucose by kidney (largely the proximal convoluted cells) is illustrated in Figure 6.39. Entry of glucose into the cell utilizes the sodium-linked glucose transporter (SGLT2) [174]. Glucose exits the cell through the basolateral membrane localized GLUT2, the same isoform of the glucose transporter found in liver, as well as pancreatic alpha and beta cells. This pathway for reclamation of glucose by the kidney is of long-standing interest in the study of diabetes. Under conditions of very high glucose concentration in the blood, the SGLT2 is saturated and can no longer remove glucose and thus glucose enters the final urine. A form of diabetes treatment is to inhibit the SGLT2, thereby lowering the saturation point and removing even more blood glucose into the urine. It should be noted that the route for glucose entry to the blood shown in Figure 6.39 is the same as that for uptake of glucose by the

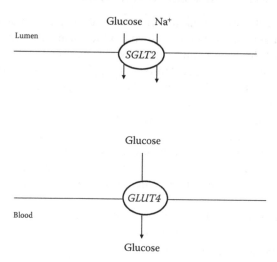

FIGURE 6.39   Two forms of glucose transport in kidney proximal tubule cells.

enterocyte, except that that the isoform for Na-dependent uptake is SGLT1; this has also been considered as a target for diabetes treatment [175].

6.6.1 KIDNEY GLUTAMINE METABOLISM: ACIDOSIS RESPONSE

The kidney plays an important role in restoring pH balance in acidosis by exporting NH_4^+ into the tubule and ultimately the final urine. The nitrogen arises from the amino acid glutamine. In response to increasing blood acidity, the proximal convoluted tubule cells induce glutaminase to greater levels so that an increased conversion of glutamine to ammonia and glutamate results [176].

The ultimate fate of the glutamate is conversion to glucose; the overall kidney pathway is

$$Glutamine \rightarrow NH_3 + glucose$$

A similar pathway was considered earlier in the chapter for the catabolism of glutamine by liver:

$$Glutamine \rightarrow urea + glucose$$

In the case of the liver, the glutamine pathway in the mitochondria involves conversion to aspartate (Figure 6.15b), which is transported into the cytosol and then condenses with citrulline. The carbon from aspartate appears as fumarate, and continues on through the gluconeogenic pathway, thus converting glutamine to glucose and urea. The pathway is largely dictated by the requirement for balancing reducing equivalents needed for gluconeogenesis and the stoichiometric requirement of equal amounts of citrulline and aspartate formation in the cytosol for the condensation reaction.

The kidney has no urea cycle; instead the end product is ammonia. In order to balance the cytosolic requirement of glucose formation for NADH in this pathway, the carbon must exit the mitochondria as malate. This in turn requires that the pathway involve 2-ketoglutarate dehydrogenase, as illustrated in Figure 6.40. The ammonium ions exit the cell and contribute to the relief of acidosis. Thus, we can view the pathway of kidney gluconeogenesis from glutamine as a metabolic pathway established for the benefit of restoration of homeostasis following acidosis. The gluconeogenic pathway is more of a metabolic necessity than a pathway contributing glucose to the whole organism as is the case in liver.

Estimates from isolated kidney tubule preparations suggest that the overall rate of kidney gluconeogenesis is about 10% that of liver [177]; however, it is not clear that the kidney actually produces net glucose. Cells of the medullary region of the kidney are largely glycolytic and could consume at least a portion of the glucose. It is clear at least in the case of glutamine gluconeogenesis that the drive for increasing this pathway is not whole-body need for glucose but rather the more specialized task of recovery from acidosis. While the mechanism of acid sensing is unknown, the recent development of a cell culture system which exhibits both ammonia formation and gluconeogenesis should contribute to that quest [178].

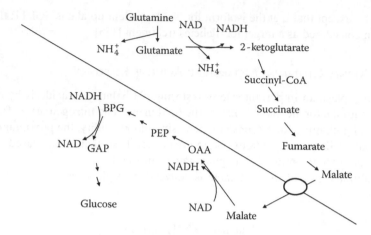

FIGURE 6.40 Glutamine metabolism in kidney during acidosis. The products are NH_4^+ exported to counteract the acidosis and glucose to balance the carbon of the substrate.

In general terms, different tissues in the body utilize distinct metabolic pathways, largely to sustain their own activities and fit their specialized purpose. For example, nerve cells have routes for the production of neurotransmitters. Energy needs are largely met by glucose, but can also be satisfied at least in part by ketone bodies. Certain cells of the immune system that produce large amounts of protein such as B cells have greatly enhanced rates of protein synthesis and thus utilize a greater portion of their energy resources for that pathway compared to other cells. Thus, no single cell type is prototypic of intermediary metabolism; each has its own modification of the broad scheme.

REFERENCES

1. Brosnan, M., Chen, L., Wheeler, C., van Dyke, T., and Koretsky, A. (1991) Phosphocreatine protects ATP from a fructose load in transgenic mouse liver expressing creatine kinase. *Am. J. Physiol.* 260, C1191–C1200.
2. Miller, K., Halow, J., and Koretsky, A. (1993) Phosphocreatine protects transgenic mouse liver expressing creatine kinase from hypoxia and ischemia. *Am. J. Physiol.* 265, C1544–C1551.
3. Koretsky, A., Brosnan, M., Chen, L., Chen, J., and van Dyke, T. (1990) NMR detection of creatine kinase expressed in liver of transgenic mice: Determination of free ADP levels. *Proc. Natl. Acad. Sci. USA* 87, 3112–3116.
4. Shearer, J., Marchand, I., Sathasivam, P., Tarnopolsky, M. A., and Graham, T. E. (2000) Glycogenin activity in human skeletal muscle is proportional to muscle glycogen concentration. *Am. J. Physiol. Endocrinol. Metab.* 278, E177–E180.
5. Melendez, R., Melendez-Hevia, E., and Cascante, M. (1997) How did glycogen structure evolve to satisfy the requirement for rapid mobilization of glucose? A problem of physical constraints in structure building. *J. Mol. Evol.* 45, 446–455.
6. Melendez, R., Melendez-Hevia, E., and Canela, E. I. (1999) The fractal structure of glycogen: A clever solution to optimize cell metabolism. *Biophys. J.* 77, 1327–1332.

7. Baskaran, S., Roach, P. J., DePaoli-Roach, A. A., and Hurley, T. D. (2010) Structural basis for glucose-6-phosphate activation of glycogen synthase. *Proc. Natl. Acad. Sci. USA* 107, 17563–17568.

8. Frame, S., and Cohen, P. (2001) GSK3 takes centre stage more than 20 years after its discovery. *Biochem. J.* 359, 1–16.

9. Cohen, P. (2002) The origins of protein phosphorylation. *Nat. Cell Biol.* 4, E127–E130.

10. Freland, L., and Beaulieu, J. M. (2012) Inhibition of GSK3 by lithium, from single molecules to signaling networks. *Front. Molec. Neurosci.* 5, 14.

11. Adeva-Andany, M. M., Gonzalez-Lucan, M., Donapetry-Garcia, C., Fernandez-Fernandez, C., and Ameneiros-Rodriguez, E. (2016) Glycogen metabolism in humans. *BBA Clin.* 5, 85–100.

12. Hems, D. A., and Whitton, P. D. (1980) Control of hepatic glycogenolysis. *Physiol. Rev.* 60, 1–50.

13. Soling, H. D., and Seck, A. (1975) Precursor specific inhibition of hepatic gluconeogenesis by glisoxepide an inhibitor of the L-aspartate/L-glutamate antiport system. *FEBS Lett.* 51, 52–59.

14. Leverve, X. M., Verhoeven, A. J., Groen, A. K., Meijer, A. J., and Tager, J. M. (1986) The malate/aspartate shuttle and pyruvate kinase as targets involved in the stimulation of gluconeogenesis by phenylephrine. *Eur. J. Biochem.* 155, 551–556.

15. Ochs, R. S., and Harris, R. A. (1980) Aminooxyacetate inhibits gluconeogenesis by isolated chicken hepatocytes. *Biochim. Biophys. Acta* 632, 260–269.

16. Ochs, R. S., and Harris, R. A. (1978) Studies on the relationship between glycolysis, lipogenesis, gluconeogenesis, and pyruvate kinase activity of rat and chicken hepatocytes. *Arch. Biochem. Biophys.* 190, 193–201.

17. Friedman, J. E., Yun, J. S., Patel, Y. M., McGrane, M. M., and Hanson, R. W. (1993) Glucocorticoids regulate the induction of phosphoenolpyruvate carboxykinase (GTP) gene transcription during diabetes. *J. Biol. Chem.* 268, 12952–12957.

18. Lamers, W. H., Hanson, R. W., and Meisner, H. M. (1982) cAMP stimulates transcription of the gene for cytosolic phosphoenolpyruvate carboxykinase in rat liver nuclei. *Proc. Natl. Acad. Sci.* 79, 5137–5141.

19. Viana, A. Y., Sakoda, H., Anai, M., Fujishiro, M., Ono, H., Kushiyama, A., Fukushima, Y., Sato, Y., Oshida, Y., Uchijima, Y., Kurihara, H., and Asano, T. (2006) Role of hepatic AMPK activation in glucose metabolism and dexamethasone-induced regulation of AMPK expression. *Diabetes Res. Clin. Pract.*

20. Newsholme, E. A., Arch, J. R. S., Brooks, B., and Surholt, B. (1983) The role of substrate cycles in metabolic regulation. *Biochem. Soc. Trans.* 11, 52–56.

21. Newsholme, E. A., and Start, C. (1973) *Regulation in Metabolism*, Wiley, London.

22. Mapes, J. P., and Harris, R. A. (1976) Inhibition of gluconeogenesis and lactate formation from pyruvate by N O-dibutyryl adenosine 3':5'-monophosphate. *J. Biol. Chem.* 251, 6189–6196.

23. Cornell, N. W., Lund, P., and Krebs, H. A. (1974) The effect of lysine on gluconeogenesis from lactate in rat hepatocytes. *Biochem. J.* 142, 327–337.

24. Nurjhan, N., Kennedy, F., Consoli, A., Martin, C., Miles, J., and Gerich, J. (1988) Quantification of the glycolytic origin of plasma glycerol: Implications for the use of the rate of appearance of plasma glycerol as an index of lipolysis in vivo. *Metabolism* 37, 386–389.

25. Vazquez, J. A., and Kazi, U. (1994) Lipolysis and gluconeogenesis from glycerol during weight reduction with very-low-calorie diets. *Metabolism* 43, 1293–1299.

26. Ochs, R. S., and Lardy, H. A. (1983) Catecholamine stimulation of hepatic gluconeogenesis at the site between pyruvate and phosphoenolpyruvate. *J. Biol. Chem.* 258, 9956–9962.

27. Whitton, P. D., Rodrigues, L. M., and Hems, D. A. (1978) Stimulation by vasopressin angiotensin and oxytocin of gluconeogenesis in hepatocyte suspensions. *Biochem. J.* 176, 893–898.

28. Kneer, N. M., Wagner, M. J., and Lardy, H. A. (1979) Regulation by calcium of hormonal effects on gluconeogenesis. *J. Biol. Chem.* 254, 12160–12168.

29. Kalhan, S. C., Bugianesi, E., McCullough, A. J., Hanson, R. W., and Kelley, D. E. (2008) Estimates of hepatic glyceroneogenesis in type 2 diabetes mellitus in humans. *Metabolism* 57, 305–312.

30. Ochs, R. S., and Harris, R. A. (1986) Mechanism for the oleate stimulation of gluconeogenesis from dihydroxyacetone by hepatocytes from fasted rats. *Biochim. Biophys. Acta* 886, 40–47.

31. Moreno, K. X., Satapati, S., DeBerardinis, R. J., Burgess, S. C., Malloy, C. R., and Merritt, M. E. (2014) Real-time detection of hepatic gluconeogenic and glycogenolytic states using hyperpolarized [2-13C]dihydroxyacetone. *J. Biol. Chem.* 289, 35859–35867.

32. Ardenkjaer-Larsen, J. H., Fridlund, B., Gram, A., Hansson, G., Hansson, L., Lerche, M. H., Servin, R., Thaning, M., and Golman, K. (2003) Increase in signal-to-noise ratio of > 10,000 times in liquid-state NMR. *Proc. Natl. Acad. Sci. USA* 100, 10158–10163.

33. Exton, J. H., and Park, C. R. (1967) Control of gluconeogenesis in liver. I. General features of gluconeogenesis in the perfused livers of rats. *J. Biol. Chem.* 242, 2622–2636.

34. Brunt, P. W. (1982) Treatment of alcohol dependence: The alcoholic patient. *Br. Med. Bull.* 38, 103–105.

35. Crow, K. E., Braggins, T. J., Batt, R. D., and Hardman, M. J. (1983) Kinetics of malate dehydrogenase and control of rates of ethanol metabolism in rats. *Pharmacol. Biochem. Behav.* 18 Suppl 1, 233–236.

36. Cornell, N. W. (1983) Properties of alcohol dehydrogenase and ethanol oxidation in vivo and in hepatocytes. *Pharmacol. Biochem. Behav.* 18 Suppl 1, 215–221.

37. Masson, S., and Quistorff, B. (1992) The 31P NMR visibility of ATP in perfused rat liver remains about 90%, unaffected by changes of metabolic state. *Biochemistry (Mosc.)* 31, 7488–7493.

38. Ouyang, X., Cirillo, P., Sautin, Y., McCall, S., Bruchette, J. L., Diehl, A. M., Johnson, R. J., and Abdelmalek, M. F. (2008) Fructose consumption as a risk factor for non-alcoholic fatty liver disease. *J. Hepatol.* 48, 993–999.

39. Brebnor, L., Phillips, E., and Balinsky, J. B. (1981) Control of urea cycle enzymes in rat liver by glucagon. *Enzyme* 26, 265–270.

40. Snodgrass, P. J. (1991) Dexamethasone and glucagon cause synergistic increases of urea cycle enzyme activities in livers of normal but not adrenalectomized rats. *Enzyme* 45, 30–38.

41. Cohen, P. P. (1981) The ornithine-urea cycle: Biosynthesis and regulation of carbamyl phosphate synthetase I and ornithine transcarbamylase. *Curr. Top. Cell. Regul.* 18, 1–19.

42. Krebs, H. A., Lund, P., and Stubbs, M. (1976) Interrelations Between Gluconeogenesis and Urea Synthesis. in *Gluconeogenesis: Its Regulation in Mammalian Species* (Hanson, R. W., and Mehlman, M. A. eds.), Wiley-Interscience, New York, pp. 269–291.

43. Ochs, R. S. (1984) Glutamine metabolism of isolated rat hepatocytes. Evidence for catecholamine activation of alpha-ketoglutarate dehydrogenase. *J. Biol. Chem.* 259, 13004–13010.

44. Martin, C., Chevrot, M., Poirier, H., Passilly-Degrace, P., Niot, I., and Besnard, P. (2011) CD36 as a lipid sensor. *Physiol. Behav.* 105, 36–42.

45. Stahl, A. (2004) A current review of fatty acid transport proteins (SLC27). *Pflügers Archiv.* 447, 722–727.

46. Hamilton, J. A., and Kamp, F. (1999) How are free fatty acids transported in membranes? Is it by proteins or by free diffusion through the lipids? *Diabetes* 48, 2255–2269.

47. Jia, Z., Pei, Z., Maiguel, D., Toomer, C. J., and Watkins, P. A. (2007) The fatty acid transport protein (FATP) family: Very long chain acyl-CoA synthetases or solute carriers? *J. Mol. Neurosci.* 33, 25–31.

48. McGarry, J. D., Leatherman, G. F., and Foster, D. W. (1978) Carnitine palmitoyltransferase I. The site of inhibition of hepatic fatty acid oxidation by malonyl-CoA. *J. Biol. Chem.* 253, 4128–4136.

49. Vance, D. E., and Vance, J. E. (2008) *Biochemistry of Lipids, Lipoproteins and Membranes*, Elsevier, Amsterdam.

50. Watmough, N. J., and Frerman, F. E. (2010) The electron transfer flavoprotein: Ubiquinone oxidoreductases. *Biochim. Biophys. Acta* 1797, 1910–1916.

51. Salvi, F., and Gadda, G. (2013) Human choline dehydrogenase: Medical promises and biochemical challenges. *Arch. Biochem. Biophys.* 537, 243–252.

52. Roe, C. R., Sweetman, L., Roe, D. S., David, F., and Brunengraber, H. (2002) Treatment of cardiomyopathy and rhabdomyolysis in long-chain fat oxidation disorders using an anaplerotic odd-chain triglyceride. *J. Clin. Investig.* 110, 259–269.

53. Vlaeminck, B., Fievez, V., Cabrita, A. R. J., Fonseca, A. J. M., and Dewhurst, R. J. Factors affecting odd- and branched-chain fatty acids in milk: A review. *Animal Feed Sci. Technol.* 131, 389–417.

54. Luo, M. J., Smeland, T. E., Shoukry, K., and Schulz, H. (1994) Delta 3,5, delta 2,4-dienoyl-CoA isomerase from rat liver mitochondria. Purification and characterization of a new enzyme involved in the beta-oxidation of unsaturated fatty acids. *J. Biol. Chem.* 269, 2384–2388.

55. McGarry, J. D., and Foster, D. W. (1980) Regulation of hepatic fatty acid oxidation and ketone body production. *Ann. Rev. Biochem.* 49, 395–420.

56. Christiansen, E. N., Thomassen, M. S., Christiansen, R. Z., Osmundsen, H., and Norum, K. R. (1979) Metabolism of erucic acid in perfused rat liver: Increased chain shortening after feeding partially hydrogenated marine oil and rapeseed oil. *Lipids* 14, 829–835.

57. Eskin, N. A. M., and McDonald, B. E. (1991) Canola oil. *Nutr. Bull.* 16, 138–146.

58. Wanders, R. J., Vreken, P., Ferdinandusse, S., Jansen, G. A., Waterham, H. R., van Roermund, C. W., and Van Grunsven, E. G. (2001) Peroxisomal fatty acid alpha- and beta-oxidation in humans: Enzymology, peroxisomal metabolite transporters and peroxisomal diseases. *Biochem. Soc. Trans.* 29, 250–267.

59. Mannaerts, G. P., Debeer, L. J., Thomas, J., and De Schepper, P. J. (1979) Mitochondrial and peroxisomal fatty acid oxidation in liver homogenates and isolated hepatocytes from control and clofibrate-treated rats. *J. Biol. Chem.* 254, 4585–4595.

60. Wanders, R. J. A., and Waterham, H. R. (2006) Biochemistry of mammalian peroxisomes revisited. *Annu. Rev. Biochem.* 75, 295–332.

61. Leighton, F., Bergseth, S., Rørtveit, T., Christiansen, E. N., and Bremer, J. (1989) Free acetate production by rat hepatocytes during peroxisomal fatty acid and dicarboxylic acid oxidation. *J. Biol. Chem.* 264, 10347–10350.

62. Williamson, D. H., Lund, P., and Krebs, H. A. (1967) The redox state of free nicotinamide-adenine dinucleotide in the cytoplasm and mitochondria of rat liver. *Biochem. J.* 103, 514–527.

63. McGarry, J. D., Mannaerts, G. P., and Foster, D. W. (1977) A possible role for malonyl-CoA in the regulation of hepatic fatty acid oxidation and ketogenesis. *J. Clin. Invest.* 60, 265–270.

64. Hawkins, R. A., Williamson, D. H., and Krebs, H. A. (1971) Ketone-body utilization by adult and suckling rat brain in vivo. *Biochem. J.* 122, 13–18.

65. Nocito, L., Kleckner, A. S., Yoo, E. J., Jones Iv, A. R., Liesa, M., and Corkey, B. E. (2015) The extracellular redox state modulates mitochondrial function, gluconeogenesis, and glycogen synthesis in murine hepatocytes. *PloS one* 10, e0122818.

66. Schutz, Y. (2004) Dietary fat, lipogenesis and energy balance. *Physiol. Behav.* 83, 557–564.

67. Conover, T. E. (1987) Textbook Error—Does citrate transport supply both acetyl groups and Nadph for cytoplasmic fatty-acid synthesis. *Trends Biochem. Sci.* 12, 88–89.

68. Conover, T. E. (1987) Reply from Conover. *Trends Biochem. Sci.* 12, 425.

69. Guthrie, G. D. (1987) The hidden counterion for the citrate antiporter. *Trends Biochem. Sci.* 12, 424.

70. Goodridge, A. G. (1973) Regulation of fatty acid synthesis in isolated hepatocytes. Evidence for a physiological role for long chain fatty acyl coenzyme and citrate. *J. Biol. Chem.* 248, 4318–4326.

71. Geelen, M. J., and Schmitz, M. G. (1993) The role of citrate in the regulation of hepatic fatty acid synthesis by insulin and glucagon. *Horm. Metab. Res.* 25, 525–527.

72. Pogson, C. I., and Randle, P. J. (1966) The control of rat-heart phosphofructokinase by citrate and other regulators. *Biochem. J.* 100, 683–693.

73. Faergeman, N. J., and Knudsen, J. (1997) Role of long-chain fatty acyl-CoA esters in the regulation of metabolism and in cell signalling. *Biochem. J.* 323, 1–12.

74. McCune, S. A., and Harris, R. A. (1979) Mechanism responsible of 5-(tetradecyloxy)-2-furoic acid inhibition of hepatic lipogenesis. *J. Biol. Chem.* 254, 10095–10101.

75. Hunt, D. W., Winters, G. C., Brownsey, R. W., Kulpa, J. E., Gilliland, K. L., Thiboutot, D. M., and Hofland, H. E. (2017) Inhibition of sebum production with the acetyl coenzyme A carboxylase inhibitor olumacostat glasaretil. *J. Invest. Dermatol.*

76. Ingebritsen, T. S., Lee, H. S., Parker, R. A., and Gibson, D. M. (1978) Reversible modulation of the activities of both liver microsomal hydroxymethylglutaryl coenzyme A reductase and its inactivating enzyme. Evidence for regulation by phosphorylation-dephosphorylation. *Biochem. Biophys. Res. Commun.* 81, 1268–1277.

77. Holland, R., Witters, L. A., and Hardie, D. G. (1984) Glucagon inhibits fatty acid synthesis in isolated hepatocytes via phosphorylation of acetyl-CoA carboxylase by cyclic-AMP-dependent protein kinase. *Eur. J. Biochem.* 140, 325–333.

78. Peng, I. C., Chen, Z., Sun, W., Li, Y. S., Marin, T. L., Hsu, P. H., Su, M. I., Cui, X. P., Pan, S. Q., Lytle, C. Y., Johnson, D. A., Blaeser, F., Chatila, T., and Shyy, J. Y. J. (2012) Glucagon regulates ACC activity in adipocytes through the CAMKK beta/AMPK pathway. *Am. J. Physiol. Endoc. M.* 302, E1560–E1568.

79. Luo, J., Hong, Y., Lu, Y., Qiu, S., Chaganty, B. K., Zhang, L., Wang, X., Li, Q., and Fan, Z. (2017) Acetyl-CoA carboxylase rewires cancer metabolism to allow cancer cells to survive inhibition of the Warburg effect by cetuximab. *Cancer Lett.* 384, 39–49.

80. Smith, J. L., and Sherman, D. H. (2008) Biochemistry. An enzyme assembly line. *Science* 321, 1304–1305.

81. Gurr, M. I., and Harwood, J. L. (1991) *Lipid Biochemistry: An Introduction*, 4th ed., Chapman & Hall, London.

82. Fernandez-Murray, J. P., and McMaster, C. (2016) Lipid synthesis and membrane contact sites: A crossroads for cellular physiology. *J. Lipid Res.*

83. Ishizuka, T., Hoffman, J., Cooper, D. R., Watson, J. E., Pushkin, D. B., and Farese, R. V. (1989) Glucose-induced synthesis of diacylglycerol de novo is associated with translocation (activation) of protein kinase C in rat adipocytes. *FEBS Lett.* 249, 234–238.

84. Zeisel, S. H. (2007) Nutrigenomics and metabolomics will change clinical nutrition and public health practice: Insights from studies on dietary requirements for choline. *Am. J. Clin. Nutr.* 86, 542–548.

85. Ameyama, M., Shinagawa, E., Matsushita, K., Takimoto, K., Nakashima, K., and Adachi, O. (1985) Mammalian choline dehydrogenase is a quinoprotein. *Agric. Biol. Chem.* 49, 3623–3626.

86. Johnson, A. R., Craciunescu, C. N., Guo, Z., Teng, Y. W., Thresher, R. J., Blusztajn, J. K., and Zeisel, S. H. (2010) Deletion of murine choline dehydrogenase results in diminished sperm motility. *FASEB J.* 24, 2752–2761.

87. Veech, R. L., Eggleston, L. V., and Krebs, H. A. (1969) The redox state of free nicotinamide-adenine dinucleotide phosphate in the cytoplasm of rat liver. *Biochem. J.* 115, 609–619.

88. Shimomura, I., Matsuda, M., Hammer, R. E., Bashmakov, Y., Brown, M. S., and Goldstein, J. L. (2000) Decreased IRS-2 and increased SREBP-1c lead to mixed insulin resistance and sensitivity in livers of lipodystrophic and ob/ob mice. *Mol. Cell* 6, 77–86.

89. Brown, M. S., and Goldstein, J. L. (2008) Selective versus total insulin resistance: A pathogenic paradox. *Cell Metab.* 7, 95–96.

90. Biddinger, S. B., Hernandez-Ono, A., Rask-Madsen, C., Haas, J. T., Alemán, J. O., Suzuki, R., Scapa, E. F., Agarwal, C., Carey, M. C., Stephanopoulos, G., Cohen, D. E., King, G. L., Ginsberg, Henry N., and Kahn, C. R. (2008) Hepatic insulin resistance is sufficient to produce dyslipidemia and susceptibility to atherosclerosis. *Cell Metab.* 7, 125–134.

91. Shimano, H. (2009) SREBPs: Physiology and pathophysiology of the SREBP family. *FEBS J.* 276, 616–621.

92. Stansbie, D., Brownsey, R. W., Crettaz, M., and Denton, R. M. (1976) Acute effects in vivo of anti-insulin serum on rates of fatty acid synthesis and activities of acetyl-coenzyme A carboxylase and pyruvate dehydrogenase in liver and epididymal adipose tissue of fed rats. *Biochem. J.* 160, 413–416.

93. Pullinger, C. R., and Gibbons, G. F. (1983) The relationship between the rate of hepatic sterol synthesis and the incorporation of [3H] water. *J. Lipid Res.* 24, 1321–1328.

94. Hashimoto, F., Taira, S., and Hayashi, H. (2000) Changes in isoprenoid lipid synthesis by gemfibrozil and clofibric acid in rat hepatocytes. *Biochem. Pharmacol.* 59, 1203–1210.

95. Schimmack, G., Defronzo, R. A., and Musi, N. (2006) AMP-activated protein kinase: Role in metabolism and therapeutic implications. *Diabetes Obesity Metab.* 8, 591–602.

96. Tesch, P. A., Thorsson, A., and Fujitsuka, N. (1989) Creatine phosphate in fiber types of skeletal muscle before and after exhaustive exercise. *J. Appl. Physiol.* 66, 1756–1759.

97. Conway, M. A., and Clark, J. F. (1996) *Creatine and Creatine Phosphate Scientific and Clinical Perspectives*. Academic Press, San Diego, CA.

98. Shulman, R. G., and Rothman, D. L. (2001) The "glycogen shunt" in exercising muscle: A role for glycogen in muscle energetics and fatigue. *Proc. Natl. Acad. Sci.* 98, 457–461.

99. Terjung, R. L., Baldwin, K. M., Mole, P. A., Klinkerfuss, G. H., and Holloszy, J. O. (1972) Effect of running to exhaustion on skeletal muscle mitochondria: A biochemical study. *Am. J. Physiol.* 223, 549–554.

100. Baldwin, K. M., Reitman, J. S., Terjung, R. L., Winder, W. W., and Holloszy, J. O. (1973) Substrate depletion in different types of muscle and in liver during prolonged running. *Am. J. Physiol.* 225, 1045–1050.

101. Fisher, J. S., Nolte, L. A., Kawanaka, K., Han, D. H., Jones, T. E., and Holloszy, J. O. (2002) Glucose transport rate and glycogen synthase activity both limit skeletal muscle glycogen accumulation. *Am. J. Physiol. Endocrinol. Metab.* 282, E1214–E1221.

102. Coulson, R. A. (1987) Aerobic and anaerobic glycolysis in mammals and reptiles in vivo. *Comp. Biochem. Physiol B Comp. Biochem.* 87, 207–216.

103. Jensen, T. E., Sylow, L., Rose, A. J., Madsen, A. B., Angin, Y., Maarbjerg, S. J., and Richter, E. A. (2014) Contraction-stimulated glucose transport in muscle is controlled by AMPK and mechanical stress but not sarcoplasmatic reticulum Ca(2+) release. *Molec. Metab.* 3, 742–753.

104. Hayashi, T., Wojtaszewski, J. F., and Goodyear, L. J. (1997) Exercise regulation of glucose transport in skeletal muscle. *AJP—Endocrinol. Metab.* 273, E1039–E1051.

105. Liang, X., Liu, L., Fu, T., Zhou, Q., Zhou, D., Xiao, L., Liu, J., Kong, Y., Xie, H., Yi, F., Lai, L., Vega, R. B., Kelly, D. P., Smith, S. R., and Gan, Z. (2016) Exercise inducible lactate dehydrogenase B regulates mitochondrial function in skeletal muscle. *J. Biol. Chem.*

106. Kaplan, N. O., and Ciotti, M. M. (1961) Evolution and differentiation of dehydrogenases. *Ann. N. Y. Acad. Sci.* 94, 701–722.

107. Fondy, T. P., and Kaplan, N. O. (1965) Structural and functional properties of the H and M subunits of lactic dehydrogenases. *Ann. N. Y. Acad. Sci.* 119, 888–904.

108. Cahn, R. D., Zwilling, E., Kaplan, N. O., and Levine, L. (1962) Nature and development of lactic dehydrogenases: The two major types of this enzyme form molecular hybrids which change in makeup during development. *Science* 136, 962–969.

109. Bloemendal, H., de Jong, W., Jaenicke, R., Lubsen, N. H., Slingsby, C., and Tardieu, A. (2004) Ageing and vision: Structure, stability and function of lens crystallins. *Prog. Biophys. Mol. Biol.* 86, 407–485.

110. Haas, R., Cucchi, D., Smith, J., Pucino, V., Macdougall, C. E., and Mauro, C. (2016) Intermediates of metabolism: From bystanders to signalling molecules. *Trends Biochem. Sci.* 41, 460–471.

111. Brooks, G. A., Dubouchaud, H., Brown, M., Sicurello, J. P., and Butz, C. E. (1999) Role of mitochondrial lactate dehydrogenase and lactate oxidation in the intracellular lactate shuttle. *Proc. Natl. Acad. Sci. USA* 96, 1129–1134.

112. Halestrap, A. P., and Denton, R. M. (1975) The specificity and metabolic implications of the inhibition of pyruvate transport in isolated mitichondria and intact tissues by alpha-cyano-4-hydroxycinnamate and related compounds. *Biochem. J.* 148, 97–106.

113. Elustondo, P. A., White, A. E., Hughes, M. E., Brebner, K., Pavlov, E., and Kane, D. A. (2013) Physical and functional association of lactate dehydrogenase (LDH) with skeletal muscle mitochondria. *J. Biol. Chem.* 288, 25309–25317.

114. Rovira, J., Irimia, J. M., Guerrero, M., Cadefau, J. A., and Cusso, R. (2012) Upregulation of heart PFK-2/FBPase-2 isozyme in skeletal muscle after persistent contraction. *Pflugers Archiv.* 463, 603–613.

115. Hue, L., Blackmore, P. F., and Exton, J. H. (1981) Fructose 2 6-bisphosphate. Hormonal regulation and mechanism of its formation in liver. *J. Biol. Chem.* 256, 8900–8903.

116. Shen, Q. W., Means, W. J., Underwood, K. R., Thompson, S. A., Zhu, M. J., McCormick, R. J., Ford, S. P., Ellis, M., and Du, M. (2006) Early post-mortem AMP-activated protein kinase (AMPK) activation leads to phosphofructokinase-2 and -1 (PFK-2 and PFK-1) phosphorylation and the development of pale, soft, and exudative (PSE) conditions in porcine longissimus muscle. *J. Agric. Food Chem.* 54, 5583–5589.

117. Minchenko, O., Opentanova, I., and Caro, J. (2003) Hypoxic regulation of the 6-phosphofructo-2-kinase/fructose-2,6-bisphosphatase gene family (PFKFB-1-4) expression in vivo. *FEBS Lett.* 554, 264–270.

118. Clem, B., Telang, S., Clem, A., Yalcin, A., Meier, J., Simmons, A., Rasku, M. A., Arumugam, S., Dean, W. L., Eaton, J., Lane, A., Trent, J. O., and Chesney, J. (2008) Small-molecule inhibition of 6-phosphofructo-2-kinase activity suppresses glycolytic flux and tumor growth. *Mol. Cancer Ther.* 7, 110–120.

119. Dickens, F., and Šimer, F. (1929) Observations on tissue glycolysis: The effect of fluoride and some other substances. *Biochem. J.* 23, 936.

120. Smolen, J. E., Kuczynski, B., Koh, E. K., Balazovich, K. J., and Woronoff, A. (1992) 'Depletion' of ATP by 2-deoxyglucose: Secretion by electroporated human neutrophils is not restored by readdition of ATP. *Biol. Signals* 1, 23–33.

121. Klaassen, M. (1996) Metabolic constraints on long-distance migration in birds. *J. Exp. Biol.* 199, 57–64.

122. Klaassen, M., Lindstrom, A., and Zijlstra, R. (1997) Composition of fuel stores and digestive limitations to fuel deposition rate in the long-distance migratory thrush nightingale, *Luscinia luscinia*. *Physiol Zool.* 70, 125–133.

123. Pimenta, A. S., Gaidhu, M. P., Habib, S., So, M., Fediuc, S., Mirpourian, M., Musheev, M., Curi, R., and Ceddia, R. B. (2008) Prolonged exposure to palmitate impairs fatty acid oxidation despite activation of AMP-activated protein kinase in skeletal muscle cells. *J. Cell. Physiol.* 217, 478–485.

124. Neely, J. R., and Morgan, H. E. (1974) Relationship between carbohydrate and lipid metabolism and the energy balance of heart muscle. *Annu. Rev. Physiol.* 36, 413–459.

125. Stanley, W. C., Recchia, F. A., and Lopaschuk, G. D. (2005) Myocardial substrate metabolism in the normal and failing heart. *Physiol. Rev.* 85, 1093–1129.

126. Menahan, L. A., and Hron, W. T. (1981) Regulation of acetoacetyl-CoA in isolated perfused rat hearts. *Eur. J. Biochem.* 119, 295–299.

127. Loiacono, F., Alberti, L., Lauretta, L., Puccetti, P., Silipigni, C., Margonato, A., and Fragasso, G. (2014) [Metabolic therapy for heart failure]. *Recenti Prog. Med.* 105, 288–294.

128. Kolwicz, S. C., Jr., Airhart, S., and Tian, R. (2016) Ketones step to the plate: A game changer for metabolic remodeling in heart failure? *Circulation* 133, 689–691.

129. Garcia, R. C., San Martin de Viale, L. C., Tomio, J. M., and Grinstein, M. (1973) Porphyrin biosynthesis. X. Porphyrinogen carboxy-lyase from avian erythrocytes: Further properties. *Biochim. Biophys. Acta* 309, 203–210.

130. Battersby, A. R., and McDonald, E. (1976) Biosynthesis of porphyrins and corrins. *Philos. Trans. R. Soc. Lond. B. Biol. Sci.* 273, 161–180.

131. Malmstrom, K., and Carafoli, E. (1976) The effect of Ca on the oxidation of beta-hydroxybutyric acid by heart mitochondria. *Biochem. Biophys. Res. Commun.* 69, 658–664.

132. Armstrong, C. T., Anderson, J. L., and Denton, R. M. (2014) Studies on the regulation of the human E1 subunit of the 2-oxoglutarate dehydrogenase complex, including the identification of a novel calcium-binding site. *Biochem. J.* 459, 369–381.

133. Pederson, B. A., Chen, H., Schroeder, J. M., Shou, W., DePaoli-Roach, A. A., and Roach, P. J. (2004) Abnormal cardiac development in the absence of heart glycogen. *Mol. Cell. Biol.* 24, 7179–7187.

134. Mora, A., Sakamoto, K., McManus, E. J., and Alessi, D. R. (2005) Role of the PDK1-PKB-GSK3 pathway in regulating glycogen synthase and glucose uptake in the heart. *FEBS Lett.* 579, 3632–3638.

135. Ameer, F., Scandiuzzi, L., Hasnain, S., Kalbacher, H., and Zaidi, N. (2014) De novo lipogenesis in health and disease. *Metabolism* 63, 895–902.

136. Hems, D. A., Rath, E. A., and Verrinder, T. R. (1975) Fatty acid synthesis in liver and adipose tissue of normal and genetically obese (ob/ob) mice during the 24-hour cycle. *Biochem. J.* 150, 167–173.

137. Solinas, G., Boren, J., and Dulloo, A. G. (2015) De novo lipogenesis in metabolic homeostasis: More friend than foe? *Mol. Metab.* 4, 367–377.

138. Simpson, I. A., and Cushman, S. W. (1986) Hormonal regulation of mammalian glucose transport. *Ann. Rev. Biochem.* 55, 1059–1089.

139. Nye, C. K., Hanson, R. W., and Kalhan, S. C. (2008) Glyceroneogenesis is the dominant pathway for triglyceride glycerol synthesis in vivo in the rat. *J. Biol. Chem.* 283, 27565–27574.

140. Bederman, I. R., Foy, S., Chandramouli, V., Alexander, J. C., and Previs, S. F. (2009) Triglyceride synthesis in epididymal adipose tissue: Contribution of glucose and non-glucose carbon sources. *J. Biol. Chem.* 284, 6101–6108.

141. Green, A., and Newsholme, E. A. (1979) Sensitivity of glucose uptake and lipolysis of white adipocytes of the rat to insulin and effects of some metabolites. *Biochem. J.* 180, 365–370.

142. Montell, E., Lerin, C., Newgard, C. B., and Gomez-Foix, A. M. (2002) Effects of modulation of glycerol kinase expression on lipid and carbohydrate metabolism in human muscle cells. *J. Biol. Chem.* 277, 2682–2686.

143. Russi, E., and Weigand, K. (1983) Analbuminemia. *Klin. Wochenschr.* 61, 541–545.

144. Hirako, S., Wakayama, Y., Kim, H., Iizuka, Y., Matsumoto, A., Wada, N., Kimura, A., Okabe, M., Sakagami, J., Suzuki, M., Takenoya, F., and Shioda, S. (2015) The relationship between aquaglyceroporin expression and development of fatty liver in diet-induced obesity and ob/ob mice. *Obes. Res. Clin. Pract.*

145. Wakayama, Y., Hirako, S., Ogawa, T., Jimi, T., and Shioda, S. (2014) Upregulated expression of AQP 7 in the skeletal muscles of obese ob/ob mice. *Acta Histochem. Cytochem.* 47, 27–33.

146. Reidy, S. P., and Weber, J. (2000) Leptin: An essential regulator of lipid metabolism. *Comp. Biochem. Physiol. A. Mol. Integr. Physiol.* 125, 285–298.

147. Martin, R. L., Perez, E., He, Y. J., Dawson, R., Jr., and Millard, W. J. (2000) Leptin resistance is associated with hypothalamic leptin receptor mRNA and protein down-regulation. *Metabolism.* 49, 1479–1484.

148. Yamauchi, T., Kamon, J., Minokoshi, Y., Ito, Y., Waki, H., Uchida, S., Yamashita, S., Noda, M., Kita, S., Ueki, K., Eto, K., Akanuma, Y., Froguel, P., Foufelle, F., Ferre, P., Carling, D., Kimura, S., Nagai, R., Kahn, B. B., and Kadowaki, T. (2002) Adiponectin stimulates glucose utilization and fatty-acid oxidation by activating AMP-activated protein kinase. *Nat. Med.* 8, 1288–1295.

149. Vu, V., Bui, P., Eguchi, M., Xu, A., and Sweeney, G. (2013) Globular adiponectin induces LKB1/AMPK-dependent glucose uptake via actin cytoskeleton remodeling. *J. Mol. Endocrinol.* 51, 155–165.

150. Koerner, A., Kratzsch, J., and Kiess, W. (2005) Adipocytokines: Leptin—The classical, resistin—The controversical, adiponectin—The promising, and more to come. *Best Pract. Res. Clin. Endocrinol. Metab.* 19, 525–546.

151. Akazawa, S., Sun, F., Ito, M., Kawasaki, E., and Eguchi, K. (2000) Efficacy of troglitazone on body fat distribution in type 2 diabetes. *Diabetes Care* 23, 1067–1071.

152. Keil, A. M., Frederick, D. M., Jacinto, E. Y., Kennedy, E. L., Zauhar, R. J., West, N. M., Tchao, R., and Harvison, P. J. (2015) Cytotoxicity of thiazolidinedione-, oxazolidinedione- and pyrrolidinedione-ring containing compounds in HepG2 cells. *Toxicol In Vitro* 29, 1887–1896.

153. Fang, X., Stroud, M. J., Ouyang, K., Fang, L., Zhang, J., Dalton, N. D., Gu, Y., Wu, T., Peterson, K. L., Huang, H. D., Chen, J., and Wang, N. (2016) Adipocyte-specific loss of PPARgamma attenuates cardiac hypertrophy. *JCI Insight* 1, e89908.

154. Tchkonia, T., Lenburg, M., Thomou, T., Giorgadze, N., Frampton, G., Pirtskhalava, T., Cartwright, A., Cartwright, M., Flanagan, J., and Karagiannides, I. (2007) Identification of depot-specific human fat cell progenitors through distinct expression profiles and developmental gene patterns. *Am. J. Physiol. Endoc. Metab.* 292, E298–E307.

155. Virtanen, K. A., Lidell, M. E., Orava, J., Heglind, M., Westergren, R., Niemi, T., Taittonen, M., Laine, J., Savisto, N. J., Enerback, S., and Nuutila, P. (2009) Functional brown adipose tissue in healthy adults. *N. Engl. J. Med.* 360, 1518–1525.

156. Ibrahim, M. M. (2010) Subcutaneous and visceral adipose tissue: Structural and functional differences. *Obes. Rev.* 11, 11–18.

157. Jensen, M. D. (2008) Role of body fat distribution and the metabolic complications of obesity. *J. Clin. Endocrinol. Metab.* 93, S57–S63.
158. Nicholls, D. G. (1979) Brown adipose tissue mitochondria. *Biochim. Biophys. Acta Rev. Bioenerg.* 549, 1–29.
159. Kitao, N., and Hashimoto, M. (2012) Increased thermogenic capacity of brown adipose tissue under low temperature and its contribution to arousal from hibernation in Syrian hamsters. *Am. J. Physiol. Regul. Integr. Comp. Physiol.* 302, R118–R125.
160. Porter, C. (2017) Quantification of UCP1 function in human brown adipose tissue. *Adipocyte* 6, 167–174.
161. Erickson, H. P. (2013) Irisin and FNDC5 in retrospect: An exercise hormone or a transmembrane receptor? *Adipocyte* 2, 289–293.
162. Zhang, Y., Xie, C., Wang, H., Foss, R., Clare, M., George, E. V., Li, S., Katz, A., Cheng, H., Ding, Y., Tang, D., Reeves, W. H., and Yang, L. J. (2016) Irisin exerts dual effects on browning and adipogenesis of human white adipocytes. *Am. J. Physiol. Endocrinol. Metab.* 311, E530–E541.
163. Lin, J. Z., and Farmer, S. R. (2016) Morphogenetics in brown, beige and white fat development. *Adipocyte* 5, 130–135.
164. Um, S. H., Frigerio, F., Watanabe, M., Picard, F., Joaquin, M., Sticker, M., Fumagalli, S., Allegrini, P. R., Kozma, S. C., Auwerx, J., and Thomas, G. (2004) Absence of S6K1 protects against age- and diet-induced obesity while enhancing insulin sensitivity. *Nature* 431, 200–205.
165. Windmueller, H. G., and Spaeth, A. E. (1974) Uptake and metabolism of plasma glutamine by the small intestine. *J. Biol. Chem.* 249, 5070–5079.
166. Windmueller, H. G., and Spaeth, A. E. (1978) Identification of ketone bodies and glutamine as the major respiratory fuels in vivo for postabsorptive rat small intestine. *J. Biol. Chem.* 253, 69–76.
167. Windmueller, H. G., and Spaeth, A. E. (1980) Respiratory fuels and nitrogen metabolism in vivo in small intestine of fed rats. *J. Biol. Chem.* 255, 107–112.
168. Watford, M. (1999) Is there a requirement for glutamine catabolism in the small intestine? *Br. J. Nutr.* 81, 261–262.
169. van de Poll, M. C. G., Ligthart-Melis, G. C., Boelens, P. G., Deutz, N. E. P., van Leeuwen, P. A. M., and Dejong, C. H. C. (2007) Intestinal and hepatic metabolism of glutamine and citrulline in humans. *J. Physiol.* 581, 819–827.
170. Hansotia, T., Maida, A., Flock, G., Yamada, Y., Tsukiyama, K., Seino, Y., and Drucker, D. J. (2007) Extrapancreatic incretin receptors modulate glucose homeostasis, body weight, and energy expenditure. *J. Clin. Invest.* 117, 143–152.
171. Nauck, M. A., Baller, B., and Meier, J. J. (2004) Gastric inhibitory polypeptide and glucagon-like peptide-1 in the pathogenesis of type 2 diabetes. *Diabetes* 53, S190–S196.
172. Cvetkovic, R. S., and Plosker, G. L. (2007) Exenatide: A review of its use in patients with type 2 diabetes mellitus (as an adjunct to metformin and/or a sulfonylurea). *Drugs* 67, 935–954.
173. Cummings, D. E., Weigle, D. S., Frayo, R. S., Breen, P. A., Ma, M. K., Dellinger, E. P., and Purnell, J. Q. (2002) Plasma ghrelin levels after diet-induced weight loss or gastric bypass surgery. *N. Engl. J. Med.* 346, 1623–1630.
174. Gilbert, R. E. (2016) SGLT2 inhibitors: Beta blockers for the kidney? *Lancet Diabetes Endocrinol.* 4, 814.
175. Song, P., Onishi, A., Koepsell, H., and Vallon, V. (2016) Sodium glucose cotransporter SGLT1 as a therapeutic target in diabetes mellitus. *Exp. Opin. Ther. Targets* 20, 1109–1125.
176. Taylor, L., and Curthoys, N. P. (2004) Glutamine metabolism: Role in acid-base balance. *Biochem. Molec. Biol. Educ.* 32, 291–304.

177. Stumvoll, M., Meyer, C., Perriello, G., Kreider, M., Welle, S., and Gerich, J. (1998) Human kidney and liver gluconeogenesis: Evidence for organ substrate selectivity. *Am. J. Physiol.* 274, E817–E826.

178. Curthoys, N. P., and Gstraunthaler, G. (2014) pH-responsive, gluconeogenic renal epithelial LLC-PK1-FBPase+cells: A versatile in vitro model to study renal proximal tubule metabolism and function. *Am. J. Physiol. Renal Physiol.* 307, F1–F11.

ENDNOTES

i   While normally absent from liver, creatine phosphokinase has been inserted genetically into liver in some studies, sometimes with intriguing results, such as increased survival from ATP depletion [1,2]. It is not clear that that normal hepatic function is retained otherwise, but it is possible to use expressed creatine phosphokinase to estimate free ADP levels [3].

ii There are also studies considering how the size of the glycogen molecule might be determined by physical considerations that are outside those of metabolic pathways and regulatory mechanisms [5,6].

iii While it may seem obvious, gluconeogenesis is unique in sharing most of its enzymes with glycolysis. The pathways run in reverse, so that steps producing ATP for glycolysis consume ATP for gluconeogenesis.

iv The hyperpolarization requires that a molecule be brought to near absolute zero, microwave irradiated, rapidly brought back to near room temperature, and the experiment conducted rapidly (within seconds) as the nuclear spin decays [32].

v   The reaction converting glyceraldehyde to glyceraldehyde phosphate is the third reaction catalyzed by the triose kinase enzyme. The most commonly considered reaction is that of glycerol to glycerol phosphate, as this is the entry of the released glycerol from adipocyte triglyceride hydrolysis. The second was the conversion of dihydroxyacetone to DHAP, which is employed for the substrate dihydroxyacetone. This relaxed specificity is similar that of hexokinase isozymes other than glucokinase.

vi In the general transamination reaction, there is a conversion of a carbonyl group to an amine, and a corresponding conversion of a separate amine to a carbonyl. For the half-reaction of a carbonyl to an amine, there is a gain of an electron pair.

vii From a metabolic standpoint, the role of CD36 is uncertain. The protein has many other roles [44] and is not the only route of fatty acid entry [45]. It is also possible that fatty acids enter cells by diffusion [46,47].

viii These bacterial also can form branched chain fatty acids which can have either a distinct metabolic pathway or else be very poorly metabolized [53].

ix The rapeseed plant was cross bred in part to minimize the content of rapeseed. The oil produced by this plant was named for the Canadian government, *canola oil* [57].

x   The abbreviation of sarcoplasmic reticulum as ER (endoplasmic reticulum) rather than SR reflects the schism between muscle aficionados and everyone else. Some use the abbreviation S/ER. The notion that *sarco* must be affixed to muscle cell biology does not mean that the sarcoplasmic reticulum is fundamentally different from the endoplasmic reticulum, nor the sarcolemma from the plasma membrane. There is also the *myokinase* as opposed to the adenylate kinase. Neurologists have a similar proclivity with *neurolemma*, but have stopped short of renaming the endoplasmic reticulum. Renaming is a concern as many form the incorrect impression that there is something special about the muscle that justifies this naming scheme. Every cell has some unique adaptation; if they were to be renamed to reflect that we would be unable to draw parallels. Hence the designation of this membrane as ER.

xi These are types in the human; however, in the well-studied murine species, there is an extra type IIx, which is a very fast muscle.

xii This is the secret to alligator wrestling: survive the short glycolytic phase of the jaw muscle until the reptile enters the gluconeogenic one at which point it will be relatively docile.

xiii This is both a Krebs cycle intermediate and a precursor to porphyrin synthesis [129,130].

xiv This was performed by the author as an experiment using frozen rat heart mitochondria prepared by the laboratory of David Green at the Enzyme Institute of the University of Wisconsin.

xv In those early heady days of its discovery, the Rockefeller University researchers were rewarded with funds of over $20 million to assign the rights of drug discoveries resulting from leptin to Amgen. However, no drugs ensued, as leptin deficiency occurs in a small number of obese patients.

7 Cellular Signaling Systems

Cellular signaling systems have by now been so thoroughly studied that summary flow-charts display a bewildering network of interactions. There are already excellent treatises of signaling systems that also consider short-term regulation, such as that of Lim et al. [1], which includes a large number of systems and biological activities. In addition, a strong summary of signaling from the standpoint of post translational protein modification is provided by Walsh [2]. The present chapter is distinct from those in providing a metabolic view of a few basic control elements. The regulatory systems discussed are those involving cAMP, Ca^{2+}, AMP, and diacylglycerol. The chapter concludes with a discussion of a few mobile cofactors proposed as regulators based on isolated kinetic analysis. Throughout the discussion, I maintain the view of evenly distributed metabolite concentration within a defined water compartment (such as cytosol) and assume a steady-state behavior. There is insight to be gained by considering signaling systems themselves as pathways. Thus, formation and destruction of signals must be accounted for, including flows across membranes as intermediary pathway steps.

7.1 CYCLIC AMP

The discovery of the first messenger molecule cyclic AMP (cAMP) by Sutherland [3] provided the first molecular insight into the action of hormones in cellular physiology. Many studies followed, implicating cAMP as the cellular intermediate formed when various extracellular hormones bind cells externally. The overall scheme of Figure 7.1 also introduces the notion of G-proteins (Gprot), which connect isolated membrane-embedded proteins [4,5].[i] G-proteins are now known to constitute a large family; the type under consideration here are the heterotrimeric G proteins that transduce signals between the external hormone and internal signal molecules, such as cAMP. All G-proteins, as indicated by their name, bind a guanidine nucleotide: either GTP or GDP: the GTP bound form is active, and the GDP bound form is inactive. All G-proteins also have a means of converting between these states. Activation involves guanine nucleotide exchange from Gprot:GDP to Gprot:GTP, catalyzed by a GEF (guanine exchange factor). As illustrated in Figure 7.1, the GEF for hormone activation is the hormone receptor (R) when it is complexed with hormone (H). Once activated, the Gprot:GTP form can dissociate from the receptor and slide along the membrane surface, just like the surface bound mobile cofactor of respiration, cytochrome c (Chapter 3). Next, Gprot:GTP can bind and activate an effector protein (E), which is also a membrane-embedded protein. In the case under consideration, E is the enzyme adenylate cyclase. This catalyzes the formation of cAMP from ATP. As shown in Figure 7.1b, G-proteins have an intrinsic GTPase activity, which means that they are continuously becoming inactive. However, it known that all G proteins have an inactivating protein that catalyzes the hydrolysis of the

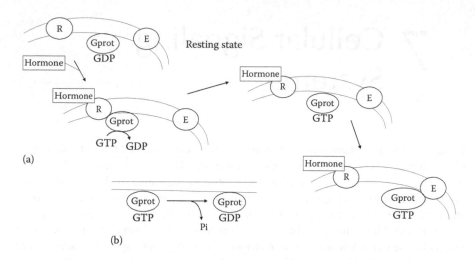

(a)

(b)

FIGURE 7.1 G proteins. G proteins (Gprot) are surface membrane-resident heterotrimers that exist bound to GDP or GTP. (a) From the resting state in which Gprot:ADP binds neither R nor E, addition of hormone forms a complex to R, which binds Gprot and catalyzes nucleotide exchange to form Gprot:GTP. This is then released from the Hormone:R complex, and can bind and active E. (b) The Gprot:GTP can be hydrolyzed to Gprot:GDP when not engaged with R or E. This is catalyzed by proteins known as RGS (receptor GTPases), a subset of the general class of GAP proteins (GTPase-activating protein) for the trimeric G proteins as illustrated and intracellular small G proteins.

GTP to GDP; these are known as GAPs (GTPase activating proteins). In the case of the heterotrimeric G-proteins, these are known as regulators of G protein signaling or RGS [6]. It is unclear what the physical arrangement is between Gprot and the RGS, but it is known that a large family of these proteins exists. Because of the ongoing presence of Gprot inactivation, restoration of the Gprot:GTP form requires the continued presence of hormone. This permits the hormone-receptor complex to act once again as a GEF and continue the activation of intracellular cAMP formation. Once the hormone has dissociated from the receptor, the system becomes inactivated.

With the discovery of cAMP, the name *second messenger* was coined, which carries out the action of the first messenger (the hormone) inside the cell. With this signaling system, a set of rules were informally compiled (e.g., ref. [7]) concerning such intracellular messengers:

1. A messenger must be produced by a large constant metabolic pool in a unidirectional fashion.
2. A messenger must be destroyed by conversion to a large metabolic pool, in a unidirectional fashion.
3. The rate of its destruction sets an upper limit to the speed of signaling.

The first two can be described succinctly as a *source* (#1) and *sink* (#2) requirement. For cAMP, the source is ATP. Since the concentration of ATP is in the millimolar range, and that of cAMP is in the micromolar range, the source is essentially unchanged in concentration. The sink is AMP, which has a (total) concentration in the hundreds of micromolar, and thus will also be unchanged by the added contribution from cAMP. Understanding these concentration values is critical to the appreciation of the notions of source and sink. Rule #3 is essentially a statement of control by the first order rate constant of the breakdown enzyme; an elaboration of this concept was established for protein steady states by Schimke [8]. For cAMP, there are distinct isoforms of phosphodiesterase that can contribute. This is of interest pharmacologically if not for cellular regulation, and unlikely to alter the treatment of cAMP levels as a steady state [9].

The major response to an elevated cAMP concentration is the activation of the enzyme now known as PKA (protein kinase A), originally called cAMP-dependent protein kinase. The enzymology and metabolic targets of this regulator are now well-known [10]; more current reviews are largely concerned with transcriptional regulation [11]. Considering a few targets in the liver illustrates how this signaling system can provide a change in metabolic climate.

In the presence of glucagon, the liver activates pathways that can be broadly considered as catabolic (Figure 7.2). Thus, glycogen breakdown to glucose, gluconeogenesis, fatty acid oxidation, and ureogenesis are activated, while glycogen synthesis, glycolysis, and fatty acid synthesis are diminished. These pathways are not independent of each other; the other constraint is the previously discussed point that only mIRR enzymes can be sites of control.

Pathway interdependence is easily observed in the case of liver, since this tissue has activate pathways for both glycolysis and gluconeogenesis, which share many steps. Control of the same enzymes can thus simultaneously affect both pathways, as evident in the Figure 7.3 for the cAMP-directed inactivation of PFK and activation of FBPase acting in part through PFK-2 formation of fructose 2,6-P_2.

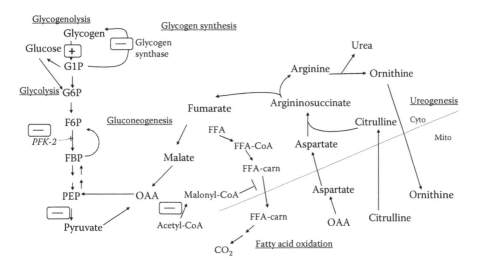

FIGURE 7.2   Glucagon effects on metabolic pathways of liver. Changes in activity of steps in the major pathways of liver are illustrated.

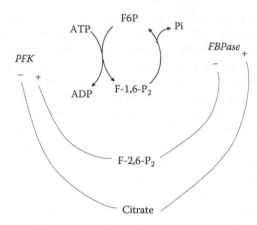

FIGURE 7.3 Control of steps in opposing directions at a substrate cycle. The formation of fructose-2,6-P_2 by PFK2 leads to activation of PFK, inhibition of FBPase, and thus the increase of glycolytic flow. Citrate produces the opposite results. Affecting both sides of the substrate cycle increases the effectiveness of the regulatory effect.

Two enzymes in these pathways are *not* direct targets of rapid cAMP regulation but are nonetheless mIRR steps: glucokinase and PEPCK. Neither enzyme is known to have any allosteric regulators. Instead, these proteins are regulated by transcriptional regulation, and thus adjust to the overall flow of metabolism on this longer time scale. In particular, PEPCK has been extensively studied and responds to a variety of hormones that regulate its transcription. As this protein has a relatively short half-life, it can dictate increases in gluconeogenesis in just a few hours [12].

For most enzymes of the pathways under consideration, the adjustments are made through near-equilibrium. It does not mean, however, that the protein levels cannot change. In fact, there are instances where near-equilibrium enzyme concentrations can be altered [13]. Why this is the case is not known. It is unlikely that such steps are usually regulatory (see Chapter 6), although it is possible that, if substantially attenuated, it is a possibility. If that is the case, measurement of the mass-action ratio for the reactions would be needed over the course of the transformation. The reason that NEQ enzymes have this status in usual situations is because of the large amount of protein the cell expresses. This ensures that the reaction steps are close to equilibrium and can be reversed under conditions when pathway flow is reversed.

7.1.1 GLUCONEOGENESIS AND cAMP

Glucagon stimulation of gluconeogenesis in the liver leads to cAMP production and PKA activation, targeting pyruvate kinase and phosphofructokinase. The increased glucose formation arises from physiological substrates such as lactate and alanine [14]. As noted in the prior chapter concerning liver gluconeogenesis, cAMP inhibition of gluconeogenesis from high concentrations of pyruvate in isolated liver cells provides clear evidence for pyruvate kinase as a cAMP target [15].

A related situation occurs in some cancer cells, as described in Chapter 10. In that case, cells with an interruption in the electron transport chain become unable to oxidize NADH, and thus are auxotrophic for the presence of pyruvate. In all cases, it is evident that metabolic pathways require a balance of mobile cofactors.

7.1.2 FEATURES OF cAMP FORMATION

The hormone-stimulated rise in cAMP is transient, lasting only several seconds. The increased cAMP concentration leads to the activation of protein kinase A, phosphorylation of target proteins such as glycogen phosphorylase kinase, and increased glycogen degradation [16]. The rise in cAMP is opposed by phosphodiesterases as mentioned above. In fact, the rate of signal destruction is instrumental in establishing how rapidly the event can take place overall [8]. There are additional termination steps for hormonal action: release of the hormone itself, destruction of the hormone, and desensitization of receptor action.

The route of cAMP formation from glucagon involves the G-protein Gs. The subscripted "s" means a stimulation of the membrane-embedded target protein adenylate cyclase. The study of G proteins has burgeoned into a large research area, of which a few general comments are in order as they relate to metabolic control.

7.2 G PROTEINS

As indicated in Figure 7.1, the standard model of G protein action is the shuttling of a mobile G protein along the inner surface of the plasma membrane, communicating between a fixed (membrane-bound) receptor and a fixed effector proteins. Of the three subunits of the trimer, the Gα is always bound to a guanine nucleotide, namely GDP in the inactive state. Upon hormone binding from the outside face of the plasma membrane, Gprot:GDP binds the occupied receptor, and Gα becomes bound to GTP instead. The now-activated Gα-GTP dissociates from the other two subunits (G$_{\beta\gamma}$), slides along the inner face of the plasma membrane and binds to its effector target. In the case of Gs, this target is adenylate cyclase.

While this is the orthodox model of G protein action, an alternative suggestion holds that the trimeric protein does not dissociate. This notion remains controversial [17], but its resolution does not preclude our ability to recognize distinct Gα subunits as the directors of different cellular control systems.

The second type of G protein discovered was known as the Gi, which stood for inhibitory G protein as it can decrease cAMP concentration provided the messenger is first activated by a separate event (e.g., ref. [18]). This mechanism was proposed when cAMP was the only molecular explanation for hormone action. If this was how Gi linked hormones act, they would always need a simultaneous Gs linked hormone to elevate the cAMP first. While it is established that Gi-linked hormones can inhibit the *in vitro* activity of adenylate cyclase [19], it is not clear how this can explain how these hormones can act on their own.

An entirely separate explanation for the mechanism of Gi-linked hormones arose from the finding that K channels activated by cholinergic muscarinic receptors of heart were activated by the $\beta\gamma$ subunits of Gi [20]. Subsequently it was shown that the $\beta\gamma$ subunits stabilize the interaction of a K channel with the membrane phospholipid phosphatidylinositol bisphosphate (PIP_2), leading to an open channel [21]. Thus, an explanation for the action of Gi-linked hormones is the stimulation of K channels, which increases cell polarization (Figure 7.4). For example, in the case of heart, vagal nerve stimulation exerts its influence by cholinergic stimulation of the pacemaker cell, enhancing polarization and thus decreasing heart rate. This is accomplished by acetyl-choline acting through a Gi protein to stimulate a K channel. The involvement of PIP_2 in this mechanism in maintaining the open state of the K channel is supported by the finding that hydrolysis of PIP_2 leads to channel inactivation [22]. A third type of G protein is involved in the hydrolysis of PIP_2 for separate purposes.

This third type of G protein—utilized by hormones such as vasopressin and the α_1-adrenergic hormones—activates plasma membrane-bound phospholipase C. The enzyme catalyzes hydrolysis of the membrane phospholipid phosphatidylinositol-bis-phosphate to inositol-1,4,5-P_3 (IP_3) and diacylglycerol (DAG). The latter is known to be an activator of protein kinase C. However, most DAG formation that leads to protein kinase C activation follows a much slower time course and accumulates to a far greater extent than the amount formed from PIP_2 (indicated by the smaller font for DAG in Figure 7.5). Thus, IP_3 formation is actually a separate signaling event from DAG formation in cells [23]. Instead, the lipid substrate is phosphatidyl choline and DAG accumulates on a slower time course. The other product of the phospholipase reaction—IP_3—is a trigger for Ca^{2+} release from the ER as described below.

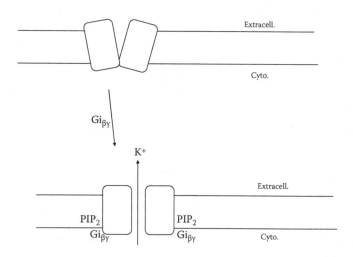

FIGURE 7.4 G_i protein activates K channels through the $\beta\gamma$-subunit. With G_i-linked hormones, the K channel is shown to be opened, leading to polarization of the cell. Note too that PIP_2 hydrolysis can also control the channel through its interaction with the channel protein.

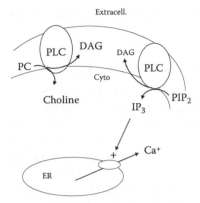

FIGURE 7.5   Two phospholipase C activities. Phospholipase C (PLC) acting on phosphatidyl choline (PLC) leads to a sustained increase in membrane diacylglycerol (DAG). On the other hand, PLC hydrolysis of phosphatidylinositol bisphosphate (PIP_2) leads to formation of water-soluble IP_3 and Ca^{2+} release from the ER. The amount of DAG formed from PIP_2 hydrolysis is small and transient.

7.3   CALCIUM SIGNALING

A rise in cytosolic Ca^{2+} is a signal for a large number of cells, triggering muscle contraction and metabolic events such as glycogenolysis. In this section, I first present mechanisms for the rise of cellular Ca^{2+} in two distinct cell types: nonexcitable and excitable cells. Second, I discuss how metabolism is affected by a rise in intracellular Ca^{2+} concentrations. The changes in Ca^{2+} itself are considered from a metabolic perspective, that is, as a pathway. Thus, the extent of entry of Ca^{2+} into a cell must be matched by its exit. A further consideration is that we are concerned largely with the *free* concentration of Ca^{2+}; most of this ion is bound to various proteins in the cytosol and to a single binding protein (calsequestrin) in the ER.

There are two viewpoints of this presentation that diverge from most treatments of Ca^{2+} events. First, the elevation of Ca^{2+} concentration in cells that leads to its signaling is considered here to arise exclusively from the internal pool of the endoplasmic reticular (ER) lumen. Second, only uniform increases in cellular spaces are considered as exerting metabolic events. This means the cell cytosol or mitochondrial matrix is treated as homogeneous. Adopting these principles avoids direct contradiction with established facts concerning Ca^{2+} regulation.

Ca^{2+} signaling in nonexcitable tissues involves an increase in the second messenger inositol trisphosphate (IP_3), somewhat analogous to cAMP. Once generated in the cytosol, IP_3 releases Ca^{2+} from the ER by binding to an ER-membrane protein that is a ligand-gated Ca^{2+} channel. The other mechanism for Ca^{2+} is found in excitable tissue, which has been well-studied in skeletal muscle. While skeletal muscle Ca^{2+} signaling historically predates those of nonexcitable tissue, the pathway for Ca^{2+} signaling is less fully understood in skeletal muscle, and even less clear in other excitable cells such as heart and neural cells. What follows is an orthodox description of the IP_3 pathway and what is known and hypothesized about excitable cell Ca^{2+} signaling.

7.3.1   CA²⁺ SIGNALING IN NONEXCITABLE CELLS

The central event leading to increased cytosolic Ca^{2+} in nonexcitable cells is the hydrolysis of the minor membrane lipid phosphatidyl inositol bisphosphate (PIP_2) by a phospholipase in the plasma membrane. This is achieved through hormone stimulus that transmits its signal (Figure 7.6) to a membrane-bound phospholipase C through G proteins such as G_P [24]. One phospholipase C product is the polar messenger molecule inositol 1,4,5 trisphosphate (IP_3) to the cytosol is a messenger. IP_3 diffuses to the ER where it binds a Ca^{2+} ion channel embedded in the ER membrane, the IP_3 receptor (IP_3R). Formation of the complex opens the channel and Ca^{2+} is released from the ER lumen to the cytosol. This resulting surge in cytosolic Ca^{2+} is terminated largely by the reuptake of Ca^{2+} back to the ER lumen, catalyzed by the pump CaATPase. The CaATPase is the major protein responsible for uptake in all cell types [24–26]. Ca^{2+} reuptake to the ER is controlled by the concentration of cytosolic Ca^{2+} itself. The operation of the release and the reuptake mechanisms provide a brief window for Ca^{2+}, on the order of a few seconds. During this time, Ca^{2+} triggers signaling events in the cytosol, such as the activation of glycogen phosphorylase kinase and thus glycogen phosphorylase and the pathway of glycogen degradation [27]. Ca^{2+} is also taken up by mitochondria, which subsequently increases the mitochondrial matrix Ca^{2+} concentration [28]. While the concentration of Ca^{2+} is greater in the mitochondrial matrix

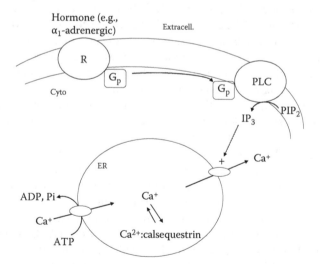

FIGURE 7.6 $Alpha_1$ adrenergic activation of Ca^{2+} and internal Ca^{2+} alterations. Activation of α_1-adrenergic receptors (R) activate the effector PLC through a G_p-type trimeric G protein, leading to IP_3 formation and Ca^{2+} release from the ER. Also shown is how Ca^{2+} is maintained at high concentration in the ER lumen through its equilibrium with calsequestrin, and the reentry of Ca^{2+} to the ER through the CaATPase.

than in the cytosol, the increase is driven by the elevation of cytosolic Ca^{2+} and the mitochondrial membrane electrical gradient. There are known targets for Ca^{2+} activation in the mitochondria, including the pyruvate dehydrogenase complex (stimulation of the pyruvate dehydrogenase phosphatase) and α-ketoglutarate DH complex [29]. These serve to increase pyruvate oxidation and the Krebs cycle, respectively.

7.3.2 Maintenance of Ca^{2+} between Extracellular and Intracellular Pools

What we have just considered is the fundamental signaling system for Ca^{2+} that occurs in a wide variety of cells. There is, however, a little more to the story, as not all the Ca^{2+} released by the ER to the cytosol is returned to the ER. This is because some Ca^{2+} is lost to the extracellular space through the action of the plasma membrane localized CaATPase [30]. Experiments conducted in isolated cells have demonstrated the relatively slow loss of Ca^{2+} from the cell when the ER CaATPase inhibited by the agent thapsigargin [31]. Over time, however, that small loss of Ca^{2+} must be replaced. Experimentally, it is possible to produce isolated cells with a deficiency in intracellular Ca^{2+} by merely removing extracellular Ca^{2+} from the medium. After several minutes, hormones that act through an increase in Ca^{2+} are no longer effective.

Under normal conditions, the small amount of Ca^{2+} that is lost is replenished by the re-entry of Ca^{2+} into the cell. This process of resupply of Ca^{2+} lost to the cell exterior has been intensely studied over the last few decades and *store-operated calcium uptake.*

7.3.3 Store-Operated Calcium Uptake (SOCE)

The pathway just described for Ca^{2+} reuptake into the cell from the exterior has the somewhat curious name of a "store-operated"[ii] Ca^{2+} entry. The process is initiated by a decrease in the ER Ca^{2+} pool, which contains a much larger Ca^{2+} concentration than the cytosol. Whereas resting cytosolic Ca^{2+} concentration is about 0.1 μM, ER Ca^{2+} is about 1 mM. This is maintained by the presence of the ER buffer protein calsequestrin [32], which has a dissociation constant for Ca^{2+} of about 1 mM.

After the ER Ca^{2+} concentration has decreased, entry of Ca^{2+} from the extracellular space is stimulated [33]. This can be achieved by release of Ca^{2+} to the cytosol due to an IP_3-linked hormone, or experimentally by removing extracellular Ca^{2+} and blocking the resupply of ER Ca^{2+} using CaATPase inhibitors like thapsigargin [34]. The most common demonstration of SOCE is illustrated in Figure 7.7. Cytosolic Ca^{2+} can be measured using fluorescent indicators that are derivatives of EGTA developed by Tsien and colleagues [35]. A common chelator is fura2 (introduced into cells as an ester that is subsequently hydrolyzed) which can assess cytosolic Ca^{2+}, as it has a binding constant of about 50 nM, close to the basal cytosolic Ca^{2+} concentration at pH 7.1. In Figure 7.7, a typical experimental Ca^{2+} time course is shown. In this setup, cells are incubated for several minutes in Ca^{2+}-free solution, and then thapsigargin (Tg) is added to deplete ER Ca^{2+}. A small rise in cytosolic Ca^{2+} appears. Next,

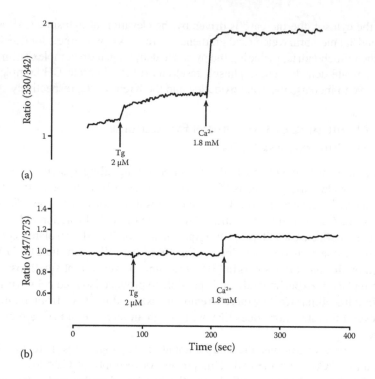

FIGURE 7.7 Cytosolic and ER Ca^{2+} during SOCE. SOCE (store-operated calcium entry) is experimentally evident by removing all extracellular Ca^{2+}, and then adding it back. (a) Cytosolic Ca^{2+} is the most common record produced to demonstrate SOCE. Cells are incubated in the absence of medium Ca^{2+}, and then thapsigargin (Tg) added to prevent any entry into the ER. There is a small rise in cytosolic Ca^{2+}. Finally, addition of millimolar Ca^{2+} to the medium shows a large increase in cytosolic Ca^{2+}. (b) ER Ca^{2+} shows little effect of Tg, but a significant increase in lumen Ca^{2+} that is more rapid kinetically than the cytosolic trace, suggesting a direct entry first into the ER and then the cytosol.

millimolar quantities of Ca^{2+} are added to the medium and a large spike in cytosolic Ca^{2+} is observed. In this experimental record, ER Ca^{2+} concentrations were also determined. Note that Tg itself has little effect on this parameter, but the addition of Ca^{2+} elevates it.

The essential data of Figure 7.7a have been recorded in many laboratories, supported by use of various cytosolic Ca^{2+} binding dyes and independent methods, such as K efflux by a Ca^{2+} dependent plasma membrane Ca^{2+} channel [36]. The most commonly accepted model is a presented as Figure 7.8. In the absence of added Ca^{2+}, the addition of thapsigargin (Tg) leads to a small rise in Ca^{2+}, explained as a release of Ca^{2+} from the ER (through the IP_3R Ca^{2+} channel, which has basal activity even in the absence of the activating ligand). Depending on the cell type, the Ca^{2+} may stay relatively high (as in Figure 7.7) or may drop towards baseline if the plasma membrane CaATPase is more active. When millimolar concentrations of extracellular Ca^{2+} are added, the Ca^{2+} enters the cytosol through the Orai1, the Ca^{2+}

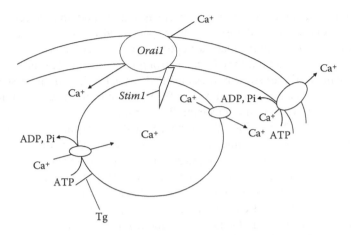

FIGURE 7.8 Orthodox view of SOCE. Ca^{2+} is postulated to enter the cytosol upon stimulation (lower ER Ca^{2+} activates Stim1, which in turn activates Orai1). When Tg is present to block CaATPase, cytosolic Ca^{2+} cannot re-enter the ER and so it accumulates in the cytosol.

channel responsible for SOCE [37]. Activation of this channel requires a protein that is embedded in the ER membrane, Stim1. The latter contains an EF hand structure within the ER lumen that senses the level of Ca^{2+} in that space. Once Ca^{2+} decreases, Stim1 binds to and activates Orai1. Ca^{2+} is postulated to enter the cytosol, but not the ER under these conditions (the presence of thapsigargin precludes entry of Ca^{2+} to the ER), and thus accumulates in the cytosol. The discoveries of Orai1 and Stim1 provided important ingredients to explain SOCE, and the model incorporates these in explaining changes in cytosolic Ca^{2+}. Note however, that the record of ER Ca^{2+} (Figure 7.7b) is not well explained by this model.

7.3.4 SOCE Is Universal

For some time, it was held that SOCE was a process limited to nonexcitable cells. The basis of this claim was an early observation of one specialized cell type, the NG115-401L, a neuroblastoma [38]. This was one of the first cells in which the effect of thapsigargin was examined, and it did not affect cytosolic Ca^{2+}. However, thapsigargin produced the now-classic SOCE signature in parotid cells, leading to the conclusion that the process of SOCE only operates on nonexcitable cells. The actual reason SOCE doesn't operate in the NG115 line is that they are deficient in Stim1. In fact, excitable cells do exhibit SOCE [39–42], indicating a wide occurrence of this mechanism.

7.3.5 Different Hypothesis for SOCE

Despite the popularity of the hypothesis just presented, there is an alternative model that historically preceded this one, which is the direct entry of Ca^{2+} from the extracellular space into the ER [43]. These investigators examined the process of uptake of extracellular calcium into a preparation of vascular smooth muscle cells. By varying Ca^{2+} concentrations in the extracellular space, and using $^{45}Ca^{2+}$ flux measurements, the authors determined that Ca^{2+} uptake was not accompanied by a change in cytosolic $[Ca^{2+}]$. In addition, they found a strong relationship between extracellular Ca^{2+} and the internal Ca^{2+} store, now established as the ER. The findings suggest that SOCE is accomplished by a direct route of Ca^{2+} entry from the extracellular space into the ER lumen without first traversing the cytosol.

As described by Putney [38], the finding that thapsigargin, an inhibitor of ER Ca^{2+} uptake, produced an accumulation of cytosolic Ca^{2+} put the direct entry model into disfavor. In the absence of thapsigargin, no change in cytosolic Ca^{2+} occurs (as estimated by lack of contraction), despite the fact that there is an influx of this ion, as measured by ^{45}Ca entry. To explain this, it was suggested that the Ca^{2+} was pumped into the ER as rapidly as it entered the cytosol. With the subsequent discovery of the two proteins involved in SOCE—Orai1 and Stim1—it seemed a fairly complete mechanism for SOCE was in place.

However, there are difficulties with two assumptions in this hypothesis. The notion that cytosolic Ca^{2+} concentration can remain constant despite a steady-state flow into the cytosol due to rapid pumping into the ER does not bear scrutiny. An increase in the rate of the ER CaATPase requires an increase in the concentration of its substrate—Ca^{2+}. This is the basis of how the ER Ca^{2+} is restored after release to the cytosol.[iii] In order to form a new steady state in which ER CaATPase pumps in more Ca^{2+}, Ca^{2+} concentration must be higher; it is not. For the CaATPase to have a higher rate at the same Ca^{2+} concentration as its unstimulated rate is a contradiction of the fundamental operation of enzyme kinetics: control by an increase in substrate. The thapsigargin data have an alternative interpretation. It is true that an inhibition of the CaATPase leads to an increase in the cytosolic Ca^{2+}, but this does not rule out direct entry into the ER and subsequent entry into the cytosol (release through the IP_3R or the ryanodine receptor, the Ca^{2+} channels of the ER). The fact that Ca^{2+} accumulates in the cytosol—for either model—means, as first anticipated by Casteels and Droogman [43], that the plasma membrane CaATPase is far less active than the ER CaATPase, hardly a controversial position today. The problem with the accepted model is that it is based entirely on measurement of cytosolic Ca^{2+} concentration, and yet the key event in SOCE involves ER Ca^{2+}. As mentioned above, the later *increases* when Ca^{2+} is added to the media back, despite the fact that the CaATPase—thought to be the only means of entry into the ER—is *inhibited*.

7.3.6   RESURRECTING THE DIRECT-ENTRY MODEL

The direct-entry model provides a simple explanation for why cytosolic Ca^{2+} does not change during SOCE: Ca^{2+} does not enter the cytosol. However, the ability of Ca^{2+} to enter the ER from the outside requires crossing two membranes at once, so that two channels would have to be connected to each other. A model is shown in Figure 7.9. It is already well established that the ER and plasma membrane have very close contact at various points, which is of course the very basis for the Orai1-Stim1 interaction. There is also a precedent for two membranes forming such a "superchannel," which is the case of gap junctions [44]. While physical proof of the interaction

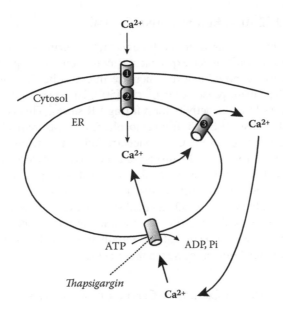

FIGURE 7.9   Direct-entry model for SOCE. Ca^{2+} enters in this model directly from the extra-cellular space into the ER lumen. This traversal across two membranes is through attached channel labeled 1 in the figure (which is the Orai1) and channel 2 (which could be either the IP_3R or the RyR). Release of Ca^{2+} to the cytosol is through channel 3, which could also be either the IP3R or the RyR. Also shown is the CaATPase in gray, inhibited by thapsigargin.

of Figure 7.9 is absent, it is clear that the plasma membrane channel for Ca^{2+} entry is Orai1 (channel 1 in the figure), and likely that the ER channel is a ryanodine receptor[iv] (channel 2 in the figure).

We have collected data supporting the direct-entry model of SOCE in two cell culture systems: the L6 muscle cells and the Jurkat T cells [46–48]. The key finding is that ER Ca^{2+}—measured in two separate ways—increases during SOCE conditions, even more rapidly than the cytosol. Moreover, blocking exit of Ca^{2+} from the ER using high concentrations of ryanodine eliminates any thapsigargin induced increase in cytosolic Ca^{2+}, and yet Ca^{2+} still enters the ER.

While the model of Figure 7.9 represents generalized SOCE, the channel used for the exit of Ca^{2+} from ER to cytosol varies with cell type: the RyR in excitable cells, and the IP_3R in others. The reason for the increase in cytosolic Ca^{2+} in the classical SOCE experiments—that is, with only the cytosolic Ca^{2+} measured—can also be interpreted as direct entry into the ER from the extracellular space followed by the exit into the cytosol.[v] As a corollary to this, recall that the earlier experiments on SOCE showed that, under conditions in which Ca^{2+} is *not* removed from the medium, that no change in cytosolic Ca^{2+} appears. It is only evident with zero external Ca^{2+}, thapsigargin treatment, and adding back Ca^{2+}, conditions which optimize expression of SOCE experimentally. Only direct entry can account for the changes in ER Ca^{2+} observed.

7.3.7   Can SOCE Also Be a Signaling System?

Many published reports, based on the finding that experimental systems show an increase in cytosolic Ca^{2+} under experimental conditions that optimize SOCE, as just discussed, have proposed that this increase in cytosolic Ca^{2+} might serve as a signal for Ca^{2+} dependent processes (e.g., ref. [50]). However, it should be recalled that SOCE is a very low flux activity for refilling ER stores. This is why the experimental setup to demonstrate SOCE from cytosolic Ca^{2+} measurements is so extreme: otherwise, no change in cytosolic Ca^{2+} concentration can be measured. In hormonally triggered cases in which cells display an increase in cytosolic Ca^{2+}, it is the intracellular pool (ER), not the extracellular one, that is responsible for activation. Excitable cells have an entirely distinct system, and while it is less well understood mechanistically, in all cases in which detailed studies exist—muscle and heart—the pool of origin for Ca^{2+} is also the ER. We consider systems of electrically induced Ca^{2+} release next, and will observe that it is equally unlikely that SOCE accounts for any stimulation of metabolic processes in these cells.

7.3.8   Introduction to Excitation-Contraction Coupling (ECC)

Excitation-contraction coupling is the process in which an electrical excitation of the plasma membrane causes a Ca^{2+} release from internal stores (ER) of the cell. The events are outlined in Figure 7.10 for the skeletal muscle, the best understood case. Initiating the event is a motor nerve, which synapses on the motor endplate region of muscle, defined by the presence of acetyl choline receptors (AchR). The depolarization propagated down the membrane as a result is represented as a squiggle. In skeletal muscle, invaginations of the membrane (T tubules) have an embedded

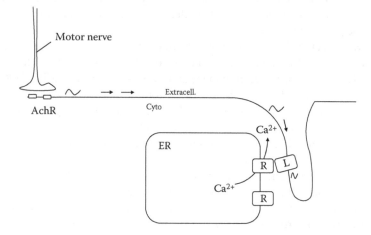

FIGURE 7.10   Excitation–contraction coupling. Nerve stimulation of muscle at the motor endplate (indicated as the region in which the acetyl choline receptors exist, or AchR) leads to a depolarization that is transmitted into the T tubule region (the dip in the membrane sketch) where the L-channel (L) is activated. This opens the ryanodine receptor (R, the ER Ca^{2+} channel) and cytosolic Ca^{2+} is produced that can stimulate contraction.

protein that responds to the depolarization by physical movement (also indicated by a squiggle in the figure). This protein is the L-channel,[vi] which is a Ca^{2+} channel, but more importantly, is located physically close to the ryanodine receptor (RyR), labelled R in the figure.[vii] The latter is a Ca^{2+} channel that can conduct Ca^{2+} from the ER lumen to the cytosol after physically being altered by the L-channel. Elevated cytosolic Ca^{2+} binds a protein attached to actin in the muscle, which exposes binding sites of actin to myosin. This initiates contraction by the sliding-filament hypothesis.

While the broad strokes of this mechanism are well worked out, there are some fine points that are unsettled, but there is one large gap: the mechanism by which contact of the L-channel with the RyR leads to activation of the latter and subsequently Ca^{2+} release.

7.3.9   WHY ECC IS CONSIDERED IN A METABOLISM TREATISE

A metabolic perspective—the consideration of how specific measurements of molecular elements is integrated into a pathway—is a means of deciding which of a variety of *in vitro* possibilities is more likely to play a part in the entire scheme. Calcium signaling itself is an essential element in regulating metabolic pathways just as it is in excitable tissues. These two considerations—that Ca^{2+} flows themselves constitute a pathway and that Ca^{2+} signaling is an important aspect of metabolic control—provide the justification for a deeper consideration of Ca^{2+} flows in excitable cells.

As a regulator of excitable tissues, Ca^{2+} assumes a primary role. For skeletal muscle, we have an extreme metabolic situation: in active muscle, virtually the entire energy consumption can be attributed to just two processes: the myosin ATPase of contraction, and the CaATPase needed to re-accumulate ER Ca^{2+} and terminate the process. The span of energy utilization between inactive muscle and very active muscle is greater than for any other tissue, and entirely under the control of Ca^{2+}. I begin with a discussion of a curious mechanism first proposed in 1970: the calcium-induced calcium release or CICR.

7.3.10   CICR

By the late 1960s, the key unsolved question in ECC was how Ca^{2+} was released from the ER to the cytosol. The relative quantities of Ca^{2+} involved in three critical spaces were well known by that point. In the extracellular space, Ca^{2+} concentration was about 1 mM. In the cell cytosol, free Ca^{2+} concentration was known to be about 0.1 µM. This very steep gradient from outside to inside the cell is striking; it would seem obvious that opening a Ca^{2+} channel would enable entry and produce the rise in cytosolic Ca^{2+}. However, it was clear that this was not the case for skeletal muscle [51]. Even with all Ca^{2+} removed from the extracellular space, the muscle could repeatedly initiate contractions upon electrical stimulation. Not only that, it was clear that the real source of Ca^{2+} for contraction is the ER.[viii]

One way of providing a rationale for the large extracellular-cytosolic Ca^{2+} gradient is the idea that it might allow a very small amount of Ca^{2+} to enter the cell, and this Ca^{2+} could itself serve to stimulate a larger release from the ER.[ix] In 1968, two laboratories independently advanced this hypothesis, known as

Calcium-Induced-Calcium-Release or CICR [52]. The experiments were conducted with isolated ER preparations, and indicated that a small amount of Ca^{2+} could itself cause a release of further Ca^{2+} from the internal stores of this organelle from skeletal muscle.

One of initial investigators (Endo) subsequently questioned the physiological significance of CICR [53,54]. For one thing, the experimental systems required conditions unlikely to prevail in intact cells: a complete absence of Mg^{2+} is required. Mg^{2+} suppresses CICR activity. For another, the concentration of Ca^{2+} required to elicit this response is far greater than observed even under conditions of active contraction. We offered an explanation for the activity measured in subcellular fractions: a Ca^{2+} activation of a protease [55].

While CICR is no longer considered a viable mechanism to explain skeletal muscle CICR, it is considered to be the mechanism for heart ECC, since that tissue does require a steady extracellular Ca^{2+} to sustain contraction. We will consider both heart and other cell types in a subsequent section. However, even for the case of skeletal muscle, the CICR hypothesis has been revisited as a rationale for the existence of a greater number of RyR over L-channel channels. The idea [56] is that the L-channel directly contacts the RyR, which releases a small, local amount of Ca^{2+} to the cytosol. This local Ca^{2+} in turn activates neighboring, unconnected RyR channels to release even greater amounts of Ca^{2+}, that is, CICR (Figure 7.11). However, it remains hypothetical, with the same difficulties associated with the original discovery. I will further explore the nature of these proposals in the following section concerning heart and smooth muscle.

A separate notion was that the same inositol phosphate system that causes release from the IP_3R as described above might also function in skeletal muscle [57]. The system does exist in the skeletal muscle, and could potentially explain ECC. However, it has not been possible to demonstrate, for example, that IP_3 can increase Ca^{2+} in skeletal muscle. The system is likely to be too slow to kinetically account for the ECC event, and may have significance in other aspects of muscle function [56].

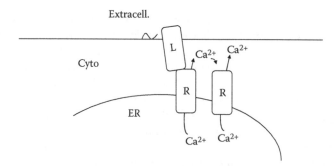

FIGURE 7.11   Model for ER Ca^{2+} release invoking CICR. In this model, Ca^{2+} release from the connected RyR is postulated to locally accumulate in high concentration near an adjacent RyR and activate it, an activity known as calcium-induced calcium release (CICR).

The bulk of current evidence is consistent with a direct physical connection between the L-channel and RyR, including both site-directed mutagenesis studies and those in which with one channel is ablated. In addition, there are studies substituting L channels and RyRs between different cell types [58,59], to demonstrate direct contact at selective regions between these two channels in the skeletal muscle. In addition, the very close overlap between the channels and the striking pattern of L-channel aligning with RyR in electron micrographs from skeletal muscle [60] argue for an important interaction. This pattern also shows unoccupied RyR at regular intervals, with every other protein unoccupied.

7.3.11   MODEL FOR HOW CA^{2+} RELEASE MIGHT BE ACHIEVED

It is established that the L-channel physically moves within the plane of the plasma membrane as the spread of depolarization reaches it. Experimental evidence of this movement is the observation of a virtual current, a membrane charge movement that results from an alteration of the disposition of the channel protein in a changing electric field [61]. This is not the result of an actual transport of an ion across the membrane, but a bobbing of the protein in response to the electric field. This movement in the L-channel physically alters the RyR embedded in the opposing, ER membrane. The ER membrane itself has no electrical potential across it [51], so no electrical signaling takes place.

A measurement of Ca^{2+} movement that occurs under depolarized conditions that is associated with the L-channel is the *excitation coupled calcium entry*, or ECCE. This acronym is unfortunately close to the one used for the overall event itself, ECC, and so they must be carefully distinguished. ECCE is the entry of Ca^{2+} during the contractile event, and this Ca^{2+} entry is a useful measure of the L-channel. The tricky thing is that the L-channel need not actually conduct *any* Ca^{2+} to trigger ECC. Thus, as discussed above, it is well known that contraction can ensue with no Ca^{2+} entry to the cell, so ECCE can be zero and the ECC can proceed for many contractile cycles.

Nonetheless, the ECCE is a useful activity that can be measured with relatively simple electrophysiological tools and provides information about the L-channel. For example, it is known that inhibitors of the channel, such as dihydropyridines and diltiazem, can suppress this current. In addition, Ca^{2+} that is elevated in the cytosol can inactivate this channel (a known property of the isolated channel as well); this is reflected in a decrease in the ECCE.

A new model for ECC is illustrated in Figure 7.12. In the resting state, the close apposition of the L-channel with the RyR occludes the RyR and it cannot release releasing Ca^{2+}. Note that a small amount of Ca^{2+} can be released to the cytosol (indicated by a dotted arrow in the figure) that represents resting Ca^{2+}; this is balanced by the CaATPase activities discussed above. The extra RyR is established: there is an approximately 2:1 ratio of RyR to L-channel in the skeletal muscle. Upon excitation of the membrane, the L-channel is displaced, so that it no longer occludes the RyR, and Ca^{2+} is released. It is likely that the opening of the RyR is transitory, briefly releasing Ca^{2+} and then becoming inactivated. The release of Ca^{2+} would require in this mechanism a continuous membrane depolarization and cycles of association

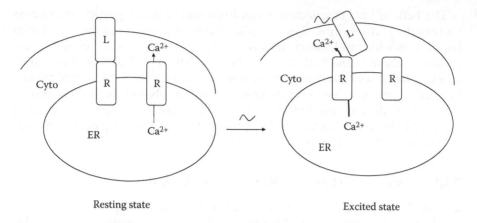

Resting state   Excited state

FIGURE 7.12 New model for ER Ca²⁺ release. In the resting state, the RyR is occluded by the L-channel. Depolarization activates the L and releases it from the R and Ca²⁺ is thereby released. Only small amounts of Ca²⁺ can be released to the cytosol in resting conditions (indicated by the small font Ca²⁺ in this condition) as the unattached channel is relatively inactive.

of the L channel with the RyR, with the assumption that this association resets the channel open. This idea is derived from the known retrograde transport of Ca²⁺ established *in vitro* as mentioned above. It is possible that the L channel could slide over to an unoccupied RyR, providing an explanation for the extra RyR.

Experimental evidence in support of this model takes advantage of the interactions between the RyR and its namesake modulator, ryanodine [62]. It is established that low concentrations of ryanodine (less than 1 μM) lock the channel into an intermediate conductance state, activating the channel and increased cytosolic Ca²⁺ is observed. Above 1 μM ryanodine, the channel is blocked, and by 10 μM essentially completely inhibited. Thus, ryanodine can be used as both an activator and inhibitor of Ca²⁺ flow through the channel, through differential binding of ryanodine.

To explore the proposed model for ECC, we titrated L6 muscle cells with ryanodine, measuring cytosolic Ca²⁺ using the indicator Fura2. The reasoning was, in the resting state, about half of the RyR would be occupied by the L-channel. Occupied RyR channels would be unable to release Ca²⁺; they would also be unable to bind ryanodine. We would anticipate that cells probed by addition of various amounts of ryanodine would show lesser binding in the resting state than in the excited state; in the latter, more RyR receptors are freed from their contact with the L-channel.

We therefore titrated muscle cells with ryanodine under conditions that should change cells from a resting to an active state. Adding the agonist Bay K-8644, a known L-channel agonist that also stimulates cells in a similar way to depolarization, was performed with various concentrations of ryanodine (Figure 7.13). We found the same shift by depolarizing the membrane (using the presence of high-medium potassium). As controls, manipulations that increase cytosolic Ca²⁺ but do not alter the conformation of L-channels, such as increasing extracellular Ca²⁺ or the

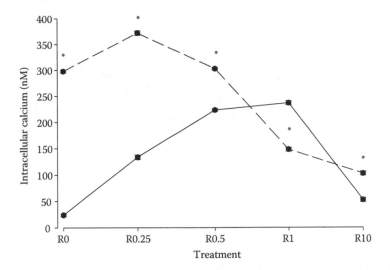

FIGURE 7.13 Ryanodine sensitivity is shifted when the L-channel is stimulated. In this experiment, the presence of the drug Bay K-8644 (which activates the L-channel and causes it to be displaced in a manner similar to membrane depolarization) causes an increase response of the RyR to ryanodine. Ryanodine (R) stimulation at low concentrations is shifted to the left (dashed line is the presence of the drug). Ryanodine is known to inhibit the RyR at high concentrations, and the sensitivity to this action is also shifted to the left. A similar result was obtained by using K depolarization in place of the drug.

direct RyR stimulant chloromethylcresol, do not affect ryanodine sensitivity [48] and did not shift the curve.

We also found that ryanodine inhibits ECCE[x] (Figure 7.14). As a control, we found no ryanodine inhibition of SOCE measured as entry currents. This result is consistent with our model in the following way. For the continued ECC, it is necessary for the L-channel to cyclically engage a RyR in order to reactivate it; the channel is known to undergo rapid inactivation following opening.[xi] Taken together, the experiments provide an explanation for ECC in skeletal muscle.

7.3.12 Ca²⁺ Signaling as a Pathway

As noted earlier, there is a distinction between the *reaction view* and the *pathway view*. Most studies of Ca^{2+} signaling take the reaction view, focusing on a single aspect of the process in order to uncover deeper details. This might be a study of ion channel properties, protein structure, or a partial process of Ca^{2+} signaling, such as SOCE or ECC. However, a pathway view is important in assessing which details make physiological sense.

If we consider Ca^{2+} as a pathway, then we are concerned with how it enters cells, and what its internal exchanges are. There must be a cellular balance of Ca^{2+} inputs with Ca^{2+} exits. In the overall pathway, Ca^{2+} signaling is initiated at the plasma membrane, either by hormonal activation (Figure 7.15a) or by electrical depolarization (Figure 7.15b). In either case, a protein channel is activated in the ER to release Ca^{2+}:

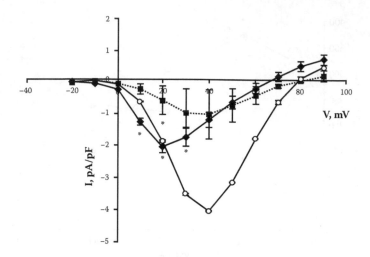

FIGURE 7.14 Ryanodine inhibits ECCE currents. The increased currents as voltage is increased (open circles) is the control signature for the L-channel, with a maximum of about +35 mV. In the presence of high ryanodine (5 μM for filled diamonds; 10 μM for filled squares) a decrease in the ECCE currents (an entry of Ca^{2+} into the cell, a measure of the activity of the L-channel) is observed. This suggests the RyR must engage the L-channel for activity; otherwise there should be no effect of inhibition of the RyR on the L-channel.

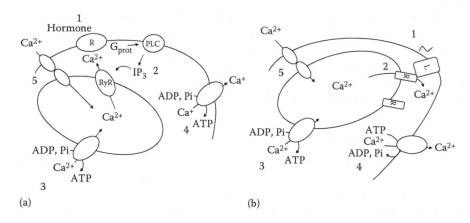

FIGURE 7.15 Ca^{2+} flows as a pathway. The entry, disposition, and exit of Ca^{2+} in nonexcitable (a) and excitable (b) cells is illustrated. (a) Hormone binding receptor (1) leads to IP_3 release (2), activating ER Ca^{2+} release, followed by CaATPase reuptake into the ER (3). A small amount of Ca^{2+} is also lost to the extracellular space by the plasma membrane CaATPase (4). This is restored by SOCE (5). (b) Excitation of the cell moves the L-channel (1) and thus stimulates the release of ER Ca^{2+} through the RyR (2). The subsequent steps (3–5) are identical to those of nonexcitable cells in a.

either by formation of the soluble ligand IP$_3$ (Figure 7.15a) or by the displacement of an occluding channel protein (Figure 7.15b). Once released to the cytosol, the rest of the pathway is similar in the two types. Most of the Ca^{2+} is taken back into the ER to terminate the signal by the CaATPase of the ER. At the same time, a small amount of Ca^{2+} is lost to the cell exterior because the elevated Ca^{2+} is also a substrate of the plasma membrane CaATPase [30]. Over time, the latter activity represents a loss of cell Ca^{2+} and would eventually deplete the store. Thus, flow through a distinct Ca^{2+} channel, the Orai1, is activated by a depletion in ER Ca^{2+}, and this flow balances that lost overall by the plasma membrane CaATPase.

Suppose it were otherwise, and that we construct a pathway out of Ca^{2+} entering directly into the cytosol. This is required if Ca^{2+} were to serve as a signal from outside the cell. In order to terminate the signal, the rate of plasma membrane CaATPase would have to be very high—in fact, it would have to achieve the rate of the ER CaATPase (which can do so by having massive amounts of this protein in the ER). Under this scenario, the ER CaATPase would not be involved in the balance of cell Ca^{2+} at all. The ER lumen is only a temporary holding space; if the Ca^{2+} flow for signaling originated from the extracellular space, its involvement would be difficult to explain. Perhaps a major reason that cells use the ER as a Ca^{2+} pool is to accommodate the massive quantities of CaATPase in the ER. It has long been established that this protein is by far the major one in muscle ER [63,64]

There is yet another factor to consider here. The amount of Ca^{2+} released into the cytosol is much larger than the rise in free Ca^{2+} [65]. Like other metabolites in the cell (e.g., ADP, AMP, NADH) the total amount of Ca^{2+} is a small fraction of the total, due to the large amount of bound Ca^{2+}. Thus, provision of Ca^{2+} by the ER to raise the cytosolic Ca^{2+} concentration by about 1 µM requires a few orders of magnitude greater release.

Similar problems exist for the mitochondria as Ca^{2+} source for the cytosol. This organelle has no high affinity binding proteins like calsequestrin in the ER, has no high capacity import pump, and with too much Ca^{2+} suffers a damage from calcium phosphate precipitation. The experiments demonstrating close tracking of mitochondrial Ca^{2+} with cytosolic rise [66] suggest instead that the mitochondria are a response organ, not a control organ for Ca^{2+}.

While the essential means of eliciting an increase in cytosolic Ca^{2+} are represented in Figure 7.15, there are variations in other cell types. This is due not only to different isoforms of the key channels and regulatory proteins involved, but also other routes, such as the Na/Ca exchanger, prominent in the heart. We consider Ca^{2+} signaling in this tissue next.

7.3.13   ECC AND THE HEART

The process of ECC in the heart is closer to skeletal muscle than to nonexcitable tissues. In heart muscle cells, the same two key proteins are involved, the L-channel and the RyR. They do have unique isoforms for the heart, and do not function when substituted in skeletal muscle cells [67]. In heart, there is also an excess of RyR over L-channel, but in this case it is even more extreme: the RyR is in about a ten-fold excess [68]. Electron micrographs also show a different arrangement for heart muscle. While there are some points of contact between the two membranes that contain these channels, there appear to be many unconnected, free RyRs and a less regular pattern than skeletal muscle.

Unlike skeletal muscle, the continuous presence of Ca^{2+} in the extracellular space is needed for contraction of heart muscle. This was first observed for the isolated frog heart in the 1880s by Ringer [69]. However, like skeletal muscle, the source of Ca^{2+} for contraction in heart is also the SR. Thus it is not surprising that the mechanism of CICR originally proposed for skeletal muscle was adopted for the heart [70]. Some studies have also suggested that Ca^{2+} regulates the RyR from the *inside* of the ER, that is, the lumen [71], making an understanding of control of Ca^{2+} release, and its separation from controlled Ca^{2+} uptake by SOCE more difficult. Since CICR requires that Ca^{2+} become localized to distinct portions of the same cytosolic water space, it further defies analysis by the simple ideas of diffusion and the usual steady state methods. Evidence is typically taken in the form of fluorescence of Ca^{2+} binding dyes which appear to distribute in specific portions of the cell in micrographs (e.g., ref. [72]). It is usually taken as an article of faith that the dye precisely reflects the free Ca^{2+} concentration, but both the time and spatial response of Ca^{2+} bound dye need not be the same as free Ca^{2+}. While this view is not universally accepted (e.g., ref. [73]), the spark evidence is commonly viewed as the key proof of CICR.

7.3.14   THE PROBLEM WITH POSITIVE FEEDBACK FOR CELLULAR CONTROL

CICR also requires that Ca^{2+} serve as its own release mechanism. The origin of this proposal, as discussed for skeletal muscle, was to justify the entry of Ca^{2+} from the cell exterior with the fact that Ca^{2+} release must originate from the cell interior (ER). This is positive feedback: as more Ca^{2+} is released from the RyR, it stimulates the RyR further, causing more release. It was this complication that probably inspired the idea that ER luminal Ca^{2+} might regulate the RyR [71], but it is not widely accepted that this can provide signal termination. There is in fact no real consensus as to how the signal might be terminated.

Positive feedback is a rare control system in physiology. When it occurs, signal termination is achieved by an irreversible and catastrophic event. For example, consider the end stage of the birthing process. Delivery of the fetus along the uterine wall is assisted by a volley of contractions of the lining muscle. The presence of the fetus itself triggers further stimulation. Thus, these contractions increase in intensity. Only the ejection of the fetus terminates these contractions. A second example is also part of reproductive biology. At the point of the estrus cycle at which the mature follicle is nearing ovulation, it secretes large concentrations of estrogens that exert a positive feedback on LH (luteinizing hormone) secretion from the pituitary. In turn, LH further stimulates follicular growth and estrogen secretion. This leads to rapid increases in both estrogens and LH until the follicle bursts (ovulation).

In these examples, there is no actual regulator that can terminate the response. Rather, there is an irreversible event that ends the positive feedback. To apply this notion to a metabolic process, we must distinguish two types of reversibility. We have already considered reversibility of a single enzymatic reaction. If we are considering a single reaction, we would classify it as NEQ. Otherwise, the reactions are mIRR, *metabolically irreversible*. A distinct form of regulation is apparent for example in phosphorylation/dephosphorylation reactions (each corresponding to different activity states). We can consider such conversions to *pathway-reversible*, as flow can be restored with one or a few reaction steps to alter the steady state. This notion was introduced in Chapter 5. Sequences that require new protein synthesis, or a series of steps that cannot be readily reversed are called *pathway-irreversible*. Under these criteria, positive feedback systems are pathway-irreversible.

It should also be clear that a positive feedback mechanism for ECC in the heart is untenable, as pathway reversibility is essential for this short term regulatory system. The need for positive feedback in a system is to provide an increasing signal for a sudden (explosive) conclusion. Termination is not just an unfortunate loose end that has yet to be explained; it is a sudden, irreversible, and critical part of positive feedback mechanisms. In fact, no new evidence in favor of CICR specifically for the heart has emerged, nor is this tissue exempt from any of the problems endemic to this mechanism [51]. To understand how the heart might function, we consider first the extracellular entry of Ca^{2+} in heart.

7.3.15   WHY Ca^{2+} ENTRY MUST OCCUR IN THE HEART

The events of cardiac contraction overall are well known [74]. The initial membrane depolarization is due to Na channel opening which subsequently activates heart-specific L channels ($Ca_v1.2$). In the case of heart, the entry of Ca^{2+} through these L channels causes a sustained depolarization. This accounts for the very long (a few hundred millisecond) contraction of the heart. Notice that the *electrical* rather than the *chemical* potential of the Ca^{2+} inflow is what is critical for the heart action potential. This suggests that the amount of Ca^{2+} entering the cell is extremely small and an ability to serve as a chemical signal is not suggested by its physiological role.

A central question is this: is Ca^{2+} entry through the L-channel also an activator of the RyR? Notice that this is separate from the physiological entry for continued electrical depolarization as just discussed. This has been addressed by Atlas and coworkers [75] by constructing a mutant L-channel that could still bind extracellular Ca^{2+} but not transport it into the cell. The result was that ER Ca^{2+} release was still stimulated. There was a requirement for Ca^{2+} on the exterior to activate the L-channel, but not for the actual entry of Ca^{2+}. These investigators also found a similar result for Ca^{2+} linked fusion of vesicles with the plasma membrane [76].

Let us use the same analysis that we used for the skeletal muscle. Entry of Ca^{2+} through that channel during excitation was not required but still occurs. The driving force must be the chemical gradient of Ca^{2+} (the calcium current entry occurs only at voltages positive to the cell interior) and happens over a narrow range. The other point to recall is the extensive buffering of Ca^{2+} in the cytosol. Like skeletal muscle, this means that massive amounts of Ca^{2+} would be required to make a substantial change in the cytosolic concentration.

7.3.16   HYPOTHESIS FOR HEART ECC

The simplest explanation for how ECC functions in heart is that it is like the skeletal muscle: physical coupling. While it is true that the heart contains other significant fluxes that shape the action potential including prolonging the contraction, these features are not part of the fundamental link between depolarization and ER Ca^{2+} release.[xii] This turns the focus on how heart and skeletal muscle are alike rather than how they are different. The unique features of the heart are unrelated to the central issue of Ca^{2+} release from the ER for signaling. Instead, they serve to alter the timing and control of the contraction.

The model shown in Figure 7.16 is similar to the skeletal muscle model of 7.12, with two variations. The first is the greater number of RyR in the heart. It is possible that the L-channel can make contact during its cycles with any of the various RyR channels, thus facilitating the ability to sustain contraction. The second feature indicated in Figure 7.16 is an extra squiggle next to the L-channel, indicating the role of L-channel Ca^{2+} entry in sustaining the depolarizing voltage across the plasma membrane. Unlike the skeletal muscle, the heart has no neural connection and is only *initiated*, and not *driven* by the Na channel depolarization. The latter arises from the pacemaker cells of the sinoatrial node, and the spread of action potential reaches the cardiomyocytes and causes this first depolarization. However, the elongation of contraction requires an elongation of the action potential; this is effected by the entry of Ca^{2+} through the L-channel. For this reason, there are *two* depolarization events symbolized in the excited state of the heart: these are labeled and enumerated in the figure. This explains why there is a need for Ca^{2+} current through the L-channel in the heart, but not in the skeletal muscle.

With skeletal muscle and heart using the L-channel:RyR interaction, and the IP_3-linked systems utilizing the second messenger, we have just two mechanisms of producing an increased Ca^{2+} for signaling. Let us consider next the question of what might be the means of Ca^{2+} signaling in other cells.

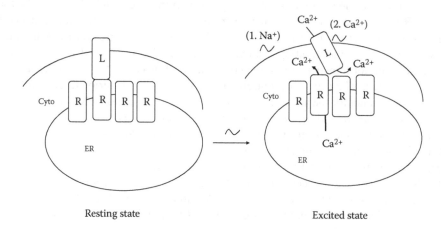

FIGURE 7.16 Excitation-stimulated Ca^{2+} release for heart. The mechanism is similar to that for skeletal muscle, except for the larger number of RyR over L-channels, and the lack of neural input to the cell. In the excited state (right-hand panel), there is (1) depolarization due to Na channels, which then excites the heart L-channel (2), which causes Ca^{2+} release, proposed to occur by a similar mechanism to skeletal muscle.

7.3.17   Triggering a Rise in Ca²⁺ in Other Cell Types

Smooth muscle has both L-channel:RyR systems as well as an IP_3-linked system [77]. Both may be used in separate conditions: in response to electrical stimulation for the first, and to hormone stimulation for the second. In smooth muscle, the isoform of the RyR is known as the RyR3, distinguishing it from the skeletal muscle (RyR1) and heart (RyR2). The RyR3 is also found in a wide variety of cells at low levels, although its function is unclear [78].

Presently, the majority opinion for smooth muscle and nerve favors extracellular Ca^{2+} entry leading to a direct increase in the cytosolic Ca^{2+}. This view is problematical for these tissues just as it is for the heart. First, as Ca^{2+} is heavily buffered in the cytosol, entry from the exterior would need to be massive to produce the expected rise to about 1 μM. Not only that, but the plasma membrane would also require massive counter ion transport to balance most of the charge of the incoming Ca^{2+}. There would also be a need for a very active export of Ca^{2+} and thus a very large quantity of plasma membrane CaATPase. Having large amounts of protein in the plasma membrane may be a problem considering the other required proteins in that membrane.

A different way of determining if intracellular Ca^{2+} rather than extracellular is the signaling ion is to determine if SOCE occurs in the cell of interest.[xiii] While it was first thought to be limited to nonexcitable cells, this process is now known to be ubiquitous. Nerve cells have the components of SOCE this activity has been shown in some studies [79,80]. A controversy remains, however, largely based on inconsistencies with the standard mode of demonstration of SOCE, which requires Ca^{2+} removal, depletion with thapsigargin, and re-addition of Ca^{2+}. Since only cytosolic Ca^{2+} is measured, it is intrinsically an indirect measurement. Nerve cells have been shown to display variable response to thapsigargin, and experiments are confounded by the presence of additional ion channels that respond to extracellular Ca^{2+} [81]. Yet the very existence of an ER Ca^{2+} store strongly suggests that even for nerve cells, a version of the skeletal muscle type Ca^{2+} signaling exists.

7.4   AMP AS A CELLULAR SIGNAL

It has long been known that AMP acts as a direct allosteric effector of enzymes of central metabolism, including glycogen phosphorylase [82], phosphofructokinase [83], and fructose bisphosphatase [84]. More recent interest is largely due to the AMP activated protein kinase, AMPK.

Assessing the signaling properties of AMP requires that we recognize that most of the AMP (like ADP) found in the cell cytosol is bound to proteins, and only a small fraction is free in solution, requiring indirect methods to measure it [85]. AMP is connected to the other adenine nucleotides, ADP and ATP, through reactions illustrated in Figure 7.17. As indicated in the diagram by the thickest arrows, the largest fluxes are the formation of ATP by the pathways of glycolysis and by mitochondria, and the conversion of ATP to ADP through cell demand (roughly, catabolism). Some reactions convert ATP to AMP and PPi, such as the activation step of fatty acids, which form acyl-CoA esters. The PPi formed in these reactions is hydrolyzed to Pi, catalyzed by the very active inorganic pyrophosphatase.[xiv] AMP can be converted

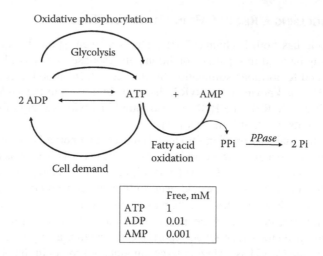

FIGURE 7.17 Cellular pathways in AMP metabolism. The central enzyme interconverting ADP with ATP and AMP is adenylate kinase. ADP can form ATP through glycolysis or oxidative phosphorylation. Usually, ADP is formed by cell demand, that is, reactions utilizing the bond energy of ATP. However, some reactions, such as fatty acid oxidation, convert ATP to AMP, with the formation of the pyrophosphate (PPi); in that case, the energy is not captured but hydrolyzed to Pi by pyrophosphatase (PPase). Also shown is the table of the approximate free concentrations of the three adenine nucleotides.

to ADP by the adenylate kinase reaction, indicated as the only NEQ reaction in the figure.

Under conditions of greater ADP formation, it is possible to drive the adenylate kinase in the direction of AMP formation. Notice that because of the relative amounts of the adenine nucleotides indicated in the bottom of the figure, conversion of some of the ATP into ADP or AMP has essentially no effect on the total ATP concentration, as amply pointed out earlier. Note, too, the much lower concentration of AMP, dictated in large part by the requirement for near-equilibrium at adenylate kinase, and the fact that the equilibrium constant for the reaction is approximately 1. Thus, a significant increase in ADP translates into a much larger increase in AMP.

To fit AMP into a classical signaling paradigm, originally established for cAMP described earlier in this chapter, we consider the following three criteria:

1. Steady source
2. Steady sink
3. Rapid rate of removal

For #1, the source is ATP, a constant and large pool (and the same source for cAMP). For #2, the sink is ADP, formed by the adenylate kinase reaction in the right-to-left direction of Figure 7.18. Total ADP is also a large sink. For #3, rapid removal is assured by the near-equilibrium nature of adenylate kinase. Thus, while these rules were not designed for such a regulatory system, it can be seen to fit if we consider of the connections and concentrations of the nucleotides in the cytosol.

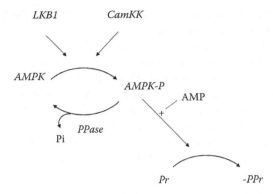

FIGURE 7.18 Regulation of AMPK. Phosphorylation of AMPK is essential for its activity. AMPK is a substrate of the protein kinases LKB1 and CamKK. AMPK is activated by the presence of AMP, which can lead to its ability to phosphorylate some cellular proteins (Pr). Not shown is the action of AMP, which is to decrease the activity of the AMPK-P phosphatase (PPase) and stimulate the action of AMPK kinase.

7.4.1   AMP-Activated Kinase (AMPK)

As recounted by Hardie and Carling [87] the AMPK was originally discovered (*sans* the AMP-activating property) as a kinase that leads to phosphorylation and inactivation of HMG-CoA reductase [88,89]. Subsequent kinetic characterization led to the key regulatory characteristics of the enzymes, illustrated as Figure 7.18. The allosteric effects include the namesake activation by AMP, as well as an inhibition by ATP. The two are frequently assembled into the ratio of AMP/ATP, and named the "energy charge" [90]. As discussed in Chapter 4, energy charge has not been a useful concept in its past incarnations [91], for two important reasons. First, ATP concentration is constant in cells. Second, the free concentrations of AMP and ADP are a small fraction of the total, so that most measurements taken to promote energy charge using total values do not reflect the steady-state values. The free conservation of AMP can regulate the AMPK (measured using indicator metabolites) can account for activation of the enzyme in some circumstances [92].

The phosphorylation status of the enzyme is a critical part of its activation, as there is very little activity in the dephosphorylated state [93,94]. The phosphorylation occurs on a threonine residue in the activation loop of the kinase. AMP leads to activation by binding AMPK and making it both a better substrate for an AMPK kinase and a poorer substrate for an AMPK phosphatase.

Two AMPK kinase activities are well established: the LKB1 and the CamKK (calmodulin-activated kinase kinase). The first of these derives its name from the liver (LK is liver kinase), and, appropriately, this is the only activity present in the hepatocyte. The LKB1 is also known as a tumor suppressor, and is found to be diminished in certain cancer states, such as breast cancer. This is likely related to its ability to phosphorylate and activate the AMPK, which is also associated with a suppression of cancer (see Chapter 10). The CamKK is present in substantial concentrations in other cell types, such as muscle. This upstream protein kinase has a calmodulin binding site and can be activated by Ca^{2+}, although the link between Ca^{2+} and AMPK activation is not established.

One hormone known to act upstream of the AMPK is adiponectin, secreted by adipocytes as a signal to increase energy expenditure [95,96]. Activation of muscle cells by adiponectin leads to an activation of AMPK through the LKB1 but it is unclear what the receptor intermediates are between the adiponectin receptor and LKB1. One metabolic action of AMPK in muscle and liver is well established, however: the phosphorylation and inactivation of acetyl CoA carboxylase [97,98]. This enzyme produces an increase in malonyl CoA as an intermediate of fatty acid biosynthesis in the liver. The same increase in malonyl CoA concentration suppresses the oxidation of long-chain fatty acids by preventing the formation of the carnitine ester that is needed for transport into the mitochondrial matrix. As malonyl CoA is both an intermediate of fatty acid biosynthesis and an inhibitor of fatty acid oxidation, it is an interesting target of the AMPK. Decreasing the cytosolic malonyl CoA concentration both directly impedes fatty acid biosynthesis and accelerates fatty acid oxidation. In muscle, only the pathway for fatty acid oxidation has significant activity, so that malonyl CoA formation is present presumably just to regulate fatty acid oxidation.

In muscle AMPK, activation also causes a stimulation of glucose transport, and this stimulation is independent of that of insulin [99]. A point of convergence between insulin and contraction is the small G protein racl, as illustrated in Figure 7.19. Studies of contraction signaling at racl, however, have not confirmed a specific connection between muscle contraction and the activation of this protein [100]. For example, some evidence suggests that AMPK can phosphorylate and active racl, but there is also evidence to the contrary: both activators of AMPK (AICAR) and suppression of AMPK activity (dominant negative expression) fail to suppress the ability of exercise to stimulate glucose transport. In all cases, racl was necessary for the exercise effect.

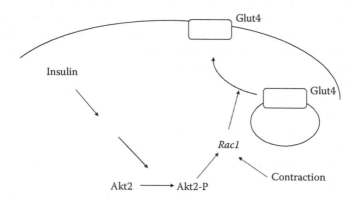

FIGURE 7.19 Insulin and contraction stimulate glucose transport in fat cells and muscle. The two effectors of glucose transport, insulin and contraction, merge at a small G protein that stimulates the vesicular movement that brings more GLUT4 to the cell surface. The two actions can affect glucose transport independently.

While it has been argued that Ca^{2+} can stimulate glucose transport, evidence for this point depends upon experiments in which glucose transport is measured in a manner that would obviate the brief rise in Ca^{2+} [101]. The measurement was performed in a "split assay:" caffeine or other agents are added to a first incubation, the media removed, and labelled 2-deoxyglucose is added in a second incubation to measure glucose transport. This separation requires that Ca^{2+}, elevated in the first incubation, survives washing and incubation in simple salts in the second. We have conducted glucose transport in this manner and demonstrate that the calcium elevation present in the first incubation in fact vanishes in the second [102]. The ability of Ca^{2+}-eliciting agents to increase glucose in this "split assay" was confirmed but it cannot be accounted for by a stimulation of glucose transport. It is not clear why this occurs, but may be the result of a depression of the cellular glycogen that was enough to increase glucose transport subsequently. Thus, Ca^{2+} is not the signal for stimulation of glucose transport either.

7.5 DIACYLGLYCEROL AND PROTEIN KINASE C

With the discovery of protein kinase C (PKC), a new regulatory molecule emerged: the lipid-soluble second messenger diacylglycerol. Its common depiction in illustrations as a molecule released from phospholipids often leaves the erroneous notion that it spills into the cytosol. This is not possible, as the molecule has essentially no water solubility. Rather, it exists entirely within the hydrophobic portion of the membrane.

Activation of PKC involves a synergistic mixture of three elements: Ca^{2+} ions, acidic phospholipids, and diacylglycerol [103].[xv] In addition, there is an experimentally observed change in the localization of the enzyme. When cells are unstimulated, PKC appears mostly in the pellet fraction, but in the presence of strong activators, such as the diacylglycerol-mimic tetradecanoylphorbol acetate (TPA), the enzyme is recovered mostly in the soluble fraction. From this evidence, the enzyme now appears to be bound to the plasma membrane. Thus, the notion that the enzyme translocates from cytosol to membrane has become widely accepted.

There is some divergence from this view. For example, analysis of the interaction of enzyme with membrane suggests simply a tighter membrane association in the presence of activators than in their absence [104]. The interaction with Ca^{2+} and negatively charged phospholipids such as phosphatidylserine may be due to Ca^{2+} serving as a bridge between PKC (in particular the C2 domain) and the phospholipid [105].

The naming of the enzyme with the letter "C" was because it appeared to be activated by the ion in cell extracts. This however was due to a Ca^{2+} stimulated protease that was present in that mixture [106]. The enzyme is known to have a "clamshell" structure in which two large portions of enzyme are separated by a hinge region [104]. In its inactive state, the two portions overlap, and the active site is occluded. In the presence of Ca^{2+}, a protease is activated in the extract and the regulatory portion is removed, providing a fragment that is always active, and immune to the regulators described above. This was called "protein kinase M"; from a metabolic perspective, a dead end.

The holoenzyme in its "clamshell" state is nonetheless of importance because the mechanism for its formation is the binding of the active site to a pseudosubstrate on the regulatory portion of the protein. Since the protein is not cleaved during activation–inactivation cycles *in vivo*, something must be able to remove the regulatory portion to achieve activation, and then restore it to return to the baseline state, and do so in conjunction with changes in the activators. Currently, activation of protein kinase is discussed in terms of specific conformational changes of the protein and assignment of domains to function, so that one domain (C1) confers DAG binding, and another (C2) interaction with phospholipids and Ca^{2+} [107]. While structural variation with different isoforms fits with distinctions in domain structure, the structure-function correlations in PKC do not explain the unusual synergy of the activators, and thus leave us without a complete mechanism for how that activation is achieved.

As discussed in Chapter 5, activation of PKC can be explained by the formation of an inverted micelle structure by DAG, inserting the regulatory lobe of PKC into the membrane. The free kinase activity thus becomes a transiently activated kinase for cytosolic substrates until the DAG is destroyed and PKC once again becomes a membrane surface-bound protein.

In the remainder of this chapter, I briefly consider molecules currently regarded by some as regulatory.

7.6 ARE ATP AND NAD REGULATORS OF METABOLISM?

As discussed previously, ATP is at high and constant concentrations, and thus unlikely to serve as a regulator or as an index of cellular energy. NAD is similarly in relatively high and unchanging concentration in cells. The partners of these cofactors—ADP and NADH—are not commonly considered metabolic regulators, although the free concentrations of these latter molecules very low and do fluctuate.

7.6.1 ATP

In Chapter 2, I have discussed at length the evidence showing ATP concentration is constant and at high concentrations in cells. Since a regulator must change in activity to exert a kinetic influence, it follows that ATP is not a regulator. Still, ATP as a regulator remains a very common assertion. This stems from very early investigations, as well as a popular treatise already cited [91], which postulated that certain adenine nucleotide ratios are analogous to a battery, explaining cellular energy metabolism. If we consider the *in vitro* findings, ATP can inhibit phosphofructokinase; perhaps glycolysis might shut down and not provide energy if ATP is abundant. However, ATP can also alter the activity of phosphoglycerate kinase. This is a near-equilibrium enzyme. When we ponder the hundreds of kinases that would be subject to regulation should ATP really change in various conditions, it becomes clear that its role as a regulator is unlikely. Moreover, it removes our understanding of mobile cofactors by making them reaction regulators instead of communication molecules between routes of metabolism.

Much has been made of ATP inhibition of K channels. For example, in the pancreatic β cell, inhibition of this channel by ATP is taken as the answer to how glucose uptake leads to insulin release [108]. But if ATP is constant in cells, then other mechanisms must be contemplated. Continuing to hold a notion that is almost certainly incorrect is not a harmless error.[xvi] There are other hypotheses for how the K channel is inactivated in the β cell, including modification by the AMPK [109].

Suppose we subject ATP to the test of intracellular signal molecule:

1. *Source—Generation from an abundant precursor.* ATP is generated from a far *less* abundant molecule, ADP, continuously.
2. *Sink—breakdown to an abundant molecule or molecules.* ATP breaks down to far *less* abundant molecules, ADP, or AMP.
3. *Rapid destruction controlling its level.* Its level is constant.

Intracellular ATP is not likely a regulator. However, it is established that *extracellular* ATP can regulate cells, and at extremely low concentrations through binding to purinergic receptors [110]. It is likely that the origin of this ATP (which is effective as a hormone at micromolar concentrations) is the release of the nucleotide from damaged cells.

7.6.2 NAD

The notion that this cofactor could also be a regulator is more recent and suggested for only a few reactions, such as the poly ADP ribosylation enzyme PARP [111] and the NAD linked deactylase sirtuin [2]. The basis of the regulatory function is essentially the argument that NAD is a substrate in these reactions, and perhaps if it is altered in concentration it could change their enzymatic rates. Beyond this, it has been argued by Walsh that protein deacetylations requiring NAD must be regulatory since it would be chemically simpler to just use water as the substrate, as in other similar reactions [2]. This is similar to the argument that multifunctional proteins must be more efficient, since they can pass the product of one reaction directly to the next. In both cases, there are surely interesting reasons for the evolutionary choice; however, neither regulation nor efficiency is a likely explanation.

Like ATP, NAD fails to satisfy any of the three criteria as a regulator. Cells rapidly change their cytosolic redox state with the introduction of different substrates of different redox level. This happens over a time scale that is far different from the deacetylation reaction, which will enable activation of DNA segments for transcription. So, while it is true that such epigenetic modifications are more rapid than the lifetime of the cell, metabolic events are much more rapid.

NAD in the PARP reaction is used as a *source* metabolic pool, not a signal. This is similar to the use of ATP in the formation of cAMP. That may be of interest in drug development. Considering NAD as regulator may obscure our understanding of the true situation and impede such advances.

7.7 OXIDANTS AND CELLULAR REGULATION

Cellular oxidants have a strong effect on metabolic pathways, and this idea has led to many studies. The idea that these exert a normal metabolic control over pathways, however [112] represents an extreme. It is difficult to rationalize how oxidative damage can serve as a selective control system, or integrate with the large number of well-known control systems, some of which have been outlined here. A more prosaic view is that oxidative damage is simply the inevitable drawback of utilizing oxygen as the electron acceptor, and systems to minimize its effects are established. The response of cells to oxidative insult does profoundly alter cellular function and its metabolism (such as the pentose phosphate pathway [113]). However, this is beyond the scope of the present study.

REFERENCES

1. Lim, W., Mayer, B., and Pawson, T. (2015) *Cell Signaling: Principles and Mechanisms*, Garland Science, New York.
2. Walsh, C. (2006) *Posttranslational Modification of Proteins: Expanding Nature's Inventory*, Roberts and Co. Publishers, Englewood, CO.
3. Sutherland, E. W. (1972) Studies on the mechanism of hormone action. *Science* 177, 401–408.
4. Gilman, A. G. (1987) G proteins: Transducers of receptor-generated signals. *Annu. Rev. Biochem.* 56, 615–649.
5. Katritch, V., Cherezov, V., and Stevens, R. C. (2013) Structure-function of the G protein-coupled receptor superfamily. *Annu. Rev. Pharmacol. Toxicol.* 53, 531–556.
6. Woodard, G. E., Jardin, I., Berna-Erro, A., Salido, G. M., and Rosado, J. A. (2015) Regulators of G-protein-signaling proteins: Negative modulators of G-protein-coupled receptor signaling. *Int. Rev. Cell Molec. Biol.* 317, 97–183.
7. Cummings, F. W. (1975) A biochemical model of the circadian clock. *J. Theor. Biol.* 55, 455–470.
8. Schimke, R. T. (1970) Regulation of protein degradation in mammalian tissues. in *Mammalian Protein Metabolism, Vol. IV* (Munro, H. N. ed.), Academic Press, New York, pp. 177–228.
9. Swillens, S., Paiva, M., and Dumont, J. E. (1974) Consequences of the intracellular distribution of cyclic 3',5'-nucleotides phosphodiesterases. *FEBS Lett.* 49, 92–95.
10. Glass, D. B., and Krebs, E. G. (1980) Protein phosphorylation catalyzed by cyclic AMP-dependent and cyclic GMP-dependent protein kinases. *Annu. Rev. Pharmacol. Toxicol.* 20, 363–388.
11. Altarejos, J. Y., and Montminy, M. (2011) CREB and the CRTC co-activators: Sensors for hormonal and metabolic signals. *Nat. Rev. Mol. Cell. Biol.* 12, 141–151.
12. Dohm, G. L., Kasperek, G. J., and Barakat, H. A. (1985) Time course of changes in gluconeogenic enzyme activities during exercise and recovery. *Am. J. Physiol.* 249, E6–E11.
13. Liang, X., Liu, L., Fu, T., Zhou, Q., Zhou, D., Xiao, L., Liu, J., Kong, Y., Xie, H., Yi, F., Lai, L., Vega, R. B., Kelly, D. P., Smith, S. R., and Gan, Z. (2016) Exercise inducible lactate dehydrogenase B regulates mitochondrial function in skeletal muscle. *J. Biol. Chem.*
14. Pilkis, S. J., Claus, T. H., and el-Maghrabi, M. R. (1988) The role of cyclic AMP in rapid and long-term regulation of gluconeogenesis and glycolysis. *Adv. Second Messenger Phosphoprotein Res.* 22, 175–191.
15. Mapes, J. P., and Harris, R. A. (1976) Inhibition of gluconeogenesis and lactate formation from pyruvate by N O-dibutyryl adenosine 3':5'-monophosphate. *J. Biol. Chem.* 251, 6189–6196.

16. Exton, J. H. (1982) *Regulation of Carbohydrate Metabolism by Cyclic Nucleotides*, Springer-Verlag, Berlin.
17. Digby, G. J., Lober, R. M., Sethi, P. R., and Lambert, N. A. (2006) Some G protein heterotrimers physically dissociate in living cells. *Proc. Natl. Acad. Sci.* 103, 17789–17794.
18. Fain, J. N., and Garcia-Sainz, J. A. (1980) Role of phosphatidylinositol turnover in alpha 1 and of adenylate cyclase inhibition in alpha 2 effects of catecholamines. *Life Sci.* 26, 1183–1194.
19. Jakobs, K. H., Aktories, K., and Schultz, G. (1979) GTP-dependent inhibition of cardiac adenylate cyclase by muscarinic cholinergic agonists. *Naunyn-Schmiedeberg's Arch. Pharmacol.* 310, 113–119.
20. Logothetis, D. E., Kurachi, Y., Galper, J., Neer, E. J., and Clapham, D. E. (1987) The beta gamma subunits of GTP-binding proteins activate the muscarinic K^+ channel in heart. *Nature* 325, 321–326.
21. Huang, C. L., Feng, S., and Hilgemann, D. W. (1998) Direct activation of inward rectifier potassium channels by PIP2 and its stabilization by Gbetagamma. *Nature* 391, 803–806.
22. Keselman, I., Fribourg, M., Felsenfeld, D. P., and Logothetis, D. E. (2007) Mechanism of PLC-mediated Kir3 current inhibition. *Channels* 1, 113–123.
23. Augert, G., Boeckino, S. B., Blackmore, P. F., and Exton, J. H. (1989) Hormonal stimulation of diacylglycerol formation in hepatocytes. Evidence for phosphatidyl choline breakdown. *J. Biol. Chem.* 264, 21689–21698.
24. Berridge, M. J. (1987) Inositol trisphosphate and diacylglycerol: Two interacting second messengers. *Annu. Rev. Biochem.* 56, 159–193.
25. Carafoli, E. (2002) Calcium signaling: A tale for all seasons. *Proc. Natl. Acad. Sci. USA* 99, 1115–1122.
26. Clapham, D. E. (1995) Calcium signaling. *Cell* 80, 259–268.
27. Assimacopoulos-Jeannet, F., Blackmore, P. F., and Exton, J. H. (1977) Studies on the a-andrenergic activation of hepatic glucose output: Studies on the role of calcium in the activation of phosphorylase. *J. Biol. Chem.* 252, 2662–2669.
28. Denton, R. M., and McCormack, J. G. (1985) Ca transport by mammalian mitochondria and its role in hormone action. *Endocrinol. Metab.* 12, 543–554.
29. McCormack, J. G., and Denton, R. M. (1985) Hormonal control of intramitochondrial Ca^{2+}-sensitive enzymes in heart, liver and adipose tissue. *Biochem. Soc. Trans.* 13, 664–667.
30. Carafoli, E. (1991) Calcium Pump of the Plasma Membrane, *Physiol. Rev.* 71, 129–153.
31. Takemura, H., Thastrup, O., and Putney, J. W., Jr. (1990) Calcium efflux across the plasma membrane of rat parotid acinar cells is unaffected by receptor activation or by the microsomal calcium ATPase inhibitor, thapsigargin. *Cell Calcium* 11, 11–17.
32. Pozzan, T., Rizzuto, R., Volpe, P., and Meldolesi, J. (1994) Molecular and cellular physiology of intracellular calcium stores. *Physiol. Rev.* 74, 595–636.
33. Putney, J. W. (1997) Capacitative calcium entry. in *Capacitative Calcium Entry* (Putney, J. W. ed.), R.G. Landes Company, Austin, Texas, pp. 53–76.
34. Thastrup, O., Cullen, P. J., Drobak, B. K., Hanley, M. R., and Dawson, A. P. (1990) Thapsigargin, a tumor promoter, discharges intracellular Ca^{2+} stores by specific inhibition of the endoplasmic reticulum Ca^{2+}-ATPase. *Proc. Natl. Acad. Sci. USA* 87, 2466–2470.
35. Tsien, R. Y. (1999) *Monitoring Cell Calcium*, Oxford University Press, New York.
36. Putney, J. W. (1986) A model for receptor-regulated calcium entry. *Cell Calcium* 7, 1–12.
37. Soboloff, J., Spassova, M. A., Tang, X. D., Hewavitharana, T., Xu, W., and Gill, D. L. (2006) Orai1 and STIM reconstitute store-operated calcium channel function. *J. Biol. Chem.* 281, 20661–20665.

38. Putney, J. W. (2011) Origins of the concept of store-operated calcium entry. *Front. Biosci. (Schol. Ed.)* 3, 980–984.

39. Kurebayashi, N., and Ogawa, Y. (2001) Depletion of Ca^{2+} in the sarcoplasmic reticulum stimulates Ca^{2+} entry into mouse skeletal muscle fibres. *J. Physiol. (Lond.)* 553 (1), 185–199.

40. Wingertzahn, M. A., and Ochs, R. S. (2001) Changes in ryanodine receptor-mediated calcium release during skeletal muscle differentiation. II Resolution of caffeine-ryanodine paradox. *Exp. Biol. Med.* 226 (2), 119–126.

41. Dirksen, R. T. (2009) Checking your SOCCs and feet: The molecular mechanisms of Ca^{2+} entry in skeletal muscle. *J. Physiol.* 587, 3139–3147.

42. Somasundaram, A., Shum, A. K., McBride, H. J., Kessler, J. A., Feske, S., Miller, R. J., and Prakriya, M. (2014) Store-operated CRAC channels regulate gene expression and proliferation in neural progenitor cells. *J. Neurosci.* 34, 9107–9123.

43. Casteels, R., and Droogmans, G. (1981) Exchange characteristics of noradrenaline-sensitive calcium stores in vascular smooth muscle cells of rabbit ear artery. *J. Physiol.* 317, 263–279.

44. Maeda, S., and Tsukihara, T. (2011) Structure of the gap junction channel and its implications for its biological functions. *Cell. Mol. Life Sci.* 68, 1115–1129.

45. Sheridan, D. C., Takekura, H., Franzini-Armstrong, C., Beam, K. G., Allen, P. D., and Perez, C. F. (2006) Bidirectional signaling between calcium channels of skeletal muscle requires multiple direct and indirect interactions. *Proc. Natl. Acad. Sci.* 103, 19760–19765.

46. Narayanan, B., Islam, M. N., Bartlelt, D., and Ochs, R. S. (2003) A direct mass-action mechanism explains capacitative calcium entry in jurkat and skeletal L6 muscle cells. *J. Biol. Chem.*

47. Islam, M. N., and Ochs, R. S. (2006) A new hypothesis for Ca^{2+} flows in skeletal muscle and its implications for other cell types. *Cell Biochem. Biophys.* 44, 251–271.

48. Pitake, S., and Ochs, R. S. (2016) Membrane depolarization increases ryanodine sensitivity to Ca^{2+} release to the cytosol in L6 skeletal muscle cells: Implications for excitation-contraction coupling. *Exp. Biol. Med.* 241, 854–862.

49. Cully, T. R., and Launikonis, B. S. (2013) Store-operated Ca^{2+} entry is not required for store refilling in skeletal muscle. *Clin. Exp. Pharmacol. Physiol.* 40, 338–344.

50. Prakriya, M., and Lewis, R. S. (2015) Store-operated calcium channels. *Physiol. Rev.* 95, 1383–1436.

51. Endo, M. (2009) Calcium-induced calcium release in skeletal muscle. *Physiol. Rev.* 89, 1153–1176.

52. Endo, M. (1977) Calcium release from the sarcoplasmic reticulum. *Physiol. Rev.* 57, 71–108.

53. Clark, M. A., Littlejohn, D., Conway, T. M., Mong, S., Steiner, S., and Crooke, S. T. (1986) Leukotriene D4 treatment of bovine aortic endothelial cells and murine smooth muscle cells in culture results in an increase in phospholipase A_2 activity. *J. Biol. Chem.* 261, 10713–10718.

54. Endo, M. (1985) Calcium release from sarcoplasmic reticulum. *Curr. Top. Mem. Transport* 25, 181–230.

55. Wingertzahn, M. A., and Ochs, R. S. (1997) Calcium mediated proteolysis enhances calcium release in skinned L6 myotubes. *Recept. Signal Transduct.* 7, 221–230.

56. Schneider, M. F. (1994) Control of calcium release in functioning skeletal muscle fibers. *Annu. Rev. Physiol.* 56, 463–484.

57. Ochs, R. S. (1986) Inositol trisphosphate and muscle. *TIBS* 11, 388–389.

58. Cherednichenko, G., Hurne, A. M., Fessenden, J. D., Lee, E. H., Allen, P. D., Beam, K. G., and Pessah, I. N. (2004) Conformational activation of Ca^{2+} entry by depolarization of skeletal myotubes. *Proc. Natl. Acad. Sci. USA* 101, 15793–15798.

59. Ahern, C. A., Sheridan, D. C., Cheng, W., Mortenson, L., Nataraj, P., Allen, P., De Waard, M., and Coronado, R. (2003) Ca^{2+} current and charge movements in skeletal myotubes promoted by the beta-subunit of the dihydropyridine receptor in the absence of ryanodine receptor type 1. *Biophys. J.* 84, 942–959.
60. Franzini-Armstrong, C., and Jorgensen, A. O. (1994) Structure and development of E-C coupling units in skeletal muscle. *Annu. Rev. Physiol.* 56, 509–534.
61. Schneider, M. F. (1981) Membrane charge movement and depolarization–contraction coupling. *Annu. Rev. Physiol.* 43, 507–517.
62. Bidasee, K. R., and Besch, Jr. (1998) Structure–function relationships among ryanodine derivatives. Pyridyl ryanodine definitively separates activation potency from high affinity. *J. Biol. Chem.* 273, 12176–12186.
63. Martonosi, A., and Halpin, R. A. (1971) Sarcoplasmic reticulum. X. The protein composition of sarcoplasmic reticulum membranes. *Arch. Biochem. Biophys.* 144, 66–77.
64. Martonosi, A., and Halpin, R. A. (1972) Sarcoplasmic reticulum. XVII. The turnover of proteins and phospholipids in sarcoplasmic reticulum membranes. *Arch. Biochem. Biophys.* 152, 440–450.
65. Royer, L., and Rios, E. (2009) Deconstructing calsequestrin. Complex buffering in the calcium store of skeletal muscle. *J. Physiol. Lond.* 587, 3101–3111.
66. Denton, R. M., and McCormack, J. G. (1985) Physiological role of Ca^{2+} transport by mitochondria. *Nature* 315, 635.
67. Nakai, J., Tanabe, T., Konno, T., Adams, B., and Beam, K. G. (1998) Localization in the II-III loop of thedihydropyridine receptor of a sequence critical for excitation–contraction coupling. *J. Biol. Chem.* 273, 24983–24986.
68. Franzini-Armstrong, C., Protasi, F., and Ramesh, V. (1998) Comparative ultrastructure of Ca^{2+} release units in skeletal and cardiac muscle. *Ann. N.Y. Acad. Sci.* 853, 20–30.
69. Miller, D. J. (2004) Sydney Ringer; physiological saline, calcium and the contraction of the heart. *J. Physiol.* 555, 585–587.
70. Fabiato, A., and Fabiato, F. (1979) Calcium and cardiac excitation–contraction coupling. *Annu. Rev. Physiol.* 41, 473–484.
71. Radwanski, P. B., Belevych, A. E., Brunello, L., Carnes, C. A., and Gyorke, S. (2013) Store-dependent deactivation: Cooling the chain-reaction of myocardial calcium signaling. *J. Mol. Cell. Cardiol.* 58, 77–83.
72. Shannon, T. R., Bers, D. M., Blatter, L. A., and Niggli, E. (2005) Confocal imaging of CICR events from isolated and immobilized SR vesicles. *Cell Calcium* 38, 497–505.
73. Copello, J. A., Zima, A. V., Diaz-Sylvester, P. L., Fill, M., and Blatter, L. A. (2007) Ca^{2+} entry-independent effects of L-type Ca^{2+} channel modulators on Ca^{2+} sparks in ventricular myocytes. *Am. J. Physiol. Cell Physiol.* 292, C2129–C2140.
74. Wohlfart, B., and Noble, M. I. M. (1982) The cardiac excitation–contraction cycle. *Pharmacol. Ther.* 16, 1–43.
75. Gez, L. S., Hagalili, Y., Shainberg, A., and Atlas, D. (2012) Voltage-driven Ca^{2+} binding at the L-type Ca^{2+} channel triggers cardiac excitation–contraction coupling prior to Ca^{2+} influx. *Biochemistry (Mosc.)* 51, 9658–9666.
76. Atlas, D. (2013) The voltage-gated calcium channel functions as the molecular switch of synaptic transmission. *Annu. Rev. Biochem.* 82, 607–635.
77. Bolton, T. B., Prestwich, S. A., Zholos, A. V., and Gordienko, D. V. (1999) Excitation–contraction coupling in gastrointestinal and other smooth muscles. *Annu. Rev. Physiol.* 61, 85–115.
78. Fill, M., and Copello, J. A. (2002) Ryanodine receptor calcium release channels. *Physiol. Rev.* 82, 893–922.

79. Steinbeck, J. A., Henke, N., Opatz, J., Gruszczynska-Biegala, J., Schneider, L., Theiss, S., Hamacher, N., Steinfarz, B., Golz, S., Brustle, O., Kuznicki, J., and Methner, A. (2011) Store-operated calcium entry modulates neuronal network activity in a model of chronic epilepsy. *Exp. Neurol.* 232, 185–194.

80. Gemes, G., Bangaru, M. L., Wu, H. E., Tang, Q., Weihrauch, D., Koopmeiners, A. S., Cruikshank, J. M., Kwok, W. M., and Hogan, Q. H. (2011) Store-operated Ca^{2+} entry in sensory neurons: Functional role and the effect of painful nerve injury. *J. Neurosci.* 31, 3536–3549.

81. Lu, B., and Fivaz, M. (2016) Neuronal SOCE: Myth or reality? *Trends Cell. Biol.* 26, 890–893.

82. Kasvinsky, P. J. (1982) The effect of AMP on inhibition of muscle phosphorylase a by glucose derivatives. *J. Biol. Chem.* 257, 10805–10810.

83. Tornheim, K., and Lowenstein, J. M. (1976) Control of phosphofructokinase from rat skeletal muscle. Effects of fructose diphosphate, AMP, ATP, and citrate. *J. Biol. Chem.* 251, 7322–7328.

84. Dang, Q., Brown, B. S., Liu, Y., Rydzewski, R. M., Robinson, E. D., van Poelje, P. D., Reddy, M. R., and Erion, M. D. (2009) Fructose-1,6-bisphosphatase inhibitors. 1. Purine phosphonic acids as novel AMP mimics. *J. Med. Chem.* 52, 2880–2898.

85. Ronner, P., Friel, E., Czerniawski, K., and Frankle, S. (1999) Luminometric assays of ATP, phosphocreatine, and creatine for estimation of free ADP and free AMP. *Anal. Biochem.* 275, 208–216.

86. Gitomer, W. L., and Veech, R. L. (1986) The accumulation of pyrophosphate by rat hepatocytes. *Toxicol. Ind. Health* 2, 299–307.

87. Hardie, D. G., and Carling, D. (1997) The AMP-activated protein kinase: Fuel gauge of the mammalian cell? *Eur. J. Biochem.* 246, 259–273.

88. Carlson, C. A., and Kim, K. H. (1973) Regulation of hepatic acetyl coenzyme A carboxylase by phosphorylation and dephosphorylation. *J. Biol. Chem.* 248, 378–380.

89. Beg, Z. H., Allmann, D. W., and Gibson, D. M. (1973) Modulation of 3-hydroxy-3-methylglutaryl coenzyme A reductase activity with cAMP and wth protein fractions of rat liver cytosol. *Biochem. Biophys. Res. Commun.* 54, 1362–1369.

90. Hardie, D. G., and Hawley, S. A. (2001) AMP-activated protein kinase: The energy charge hypothesis revisited. *Bioessays* 23, 1112–1119.

91. Atkinson, D. E. (1977) *Cellular Energy Metabolism and Its Regulation*, Academic Press, New York.

92. Ouyang, J., Parakhia, R. A., and Ochs, R. S. (2011) Metformin activates AMP kinase through inhibition of AMP deaminase. *J. Biol. Chem.* 286, 1–11.

93. Davies, S. P., Helps, N. R., Cohen, P. T., and Hardie, D. G. (1995) 5'-AMP inhibits dephosphorylation, as well as promoting phosphorylation, of the AMP-activated protein kinase. Studies using bacterially expressed human protein phosphatase-2C alpha and native bovine protein phosphatase-2AC. *FEBS Lett.* 377, 421–425.

94. Hawley, S. A., Pan, D. A., Mustard, K. J., Ross, L., Bain, J., Edelman, A. M., Frenguelli, B. G., and Hardie, D. G. (2005) Calmodulin-dependent protein kinase kinase-beta is an alternative upstream kinase for AMP-activated protein kinase. *Cell Metab.* 2, 9–19.

95. Minokoshi, Y., Shiuchi, T., Lee, S., Suzuki, A., and Okamoto, S. (2008) Role of hypothalamic AMP-kinase in food intake regulation. *Nutrition* 24, 786–790.

96. Imai, K., Inukai, K., Ikegami, Y., Awata, T., and Katayama, S. (2006) LKB1, an upstream AMPK kinase, regulates glucose and lipid metabolism in cultured liver and muscle cells. *Biochem. Biophys. Res. Commun.* 351, 595–601.

97. Velasco, G., Geelen, M. J., and Guzman, M. (1997) Control of hepatic fatty acid oxidation by 5'-AMP-activated protein kinase involves a malonyl-CoA-dependent and a malonyl-CoA-independent mechanism. *Arch. Biochem. Biophys.* 337, 169–175.

98. Smith, A. C., Bruce, C. R., and Dyck, D. J. (2005) AMP kinase activation with AICAR further increases fatty acid oxidation and blunts triacylglycerol hydrolysis in contracting rat soleus muscle. *J. Physiol. Lond.* 565, 547–553.

99. Burcelin, R., Crivelli, V., Perrin, C., Da Costa, A., Mu, J., Kahn, B. B., Birnbaum, M. J., Kahn, C. R., Vollenweider, P., and Thorens, B. (2003) GLUT4, AMP kinase, but not the insulin receptor, are required for hepatoportal glucose sensor-stimulated muscle glucose utilization. *J. Clin. Invest.* 111, 1555.

100. Sylow, L., Moller, L. L., Kleinert, M., Richter, E. A., and Jensen, T. E. (2014) Rac1—A novel regulator of contraction-stimulated glucose uptake in skeletal muscle. *Exp. Physiol.* 99, 1574–1580.

101. Youn, J. H., Gulve, E. A., and Holloszy, J. O. (1991) Calcium stimulates glucose transport in skeletal muscle by a pathway independent of contraction. *Am. J. Physiol.* 260, C555–C561.

102. Balu, D., Ouyang, J., Parakhia, R.A., Pitake, S., Ochs, R.S. (2016) Ca^{2+} effects on glucose transport and fatty acid oxidation in L6 skeletal muscle cell cultures. *Biochem. Biophys. Rep.* 5, 365–373.

103. Kikkawa, U., Kishimoto, A., and Nishizuka, Y. (1989) The protein kinase C family: Heterogeneity and its implications. *Annu. Rev. Biochem.* 58, 31–44.

104. Bell, R. M., and Burns, D. J. (1991) Lipid activation of protein kinase C. *J. Biol. Chem.* 266, 4661–4664.

105. Conesa-Zamora, P., Lopez-Andreo, M. J., Gomez-Fernandez, J. C., and Corbalan-Garcia, S. (2001) Identification of the phosphatidylserine binding site in the C2 domain that is important for PKC alpha activation and in vivo cell localization. *Biochem* 40, 13898–13905.

106. Takai, Y., Kishimoto, A., Iwasa, Y., Kawahara, Y., Mori, T., and Nishizuka, Y. (1979) Calcium-dependent activation of a multifunctional protein kinase by membrane phospholipids. *J. Biol. Chem.* 254, 3692–3695.

107. Newton, A. C. (1997) Regulation of protein kinase C. *Curr. Opin. Cell Biol.* 9, 161–167.

108. Weiss, J. N., and Lamp, S. T. (1987) Glycolysis preferentially inhibits ATP sensitive K^+ channels in isolated guinea pig cardiac myocytes. *Science* 238, 67–69.

109. Andrey, S., Sofija, J., Qingyou, D., Grant, B., Allyson, K. C., Mei, S., Kei, S., Rong, T., and Aleksandar, J. (2006) AMP-activated protein kinase mediates preconditioning in cardiomyocytes by regulating activity and trafficking of sarcolemmal ATP-sensitive K^+ channels. *J. Cell. Physiol.* 210, 224–236.

110. Woo, K., Sathe, M., Kresge, C., Esser, V., Ueno, Y., Venter, J., Glaser, S. S., Alpini, G., and Feranchak, A. P. (2010) Adenosine triphosphate release and purinergic (P2) receptor-mediated secretion in small and large mouse cholangiocytes. *Hepatology* 52, 1819–1828.

111. Zhang, D. X., Zhang, J. P., Hu, J. Y., and Huang, Y. S. (2016) The potential regulatory roles of NAD(+) and its metabolism in autophagy. *Metabolism* 65, 454–462.

112. Chandel, N. S. (2015) *Navigating Metabolism*, Cold Spring Harbor Laboratory Press, Cold Spring Harbor, New York.

113. Stincone, A., Prigione, A., Cramer, T., Wamelink, M. M., Campbell, K., Cheung, E., Olin-Sandoval, V., Gruning, N. M., Kruger, A., Tauqeer Alam, M., Keller, M. A., Breitenbach, M., Brindle, K. M., Rabinowitz, J. D., and Ralser, M. (2015) The return of metabolism: Biochemistry and physiology of the pentose phosphate pathway. *Biol. Rev. Camb. Philos. Soc.* 90, 927–963.

ENDNOTES

i. The more recent studies on G proteins are largely structural, and while useful for drug design and mechanistic detail, is less directly informative for metabolic analysis.

ii. In earlier literature, this process of ER Ca^{2+} content controlling extracellular Ca^{2+} uptake was called *capacitive* Ca^{2+} uptake, drawing on the analogy of an electrical capacitor discharge initiating a current.

iii. While it is possible that CaATPase can be allosterically regulated as in the heart, there is no evidence to suggest that this occurs during SOCE.

iv. It is also possible that the ER entry channel is the IP_3R. However, all cells are known to also have RyR (albeit a separate isoform from excitable cells), and the RyR has been well established to enable Ca^{2+} flow in either direction [45]. This property is needed to explain Ca^{2+} entry into the ER during SOCE and release to the cytosol; these are opposing directions.

v. Yet another alternative view is that muscle in particular has no SOCE, based on the argument that use of CaATPase inhibitors is too extreme [49]. This is a return to the "nonexcitable cell only" notion, but with newer experiments. Yet, despite their insistence on more physiological models, avoiding extreme situations, they offered their own: a "skinned muscle" preparation, which removes the plasma membrane. While some issues of this study remain unsettled, in the absence of SOCE for any cell type, the overall flow of Ca^{2+} as a pathway is incomplete as the very small loss of Ca^{2+} with each contraction must be replaced.

vi. This is named as a description of the currents produced by the channel studied in isolation; these are *long* as opposed to other types, such as the T channel (*tiny*). The channel is also known as the L-channel or dihydropyridine receptor, in this case named because of certain dihydropyridine drugs that bind it. Formally, it is classified by a third name, $Ca_{v1.1}$.

vii. The drawing indicates two RyR for each L-channel, which is the approximate ratio in skeletal muscle; the reason for this intriguing result is unknown. Not shown in the figure is another ER vesicle also connected to the T tubule. It is commonly observed in electron micrograph sections that the T-tubule has contact with an ER on both sides; together this is called the *triad*.

viii. The terminology used here for membranes and spaces is the same for all cell types. For muscle, the ER is often referred to as the SR (sarcoplasmic reticulum), and the plasma membrane as the sarcolemma. In nerve, the plasma membrane is renamed as neurolemma. Fortunately, this practice has not spread to other cell types.

ix. This presumes that there is some Ca^{2+} to enter. As just noted, muscle contraction can proceed in the complete absence of extracellular Ca^{2+} for many contraction cycles. Early proponents of the hypothesis thus imagined that there are local pools of Ca^{2+} that provided the source; this has not been borne out.

x. The experiments presented here use patch clamp electrophoresis of L6 cells in "whole-cell" configuration. That means that a glass pipette of about 1 micron in diameter is placed on the cell membrane, sealed by suction, and then ruptured by increased suction to create an opening of the pipette interior into the cell interior. A controlled amplifier can apply constant voltage of various values and the current measured. By convention, inward current has a negative value.

xi. As mentioned above, Ca^{2+} itself rapidly inactivates the channel; this is a common feature of Ca^{2+} channels.

xii. The shaping of action potentials in heart is strongly affected by other channels, including multiple K channels specific to heart ventricular cells.

xiii. The existence of SOCE demonstrates that ER is the Ca^{2+} store, as this activity is used to restore the small amount of Ca^{2+} lost to the cell through the plasma membrane CaATPase. In fact, if extracellular Ca^{2+} could be a signal it would be directly exported through the plasma membrane CaATPase as well, with no utilization of the ER.

xiv. In an imaginative series of studies, the laboratory of Veech once proposed that mammalian pyrophosphatases might not be very active, and leave high concentrations of pyrophosphate to serve as phosphate bond energy donors, as they do in bacteria. However, this was ultimately rescinded [86] as in Section 2.11.

xv. Most studies of PKC in recent years focus on the large number of new isozymes that have been discovered, some of which do not display the dependence on some or even any of the activators of the older, "classical" enzymes. It remains unclear how some forms—particularly those which none of the activators—are controlled but may be important in non-metabolic roles, as some appear in the nucleus.

xvi. The notion of harmless error may seem overly philosophical, but the fact is there are such things. None of us, nor any finite group of us are omniscient. Thus, it is inevitable that in any field of investigation it is not possible for us to know even the most elementary ideas of certain fields that are likely to be relevant to our study. It is certain that we are entirely wrong about at least some aspects of our understanding. Despite that, we can make conclusions. This is akin to the legal use of "harmless error" not affecting the outcome of a decision. In a sense, the harmless error in scientific hypothesis is the obverse of Occam's Razor. In Occam's Razor, we are advised to take the simplest mechanism, since adding more notions to our construct obfuscates the result. In harmless error, we add extraneous notions that are without value to the conclusion.

8 Pathways and Energy Contributions

Cellular energy is a common thread in metabolism. It is a holistic view, a means of summarizing and organizing our ideas. Energy supply or demand is also sometimes invoked to suggest the significance of a process in the cell. For example, while protein synthesis is not usually considered to be a significant energy drain for the cell, it is very active in lymphocytes [1]. Still, most cell operations that involve activities such as transcription, translation, or acetylation operate over such a long time scale that they are insignificant in terms of the utilization of energy. This chapter will focus on a few issues of metabolism that connect to cellular energy.

8.1 CREATINE PHOSPHOKINASE SYSTEM

One way of summarizing cellular energy pathways is to list three means of ATP production:

1. Creatine phosphokinase
2. Glycolysis
3. Krebs cycle

They can be ranked: Creatine phosphokinase is very rapid, supplying ATP within seconds, from a single pool somewhat larger than ATP itself. Next is the slower glycolytic pathway, and finally the much slower Krebs cycle, coupled to oxidative phosphorylation.

This summary must be tempered by the fact that creatine phosphokinase (CPK) is present only in a few cells, such as skeletal muscle, heart, and nerve. More significantly, this single reaction is not really an energy supply pathway: unlike glycolysis or the Krebs cycle, it does not draw energy from nutrients, but rather stores a modest amount of bond energy under conditions of rest. It is best viewed as an "ATP buffer" to ensure a constant concentration of ATP.

The formation of creatine involves multiple organs [2]. Most of the compound is synthesized by liver, or ingested in the diet, transported to muscle (where most of it is found), and finally degraded to creatinine and excreted by the kidney (Figure 8.1). The precursor of creatinine is phosphocreatine [3], so this excretion product is formed at a higher rate in the resting condition, as this is the state of higher concentrations of phosphocreatine.

Most CPK activity is cytosolic. There is an isozyme of CPK in the mitochondria, although its function is unclear, as there is no need for buffering within the mitochondria [2]. Nonetheless, Miller et al. [4] expressed this isoform in the liver, a tissue that normally has no CPK. While this enzyme does not normally exist in

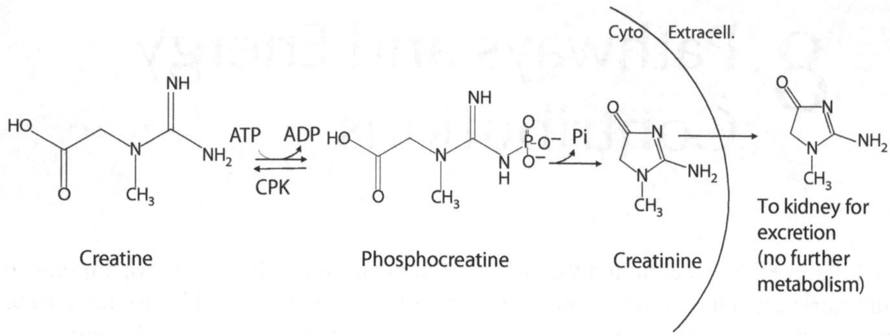

FIGURE 8.1 Creatine and creatine phosphate metabolism.

liver, one study suggested a form of this enzyme may appear in hepatic cancer [5]. An experimental use of hepatocytes expressing cytosolic CPK was to test the notion that hepatic ADP concentration was mostly bound. The investigators confirmed this, finding a free concentration to be in the range of the tens of micromolar [6]. This is also known from nmr measurements [7]; this puts the resulting ATP/ADP ratio at about 100.

8.2 USING THE CREATINE PHOSPHATE REACTION *IN VIVO* AND IN CELLS: CREATINE PHOSPHATE RECOVERY

A method that takes advantage of the creatine phosphokinase equilibrium in muscle has been used *in vivo* to measure not simply the concentration of ATP, but rather the dynamics of cellular ATP production. This is known as *phosphocreatine recovery* [8–10].

The idea behind this approach is that the CPK equilibrium is shifted in the direction of creatine during energy depletion, and then towards creatine phosphate following energy restoration. Creatine phosphate is measured *in vivo* using ^{31}P-NMR. Energy depletion is established with an arm cuff; energy restoration is enabled by simply removing the cuff. As blood flow returns to the muscle tissue, the time course of appearance of the creatine phosphate resonance is taken as the formation of ATP by mitochondria. This conceptually simple experimental is thus an *in vivo* measurement of mitochondrial ATP synthesis.

There are a few assumptions made in this measurement system. First, the NMR system can measure only creatine-P, and not creatine. The assumption is that total creatine pool is constant. While this is probably based on the concept that creatinine production by the body is relatively constant, it is not necessarily the case that this is true for the specific capillary bed in question. In particular, since the creatinine arises only from creatine phosphate, it is possible that some error is introduced in assuming a constant total creatine pool over the time course of the measurement. A second assumption is that glycolysis contributes negligibly to ATP formation. This one was assessed in the cited study by measuring lactate production across the muscles of the forearm. This was shown not to account for the bulk of the ATP formation, which is reasonable considering that the stimulus is reoxygenation.

We adapted the system used for *in vivo* study of isolated cell cultures [11]. In order to do so, a substitution was needed for anoxia as this is experimentally cumbersome in cell culture (rapid switching to oxygen-replete conditions is difficult to achieve). Instead, we used electron transfer inhibition (with azide, the complex IV inhibitor) or uncoupling (with DNP, dinitrophenol). Control experiments were necessary to show that the perturbants could block the flow of energy temporarily but not destroy the integrity of the cells. This would reflect the situation that occurs in the arm cuff *in vivo* experiment. We found conditions for which about 15 min of incubation of inhibitor or uncoupler (depending on concentration) was sufficient to lower the total ATP temporarily, but enable its restoration when the media was changed to remove the agents.

Of key importance in these experiments are the tests to ensure the integrity of the cell. This is the most fundamental issue of cellular energy evaluation: the survival of the cell. There were surprises in store.

We used three assays of cellular vitality:

1. Total ATP content
2. Trypan blue
3. MTT

ATP content is a sensitive method of analysis for cell viability, as demonstrated for isolated hepatocytes [12]. As mentioned in prior chapters of this work, the level of ATP is constant in cells, although the value of total ATP is known to vary between different cells.

We found, however, that ATP concentration could drop in cell culture, and did so over the period of time that an agent interrupting mitochondrial energy formation was present. This means that the measurement of total ATP as a reporter of cell viability is not absolute: under some conditions, at least as observed in isolated cell cultures, the ATP level can decrease. The reason for the drop is likely that the interruption of electron flow by azide, or the uncoupling of oxidative phosphorylation by DNP prevents the mitochondria from converting ADP to ATP in sufficient quantities to maintain a resting ATP level. It is not known if there is a physiological correlate to these experiments, although it is established that there are physiological states of uncoupling, as there is a known family of uncoupling proteins in some cells (such as the brown fat cell [13,14]). Still, in the context of the experiment we are interested in here, it is clear that the viability of the cell is maintained as the ATP level is restored once the inhibitor or uncoupler is removed. This provides a caveat to the use of ATP as a reporter of cell viability.

Trypan blue, an example of dye exclusion analysis for cell viability, was unaffected by the temporary presence of energy interruption; it alone correctly reported ongoing cell viability. The disadvantage of using Trypan blue (and similar agents that can enter only cells with incomplete cell membrane barriers) is that it is less sensitive than other measurements.

Finally, the results using the MTT assay were unusual. This agent does not actually measure the viability of cells, but rather is a measurement of mitochondrial electron flow, as this is a tetrazolium salt that accepts electrons from cytochrome c of the respiratory chain [15]. The tacit assumption of users of this method is that it reports functioning mitochondria and thus reflects cell viability. In many cases, this is indeed the case. However, we found conditions in which cell viability is not associated with electron transfer to MTT.

First, the MTT assay showed a decrease when energy interruption was imposed [11]. This follows from the experimental protocol, but it also means that it no longer tracks cell viability. The problem is the same as that of total cellular ATP: in the initial period, ATP also decreased, but could return to the initial level if the perturbants were removed after a brief exposure. MTT reduction also returned to normal, and, as pointed out above, trypan blue exclusion was unaffected. However, a second result was perhaps even more surprising. We were investigating the action of the drug metformin on mitochondrial respiration in intact cells (the reason for developing this system in the first place). We found that phosphocreatine recovery in the cell cultures corresponds to an increased MTT reduction. Interpreted as a measure of living cell numbers, it would have to mean that the cells were more than 100% viable. Clearly, MTT instead is measuring an increase of electron flow that is greater than baseline (i.e., in the absence of metformin). We also found that metformin stimulates ATP formation by mitochondria in the phosphocreatine recovery assay, which led us to conclude that metformin activates mitochondria—the opposite of a general view that metformin inhibits this process. I will discuss metformin itself in more detail in Chapter 10.

8.3   ENERGY ESTIMATES USING ISOLATED MITOCHONDRIA

Studies of isolated mitochondria between the middle to late decades of the twentieth century[i] culminated in the Mitchell hypothesis, the mechanism by which mitochondria produce ATP. This hypothesis, now universally accepted, is that a proton gradient formed by the respiratory chain across the inner mitochondrial membrane is the "energy intermediate" for the formation of ATP, achieved by its re-entry through the ATP synthase complex [18–20]. Generally speaking, these experiments determined the *potential* for energy production, and did not purport to estimate the actual amount of energy that cells made under specific circumstances based upon this subcellular fragment.

In the 1980s, some studies of hormone action were conducted by injecting animals with drugs or hormones, followed by isolation from a target tissue—often liver, for which mitochondrial preparation is technically simple. Jensen et al. [21] summarized various investigations of their lab and others on metabolic changes evident in isolated mitochondria following *in vivo* injection of rats with the hormones cortisol, glucagon, and epinephrine. One consistent "stable change" (i.e., an alteration that persisted after the experimental procedure) was a rise in mitochondrial ATP content. The complication in such investigations is that very rapid exchanges across membranes occur once cells have been opened, so that the results are a very indirect reflection of metabolic events. While it suggests some energetic consequence of these hormones, the connections remain unclear.

Similar complications occur when isolated cells are treated with detergent to assess respiration (e.g., ref. [22]). This does not mean that the isolated mitochondria model is not useful. For example, Davis and coworkers [23,24] showed that isolated mitochondria in the presence of an artificial "cell demand" established by titration with purified exogenous F_1ATPase stimulate respiration over a range of ATP/ADP ratios that are very close to those determined by the near-equilibrium indicator methods of adenylate kinase and creatine phosphate kinase discussed above, that is, on the order of 50–100. This puts free ADP at about 1% of the total concentration, in close agreement with other estimates. What such experiments cannot do is provide steady state estimates of actual energy utilization in intact cells.

8.4 ENERGY ESTIMATES BY DIRECT OXYGEN CONSUMPTION IN INTACT CELLS

One means of providing energy utilization estimates in intact cells is the analysis of oxygen consumption in intact cells. This is not in itself a new idea; the standard oxygen (Clark) electrode has been used with cell suspensions in the past. The introduction of the Seahorse instrument, however, extends measurements to cell cultures; this was discussed in Chapter 3.

The use of the oxygen analytic device to measure respiration was extended by a protocol developed by Nicholls (author with Ferguson of the monograph *Bioenergetics* [25]), in which a series of inhibitors and uncouplers are added and the oxygen consumption given interpretations. A typical trace of oxygen formation (OCR, or oxygen consumption rate) is displayed next to a sketch of expected results from isolated mitochondria in Figure 8.2. In the mitochondrial trace (Figure 8.2a), oxygen consumption is related to the slope of the lines, as the ordinate is the oxygen concentration (the maximum amount is just the amount of oxygen saturating the buffer; the minimum is established by adding the reductant dithionite at the end of the experiment). The experiment is usually conducted over a period of about 5–10 min, far shorter than the OCR assay (Figure 8.2b). In the case of mitochondria, the first addition represents the mitochondria, and media containing the substrates, buffer, and inorganic phosphate. Two additions of limiting amounts of ADP are added; in each case, respiration increases and then levels off. As discussed in Chapter 3, the faster rate is the "state 3" rate, and the slower, the "state 4." After addition of oligomycin (the F_1F_oATP synthetase inhibitor), further addition of ADP has no effect. Uncoupler (F, for FCCP) rapidly releases respiration, which is halted by addition of rotenone (R).

The OCR protocol is different, as no direct ADP can be added (as a charged hydrophilic molecule cannot penetrate the plasma membrane), nor can intracellular substrate concentrations be precisely controlled. Each point in the figure represents a rate of oxygen utilization measured over a period of several seconds. The initial, baseline rate is first lowered by the addition of oligomycin. To replicate that in the isolated mitochondria, it would be necessary to add oligomycin before the mitochondria had depleted its ADP. In cells in a steady state, the ADP is at a constant level. Subtraction of the oligomycin rate from the baseline estimates cell respiration. After uncoupler addition (F), a high rate of respiration ensues, followed by rotenone to reduce the OCR to effectively zero. Note in this case, antimycin A (AA) caused a slight further decrease. While the uncoupler-induced rate is suggested to be taken as the "maximum respiration," it is greater than the maximal respiration that can be achieved with ADP-released respiration. In studies of isolated mitochondria, the uncoupler-stimulated rate is far greater than that achieved by maximum ADP levels. While it could be argued that this might apply in cells that contain uncoupling protein, this is not the usual situation. A further distinction is that intact cells have a glycolytic pathway that is operating at the same

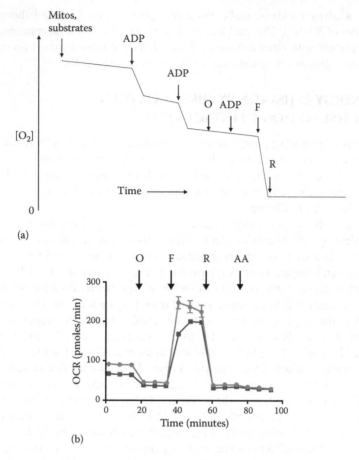

(a)

(b)

FIGURE 8.2 Mitochondrial and intact cell respiration. Oxygen consumption records of (a) isolated mitochondria and (b) cell cultures. (a) Mitochondrial respiration is measured over a time course of several minutes using a Clark electrode. Additions are ADP, oligoymcin (O), FCCP (F), and rotenone (R). (b) Cell cultures (circles, control; squares, in the presence of 10-mM metformin) were incubated in a Seahorse instrument. Abbreviations for additions are the same as in a except for antimycin A (AA).

time, and its alterations (which can also be assessed at least indirectly by simultaneous pH measurement) are influenced by alterations in the mitochondria.

A separate complication is the lack of bicarbonate in the medium, which means cells will be deprived of any carboxylation reactions which typically have very high, millimolar Km values for HCO_3^-. This can be serious in certain types of cells, such as kidney and liver, in which pyruvate carboxylase is essential for the continuous replenishment of the Krebs cycle. On the other hand, other cells are not greatly affected; for example, we have found no difference if bicarbonate is added in muscle L6 cell cultures, although the issue has not been thoroughly investigated. The technique is in use in an increasing number of laboratories; many of these are archived by the company manufacturing the instrument itself [26].

As an example of using oxygen consumption as an index of cellular energy, Princiotta et al. [1] performed an energy accounting in a fibroblast cell line. The authors found a decrease in the presence of cycloheximide of about one third of the total oxygen consumption. Based on their finding that oligomycin caused a 75% decrease in oxygen consumption, they concluded that total energy consumption devoted to protein synthesis (based on cycloheximide) is 45%.

On the other hand, cycloheximide had no measurable effect at all on liver slices despite an 84% inhibition of protein synthesis [27]. It is possible that the extent of protein synthesis varies with cells; liver protein synthesis is a very small portion of energy utilization, whereas in fibroblasts, and likely immune cells, it is a major one. In a similar way, the amount of energy contributed by the Na pump in response to thyroid hormone was demonstrated to vary with cell type, being considerable in liver and kidney and marginal in cerebellum [28].

Statements on the percentage of energy utilization based solely on an inhibitor result can also be misleading: it is not always clear that inhibition of the target leaves all other pathways unaffected. It is often the case that conclusions on energy utilization are based on intuition rather than any measurement at all. For example, a recent book on cell signaling [29] states that protein synthesis and transcription are "energetically costly," comparing these processes to the "energetically cheap" processes of short-term regulation (p. 9 and beyond). This idea is probably derived from considering the number of ATP molecules used in forming a protein, but not taking into account the rates of protein synthesis compared to the rates of metabolic pathways. Measurement of energy utilization, while imperfect, it at least an indication of the overall rates.

8.5   OVERLAP OF ENERGY AND SIGNALING

We generally assume that signaling systems must not use a significant amount of cellular energy. If they did, the signal itself would be a major consumer of cell demand, and it in turn would have to be regulated! Still, there are some exceptions. For example, we can anticipate that a major portion of metabolism of nerve cells and glia is devoted to neurotransmitters, such as glutamate formation [30]. In some cases, enzymes that are major components of metabolic pathways in one cell are used as signaling pathways in others, such as the enzyme acetyl CoA carboxylase. This is important for fatty acid biosynthesis in liver and fat cells, but exists for the purpose of forming the signal malonyl CoA in muscle [31,32] and certain neurons [33] just to regulate fatty acid oxidation.

One other curious overlap of signaling and metabolic activity is found in the beta cells of the pancreas, devoted to the secretion of insulin. It is the metabolism of glucose (and other metabolites) that leads to the triggering of insulin release. The ratio of ATP/ADP can be monitored by a genetically inserted fluorescent protein and it is evident that a strong correlation between metabolic demand, K channel, and insulin release exists [34,35]. What is distinct in this cell type is that the mobile cofactor turnover (ATP and ADP) is not merely the intermediate that is important to supply cell synthesis, but that it is key to the very function of the cell: to release insulin. While the mechanism is not fully elucidated, it centers on the depolarization of the cell by inhibiting a K channel, leading to vesicular fusion and release of insulin. What is clear is that the K channel—the ATP-dependent K channel—is not regulated by ATP *per se* but by the metabolism of the cell itself, directed by external nutrient supply.

8.6 CATABOLISM AND ANABOLISM RECONSIDERED

The terms catabolism and anabolism are useful classifications in broadly dividing metabolic sequences between breaking down nutrients for their energy (catabolism) and building or maintaining basic cell structures (anabolism). To put it more succinctly:

$$\text{Catabolism} \rightarrow \text{ATP} \rightarrow \text{Anabolism}$$

Still, it is also clear that this pair of terms is more of a convenience than an absolute rule. For example, fatty acid biosynthesis would seem to be clearly classified as anabolic. Yet, Flatt [36] showed in 1970 that fatty acid biosynthesis in adipocytes is an energy yielding rather than an energy utilizing pathway. This is based upon analysis of converting glucose to fatty acids, and considering the amount of NADPH that can be formed from the pentose phosphate shunt and the malic enzyme step. Since the conversion also involves NADH formation as pyruvate is converted to acetyl CoA in the mitochondria, the overall balance of the pathway is an excess of reducing equivalents beyond those required for the synthesis of fatty acids. Flatt further proposed that this pathway is "self-limiting," as it produces, rather than utilizes, energy during biosynthesis. This is another way of saying that if there is no other means of utilizing mobile cofactors (NADH, ATP), the pathway cannot proceed.

It should also be noted that the pathway of gluconeogenesis and ureogenesis from glutamine (Chapter 6) produces much of the energy needed by virtue of the fact that carbon from glutamine traverses part of the Krebs cycle between α-ketoglutarate and oxaloacetate (Figure 6.15). Other tissues can of course utilize glutamine as an energy source, such as the intestinal epithelia [37].

8.7 ρ^0 CELL: A CELL ENGINEERED TO HAVE NO MITOCHONDRIAL ENERGY FORMATION

As a final issue concerning energy in metabolism, consider the ρ^0 cell, a cell engineered to have no respiratory chain by sustained incubation with ethidium bromide [38]. This causes a destruction of mitochondrial DNA, which codes for several subunits of the mitochondrial respiratory chain. These cells nonetheless maintain a membrane potential, which is established by the reverse direction of the F_0F_1ATP synthetase [39]. It is of interest that these mitochondria are able to import ATP for intramitochondrial hydrolysis. The proteins of the inner membrane have lost some of the mitochondrially encoded subunits. While the F_0F_1ATP synthetase and the adenine nucleotide translocase have activity, it is in the opposite direction from normal mitochondria. The pathway for establishing a membrane potential in these cells is illustrated in Figure 8.3. It was estimated that about 10% of the glycolytic flux in these cells was needed to produce the ATP that is used to drive the membrane potential [39]. There has been further interest in this model, including speculation on why the cells maintain a membrane potential at all. One study suggests that membrane potential is a "reliable indicator" of mitochondrial function [40]. This seems unlikely; after all, these mitochondria aren't functioning at all, as they are

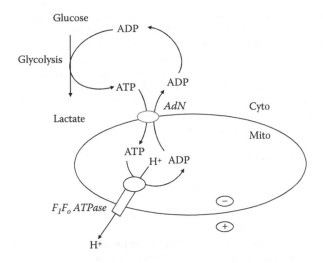

FIGURE 8.3 Membrane potential development in Rho-zero cells. Cells deprived of fully functioning mitochondria by selection after ethidium bromide treatment maintain a membrane potential by supplying ATP through glycolysis to the mitochondria. The activity of the adenine nucleotide translocase (AdN) runs in reverse to that of normal cells due to loss of some of the subunits of the complex. Similarly, the exit of protons through F_1F_0ATPase is possible because some of those proteins are also lost in the modified cells. The amount of membrane potential is less than half that achieved in normal cells, but enough to allow cell survival.

consuming rather than forming ATP. A simple explanation for why these cells maintain a membrane potential may simply be the fact that this enables cells to survive the ethidium bromide selection, providing a means to import nuclear proteins and establish pathways essential to cell survival. While the ρ^0 cell is an intriguing model for some purposes, such as genetics or mitochondrial dynamics, its connection to metabolism may well be more significant in the analysis of disease states [41].

REFERENCES

1. Princiotta, M. F., Finzi, D., Qian, S.-B., Gibbs, J., Schuchmann, S., Buttgereit, F., Bennink, J. R., and Yewdell, J. W. (2003) Quantitating protein synthesis, degradation, and endogenous antigen processing. *Immunity* 18, 343–354.
2. Wyss, M., and Kaddurah-Daouk, R. (2000) Creatine and creatinine metabolism. *Physiol. Rev.* 80, 1107–1213.
3. Iyengar, M. R., Coleman, D. W., and Butler, T. M. (1985) Phosphocreatinine, a high-energy phosphate in muscle, spontaneously forms phosphocreatine and creatinine under physiological conditions. *J. Biol. Chem.* 260, 7562–7567.
4. Miller, K., Sharer, K., Suhan, J., and Koretsky, A. P. (1997) Expression of functional mitochondrial creatine kinase in liver of transgenic mice. *Am. J. Physiol. Cell Physiol.* 272, C1193–C1202.
5. Enooku, K., Nakagawa, H., Soroida, Y., Ohkawa, R., Kageyama, Y., Uranbileg, B., Watanabe, N., Tateishi, R., Yoshida, H., Koike, K., Yatomi, Y., and Ikeda, H. (2014) Increased serum mitochondrial creatine kinase activity as a risk for hepatocarcinogenesis in chronic hepatitis C patients. *Int. J. Cancer* 135, 871–879.

6. Brosnan, M., Chen, L., Wheeler, C., van Dyke, T., and Koretsky, A. (1991) Phosphocreatine protects ATP from a fructose load in transgenic mouse liver expressing creatine kinase. *Am. J. Physiol.* 260, C1191–C1200.

7. Jeneson, J. A., Westerhoff, H. V., and Kushmerick, M. J. (2000) A metabolic control analysis of kinetic controls in ATP free energy metabolism in contracting skeletal muscle. *Am. J. Physiol. Cell Physiology* 279, C813–C832.

8. Walter, G., Vandenborne, K., McCully, K. K., and Leigh, J. S. (1997) Noninvasive measurement of phosphocreatine recovery kinetics in single human muscles. *Am. J. Physiol. Cell Physiol.* 272, C525–C534.

9. Haseler, L. J., Hogan, M. C., and Richardson, R. S. (1999) Skeletal muscle phosphocreatine recovery in exercise-trained humans is dependent on O2availability. *J. Appl. Physiol.* 86, 2013–2018.

10. Forbes, S. C., Paganini, A. T., Slade, J. M., Towse, T. F., and Meyer, R. A. (2009) Phosphocreatine recovery kinetics following low- and high-intensity exercise in human triceps surae and rat posterior hindlimb muscles. *Am. J. Physiol. Regul. Integr. Comp. Physiol.* 296, R161–R170.

11. Vytla, V. S., and Ochs, R. S. (2013) Metformin increases mitochondrial energy formation in L6 muscle cell cultures. *J. Biol. Chem.* 288, 20369–20377.

12. Cornell, N. W. (1983) Evaluation of hepatocyte quality: Cell integrity and metabolic rates. in *Isolation, Characterization, and Use of Hepatocytes* (Harris, R. A., and Cornell, N. W. eds.), Elsevier Biomedical, New York, pp. 11–20.

13. Keipert, S., and Jastroch, M. (2014) Brite/beige fat and UCP1—Is it thermogenesis? *Biochim. Biophys. Acta* 1837, 1075–1082.

14. Virtanen, K. A., Lidell, M. E., Orava, J., Heglind, M., Westergren, R., Niemi, T., Taittonen, M., Laine, J., Savisto, N. J., Enerback, S., and Nuutila, P. (2009) Functional brown adipose tissue in healthy adults. *N. Engl. J. Med.* 360, 1518–1525.

15. Vistica, D. T., Skehan, P., Scudiero, D., Monks, A., Pittman, A., and Boyd, M. R. (1991) Tetrazolium-based assays for cellular viability: A critical examination of selected parameters affecting formazan production. *Cancer Res.* 51, 2515–2520.

16. Racker, E. (1976) *A New Look at Mechanisms in Bioenergetics*, Academic Press, New York.

17. Blankenship, R. E. (2014) *Molecular Mechanisms of Photosynthesis*, 2nd edition. John Wiley and Sons, West Sussex, UK.

18. Boyer, P. D., Chance, B., Ernster, L., Mitchell, P., Racker, E., and Slater, E. C. (1977) Oxidative phosphorylation and photophosphorylation. *Annu. Rev. Biochem.* 46, 955–966.

19. Skulachev, V. P., Hinkle, P. C., and Mitchell, P. D. (1981) *Chemiosmotic Proton Circuits in Biological Membranes*, Addison-Wesley, Advanced Book Program/World Science Division, Reading, MA.

20. Slater, E. C. (1981) *Chemiosmotic Proton Circuits in Biological Membranes: In Honor of Peter Mitchell*, Addison-Wesley, Advanced Book Program/World Science Division, Reading, MA.

21. Jenson, C. B., Sistare, F. D., Hamman, H. C., and Haynes, R. C. (1983) Stimulation of mitochondrial functions by glucagon treatment. Evidence that effects are not artifacts of mitochondrial isolation. *Biochem. J.* 210, 819–827.

22. Owen, M. R., Doran, E., and Halestrap, A. P. (2000) Evidence that metformin exerts its anti-diabetic effects through inhibition of complex 1 of the mitochondrial respiratory chain. *Biochem. J.* 348 Pt 3, 607–614.

23. Davis, E. J., Lumeng, L., and Bottoms, D. (1974) On the relationships between the stoichiometry of oxidative phosphorylation and the phosphorylation potential of rat liver mitochondria as functions of respiratory state. *FEBS Lett.* 39, 9–12.

24. Davis, E. J., and Lumeng, L. (1975) Relationships between the phosphorylation poten-
    tials generated by liver mitochondria and respiratory state under conditions of adenos-
    ine diphosphate control. *J. Biol. Chem.* 250, 2275–2282.
25. Nicholls, D. G., and Ferguson, S. J. (2013) *Bioenergetics.* 4th edition. Ed.
26. Technologies, A. (2016) Publications with Seahorse XF Data.
27. Ismail-Beigi, F., Dietz, T., and Edelman, I. S. (1976) Thyroid thermogenesis: Minimal
    contribution of energy requirement for protein synthesis. *Mol. Cell. Endocrinol.*
    5, 19–22.
28. Ismail-Beigi, F., and Edelman, I. S. (1971) The mechanism of the calorigenic action of
    thyroid hormone. Stimulation of Na plus + K plus-activated adenosinetriphosphatase
    activity. *J. Gen. Physiol.* 57, 710–722.
29. Lim, W., Mayer, B., and Pawson, T. (2016) *Cell Signaling: Principles and Mechanisms.*
30. Westergaard, N., Sonnewald, U., and Schousboe, A. (1995) Metabolic traffick-
    ing between neurons and astrocytes: The glutamate/glutamine cycle revisited. *Dev.
    Neurosci.* 17, 203–211.
31. Spriet, L. L. (1998) Regulation of fat/carbohydrate interaction in human skeletal muscle
    during exercise. *Adv. Exp. Med. Biol.* 441, 249–261.
32. Saggerson, D. (2008) Malonyl-CoA, a key signaling molecule in mammalian cells.
    Annu. Rev. Nutr. 28, 253–272.
33. Lopaschuk, G. D., Ussher, J. R., and Jaswal, J. S. (2010) Targeting intermediary metab-
    olism in the hypothalamus as a mechanism to regulate appetite. *Pharmacol. Rev.* 62,
    237–264.
34. Tarasov, A., Dusonchet, J., and Ashcroft, F. (2004) Metabolic regulation of the pancre-
    atic beta-cell ATP- sensitive K^+ channel - A pas de deux. *Diabetes* 53, S113–S122.
35. Tarasov, A. I., and Rutter, G. A. (2014)—Use of genetically encoded sensors to monitor
    cytosolic ATP/ADP ratio in living cells. Chapter 15. in *Methods Enzymology* (Lorenzo,
    G., and Guido, K. eds.), Academic Press, pp. 289–311.
36. Flatt, J. P. (1970) Conversion of carbohydrate to fat in adipose tissue: An energy-
    yielding and therefore self-limiting process. *J. Lipid Res.* 11, 131–143.
37. Martin, G., Ferrier, B., Conjard, A., Martin, M., Nazaret, R., Boghossian, M., Saadé, F.,
    Mancuso, C., Durozard, D., and Baverel, G. (2007) Glutamine gluconeogenesis in the
    small intestine of 72 h-fasted adult rats is undetectable. *Biochem. J.* 401, 465–473.
38. Morais, R., Zinkewich-Peotti, K., Parent, M., Wang, H., Babai, F., and Zollinger, M.
    (1994) Tumor-forming ability in athymic nude mice of human cell lines devoid of mito-
    chondrial DNA. *Cancer Res.* 54, 3889–3896.
39. Buchet, K., and Godinot, C. (1998) Functional F1-ATPase essential in maintaining
    growth and membrane potential of human mitochondrial DNA-depleted rho degrees
    cells. *J. Biol. Chem.* 273, 22983–22989.
40. Bagkos, G., Koufopoulos, K., and Piperi, C. (2014) A new model for mitochondrial
    membrane potential production and storage. *Med. Hypotheses* 83, 175–181.
41. Cook, C. C., and Higuchi, M. (2012) The awakening of an advanced malignant can-
    cer: An insult to the mitochondrial genome. *Biochim. Biophys. Acta-Gen. Subj.* 1820,
    652–662.

ENDNOTE

i. These were years of intense study in which many biochemists were working on the
   goal of mitochondrial energy, best summarized by Racker [16]: "Anyone who is not
   thoroughly confused just doesn't understand the problem"[17].

9 Computer Modeling Studies in Metabolism

Two types of metabolic analysis have been performed with modern computational methods. The first originated with the mapping of metabolic pathways, in the form of wall charts and book compilations (e.g., ref. [1]). These have progressed to viewable PDF files [2] and several searchable databases. The second is modeling and flux analysis. In more recent years, measurement of large numbers of metabolites involving increasing technical sophistication has created a new approach: metabolomics. In this chapter I will provide a summary of these topics, which hold the potential to greatly expand both our understanding of metabolism and future approaches to it.

9.1 DATABASES

The conversion of metabolic pathway representation from chart to computer form has led to several separate databases, mostly in the public domain. In the early days of this effort, metabolic databases were scarce as most were developing representations of molecular biology. For example, a conference in 2000 had a section on metabolism, which was represented by just one paper describing a metabolic database among about 70 others on the topics such as genomics, evolution, transcriptional signaling, inborn errors of metabolism, polymorphisms, and cell cycles [3]. Subsequently, several databases containing metabolic information have emerged. The original dataset of enzymatic reactions for most of these was derived from the Enzyme Commission listings first compiled by Bairoch [4]. Some of these databases are listed in Table 9.1.

While they are not the only metabolic compilations in existence, the list includes some of the largest (KEGG), and illustrates the fact that many are specialized databases, such as for a particular species (E. Coli) or metabolite type (lipids). The major use of metabolic databases has been a lookup function, either to assign genetic function, or to collect specific information about a reaction. For specific communities, for example, Drosophila researchers or those interested in specific bacteria, the idea of a specialized database has clear value. What is missing is the ability to utilize databases for analysis of metabolic interrelationships, as the pathways are typically predefined (as they would be in a textbook).

I have previously outlined a future direction for metabolic pathway representation [5,6], although the full construction of these notions has yet to be implemented. What is required is an integration of multiple existing databases. The essential idea is fairly simple. Rather than represent pathways as stagnant preconstructions, the pathway itself needs to be variable. This enables broad connectivities. In addition, making transporters equivalent to enzymes (they become a catalysis of cell space change) would enable both realistic intracellular as well as interorgan pathways.

TABLE 9.1
Metabolic Databases

Database	Description
BioCarta	Broad pathway collections, as relates to molecular biology.
BRENDA	More specific enzyme information database. Includes search facilities by ligand, by enzyme, and by prestructured pathway sequences.
ECMDB	Metabolic information specific to *E. coli*. Includes other details of metabolites, such as spectra, useful in metabolomics.
EXPASY	Broad informatics database containing a wall chart map of pathways converted to PDF.
FlyBase	Drosophila database with some enzyme information, mostly related to biology of this species and its genetics.
IUBMB ExplorEnz	Reaction database enabling search by enzyme name or EC number, focused on enzyme information as the output.
KEGG	A combined database repository including metabolic reactions, with the intent of correlating various biological activities and disease states. Includes search of reactions and metabolites, as well as some pathways.
LmSmdB	A "one-stop shop" for all information related to both Leishmania major and *Schistosoma mansoni*; metabolic data is included.
MetaCyc	A database of reactions and pathways originally devoted to *E. coli* and then expanded to add some tissue type information. Provides fixed pathways, reaction information, and links to genes.

A full implementation must account for species, cell space, and cell type. The metabolic notions that have been outlined in prior chapters could be incorporated. For example, a reaction pathway as constructed has pathway intermediates and these intermediates typically do not include all reactants. That is, there is a pathway view and a reaction view. Existing databases of metabolic reactions are squarely in the camp of exclusively representing the reaction view. There is no separation of mobile cofactors and fixed cofactors, no enforcement of the cell space or cell type for the reaction; no designation of near-equilibrium or metabolically irreversible status of the reaction. Such qualities would transform representations that currently hold a great deal of data into a true metabolic tool.

Essentially, the existing databases of metabolic information serve as repositories rather than as working tools to extend metabolic analysis. A direct use is the correlation of gene changes to metabolite changes, such as a study in which alterations effected by making mice transgenic for the control protein PGC-1α in muscle causes genetic changes in several pathways [7]. In this example, there were some wholesale changes in one pathway: genes for the Krebs cycle enzymes, and corresponding metabolites, all increased about two-fold. The other pathway metabolites (including some amino acids, nucleotides, and pentose phosphate pathway intermediates) were more complex. This is because metabolites intersect between pathways, and because—unlike genes—they are so context-dependent (time, substrates, hormones) so correlations are difficult to achieve.

9.2   FLUX MODELING OF PATHWAYS AND METABOLIC CONTROL ANALYSIS

The use of various forms of computational modeling have long been used to both assist understanding and to advance it into areas that could not otherwise be reached. As a simple example of the first, consider the novel idea of using simultaneous equations for acid base reactions in the body rather than using simplifying assumptions, avoiding even the use of pH in favor of simply [H^+]. The calculations, while modest, still require a computer and can provide insight into acid base balance that is otherwise difficult using simplifications such as the Henderson–Hasselbach equation [8]. One important result of that effort is the recognition that concentrations of buffer components are themselves less important that the *strong ion difference*, the concentration of strong ion cations minus strong ion anions. These have found clinical use [9].

A long tradition of computers used to solve simultaneous differential equations in metabolism has provided models of major routes of intermediary metabolism [10]. Still, much of this effort has been trained on microbial metabolism (e.g., ref. [11]) in part because of its relative metabolic simplicity but also because of the interest in using microbes as a synthetic factory [12]. Several such strategies were outlined by Bailey [13].

Modeling studies themselves offer the apparent objectivity in removing the accumulated "system rules" in favor of mathematical rigor. Still, the model itself must be selected on the basis of individual constraints of the investigator. For example, a study of fatty liver disease [14] concluded that zonation of the liver was an important feature in the accumulation of fatty deposits. Of course, the model built in such zonation features and biochemical differences between them (the periportal, near the entry of blood to the organ and the pericentral, near the drain) have been routinely observed in the past. Thus, the quantitation does not actually test the significance of zonation. A complication with the idea is that connected pathways would no longer function if they were constrained to distinct cell populations.

9.3   METABOLIC CONTROL ANALYSIS

One type of metabolic analysis has its origins in an idea first advanced for enzyme mechanisms [15] in which the rate-limiting step may vary with conditions. In a similar way, the rate-limiting step of a pathway can shift as alterations are made to different enzymatic steps, such as increased protein synthesis. It is common that more than one pathway step has the potential to be rate-limiting. For example, there are at least three in glycolysis: hexokinase, phosphofructokinase, and pyruvate kinase. However, it still must be recognized that not *all* the steps have this potential. As we have seen in the prior chapters, most enzymes achieve near-equilibrium in cells, and thus are unlikely to serve as control points. This gives us a means to determine which of a vast number of findings might be worth pursuing. For example, the view that near-equilibrium enzymes like glyceraldehyde phosphate dehydrogenase is rate-limiting [16] is unlikely, despite elegant metabolic control analysis.

The technique itself [17] can be an effective tool. Metabolic control analysis is operationally complex, involving mathematics that is beyond what most biological investigators are comfortable with. In its simplest form, it requires a measurement of total flux of a pathway, and the availability of a specific inhibitor of one of the steps. Titration with that inhibitor has a measurable effect on the target enzyme itself as well as the pathway; the comparison provides information as to how much of the pathway flux is controlled by that step. One of the more intriguing applications is an investigation of the relationship between distinct mitochondrial mutations and fundamental parameters of oxidative phosphorylation (e.g., the adenine nucleotide translocase) [18]. The study confirmed prior expectations, suggesting that with accumulated mutations, a rate limitation hits a threshold with thereby can compromise mitochondria function.

9.4   METABOLOMICS

Prior to about 2000, when much of the biological community and the public alike were excited about the determination of the entire sequence of human DNA, a word was coined: the *genome*. The excitement was based on sound reasons: having this information provides enormous insight into virtually every aspect of biology [19].

By the early 2000s, the next phase of massive data collection, dubbed *proteomics*, was already a decade old [20]. Heavily technique-oriented (largely mass spectroscopy), proteomics resembled genomics in its intense focus on methodology and bright future for applications. Proteomics is closer to cellular function.

The success of the first two inexorably led to a third effort: *metabolomics*.[i] Cui et al. [21] suggest this approach goes back decades, citing a report of Pauling et al. [22] from 1971, in which 250 compounds were measured in body fluids by gas chromatography. Such analysis bears striking similarity in that the report was a technical tour de force, and had clear potential for the clinical as well as for research. Still, metabolomics is distinct from the ability to measure a large number of compounds in a body fluid. The difference is evident if we revisit the sequence from DNA to metabolites. In metabolomics, knowing the DNA sequence gives us the completeness of the gene with all the possibilities of our nature. In proteomics, all of the possible protein sequences provides the full potential that can be expressed by genes. In accordance with this trend, in metabolomics the goal is completeness: not just collecting, but collecting *everything*.

Another distinction between the newer efforts of metabolomics and the earlier multiple measurement studies is the use of the database and analytic developments that have been outlined in the prior sections. The notion that massive data sets can be collected and appropriately massaged by computer implies large advances on older and simpler types of analysis.

The excitement stems from the fact that typically metabolites are measured a few at a time and the available information is thereby somewhat limited in scope. The heady prospect of massive numbers of measurements might produce insights that are entirely unexpected and go far beyond what can be decided from focused analysis on relatively few compounds.

9.4.1 METABOLITE DATASET

A seemingly simple question to ask of metabolomics is what is the number of metabolites in a cell? By one estimate, the number of metabolites known to occur in yeast and some other organisms is between 10- and 100-fold lower than the number of genes and proteins [23]. This estimate was drawn from three citations. The study [24] compared the number of open reading frames, reactions listed in the KEGG database, textbooks, and the Enzyme Commission database (Bairoch's source, as above), are arrived at

700 ORFs
580 metabolites
1175 reactions

A second source [25] using the KEGG database cited

110,018 genes
3705 enzymes
5645 compounds
5207 reactions

That may be the origin of the 10- to 100-fold distinction. Of course, this is not the actual number of genes nor it is the actual number of compounds. The 110,000 genes are now recognized to be a gross overestimate. Note that the enzymes, reactions, and compound numbers were similar.

A separate citation [26] put the number of genes for yeast at 6,000 (about the number of compounds in the KEGG database). More interestingly, this was cited as an example of the use of metabolomics for genomics. This study points out two issues that are not usually brought forward. First, like proteomics, metabolomics is context-dependent, meaning various states of change in the organism alter their measurement. Second, also like proteomics, metabolomics reflect function rather than potential.

More recently, new estimates have gone in the other direction. For example, Fessenden [27] cites the Human Metabolome Database as having over 42,000 metabolites, and suggests this is a conservative estimate. This is cited in the context of a succinct review of the new field of single-cell metabolomics. This is a complexity of a new order; it is intriguing both from a theoretical and technical viewpoint to determine such distinctions.

One thing is certain about the number: it is larger than the total number of molecules that can be resolved in present-day analysis. Either a defined subset is considered (perhaps those of well-known pathway intermediates of ample concentration in cells), or a sample is analyzed by a method and all compounds that can be resolved are analyzed (the "untargeted" approach).

9.4.2 RESULTS FROM COMMERCIAL VENDORS

One way of performing metabolomics is to simply send samples to a commercial enterprise that performs the analysis and sends back the data. As an example, the company Metabolon (www.metabolon.com) provides this service, sending the investigator data in the form of relative numbers. The data is displayed graphically as a heat map, with separate colors representing different relative concentrations. Since there are hundreds of metabolites measured, it is sometimes just a pattern distinction, say between cells taken from cancer or normal tissues that is apparent.

The company also provides a listing of publications resulting from investigators using the service. As one example, Fu et al. [28] found a defect in the formation of phosphatidylcholine from phosphatidylethanolamine in obesity from the metabolite patterns. Other companies also exist that do similar measurements, although comparisons are not precise because the list of analytes are not made public.

9.4.3 INDEPENDENT LABORATORIES

Independent laboratories focusing on metabolomic style investigations have a similar large investment in technical expertise to conduct large numbers of measurements. One form of analysis is an *untargeted* approach, which has some similarities to the early studies from Pauling's laboratory mentioned above. Rather than having a set number of known targets to analyze, as many peaks as possible are obtained for a particular treatment, and analysis ensues. As an example of the success of this method, Witherspoon et al. [29] found that treatment of colon cancer with difluomethylornithine (an ornithine decarboxylase inhibitor) resulted from a decline in thymidine concentration. While further investigation is needed to explain why thymidine itself was decreased, it is clear that metabolomics provided the lead for this investigation.

It is clear from just these few examples that the current and potential applications of metabolomics can explain the interest in the method. As a caveat, it is also clear that despite the idea of measuring everything, there is no guarantee that the investigator will understand everything about the particular problem. Having thousands of measurements does not mean that we can ignore fundamental principles, such as assigning regulatory properties to near-equilibrium enzymes, or taking total concentrations of metabolites such as ADP, AMP, and NADH that are known to be a small fraction of the measured total. An endemic limitation to the analysis is the difficulty in measuring a time course or varying conditions which is possible for one or a few compounds. Thus, the exploratory work is typically performed on the basis of a limited parameters. In this sense, the *omics* power is greatest for DNA, which is fixed, less so for proteins, which vary but on a smaller scale, and worst for metabolites, which can vary on a time scale of seconds to minutes, and strikingly between treatments. Thus, while the methodology has obvious power, it is equally important to recognize its limitations.

9.4.4 PROTEIN–PROTEIN INTERACTIONS

Of other available databases, there are collections of proteins interacting with other proteins, such as STRING (string-db.org). Much of this data is collected using the yeast 2-hybrid cross, which can detect any interaction between proteins stable enough to survive the experimental conditions. Prior to becoming its own field, Srere [30] suggested that proteins in metabolic pathways actually stuck together. This contrasts with most thinking that the connections are between metabolites, which is in accord with metabolic ideas, such as the matching of intermediates to the Km values of their enzymes [31], as well as other kinetic and thermodynamic analysis. Presented with a large catalog of protein interactions makes it difficult to discriminate between them, and some interactions that do occur may not be observed [32]. For example, the Orai1-IP$_3$R interaction discussed in Chapter 7 would not be observed under most conditions of measurement since this only happens in Ca^{2+} depletion conditions. Thus, even known interactions depend on conditions and may not be observed.

In summary, there is a constant and growing effort to catalog data about metabolism. Some of it is driven by the feeling that it represents a completeness. Still there is no doubt that much of the collection process has clear value to metabolism, and future improvements and extension in the joining of such data will prove valuable.

REFERENCES

1. Salway, J. G. (2004) *Metabolism at a Glance*, 3rd ed., Blackwell Pub., Malden, MA.
2. Biochemicals Pathway Maps (2016) http://web.expasy.org/pathways/.
3. Mendes, P., Bulmore, D. L., Farmer, A. D., Steadman, P. A., Waugh, M. E., and Wlodek, S. T. (2000) PathDB: A second generation metabolic database. In *Bioinformatics of Gene Regulation, Gene Networks and Metabolic Pathways* (Kolchanov, N. ed.), Russian Academy of Sciences, Novosibirsk, Russia.
4. Bairoch, A. (1994) The ENZYME data bank. *Nucleic Acids Res.* 17, 3626–3627.
5. Ochs, R. S., and Conrow, K. (1991) A computerized metabolic map. *J. Chem. Inf. Comput. Sci.* 31, 132–137.
6. Ochs, R. S., Qureschi, A., Sycz, A., and Vorbach, J. (1996) A computerized metabolic map. II. Relational structure, extended modeling, and a graphical interface. *J. Chem. Inf. Comput. Sci.* 36, 594–601.
7. Hatazawa, Y., Senoo, N., Tadaishi, M., Ogawa, Y., Ezaki, O., Kamei, Y., and Miura, S. (2015) Metabolomic analysis of the skeletal muscle of mice overexpressing PGC-1alpha. *PloS One* 10, e0129084.
8. Stewart, P. A. (1981) *How to Understand Acid-Base*, Elsevier, New York.
9. George, Y. W. H. (2015) *Easy Way to Understand Stewart's Acid-Base*.
10. Garfinkel, D. (1985) Computer modeling of metabolic pathways. In *Metabolic Regulation* (Ochs, R. S., Hanson, R. W., and Hall, J. eds.), Elsevier, New York, pp. 20–26.
11. Edwards, J. S., Covert, M., and Palsson, B. (2002) Metabolic modelling of microbes: The flux-balance approach. *Environ. Microbiol.* 4, 133–140.
12. Zhang, F., Rodriguez, S., and Keasling, J. D. (2011) Metabolic engineering of microbial pathways for advanced biofuels production. *Curr. Opin. Biotechnol.* 22, 775–783.
13. Bailey, J. E. (1991) Toward a science of metabolic engineering. *Science* 252, 1668–1674.
14. Ashworth, W. B., Davies, N. A., and Bogle, I. D. (2016) A computational model of hepatic energy metabolism: Understanding zonated damage and steatosis in NAFLD. *PLoS Comput. Biol.* 12, e1005105.
15. Cleland, W. W. (1975) What limits the rate of an enzyme-catalyzed reaction? *Acc. Chem. Res.* 8, 145–151.
16. Shestov, A. A., Liu, X., Ser, Z., Cluntun, A. A., Hung, Y. P., Huang, L., Kim, D., Le, A., Yellen, G., Albeck, J. G., and Locasale, J. W. (2014) Quantitative determinants of aerobic glycolysis identify flux through the enzyme GAPDH as a limiting step. *Elife* 3.
17. Cornish-Bowden, A. (1995) Metabolic control analysis in theory and practice. In *Advances in Molecular and Cell Biology* (Bittar, E. E. ed.), Elsevier, pp. 21–64.

18. Mazat, J. P., Letellier, T., Bedes, F., Malgat, M., Korzeniewski, B., Jouaville, L. S., and Morkuniene, R. (1997) Metabolic control analysis and threshold effect in oxidative phosphorylation: Implications for mitochondria pathologies. *Mol. Cell. Biochem.* 174, 143–148.
19. Collins, F. S., Morgan, M., and Patrinos, A. (2003) The Human Genome Project: Lessons from large-scale biology. *Science* 300, 286–290.
20. Patterson, S. D., and Aebersold, R. H. (2003) Proteomics: The first decade and beyond. *Nat. Genet.* 33, S311–S323.
21. Cui, Q., Lewis, I. A., Hegeman, A. D., Anderson, M. E., Li, J., Schulte, C. F., Westler, W. M., Eghbalnia, H. R., Sussman, M. R., and Markley, J. L. (2008) Metabolite identification via the Madison Metabolomics Consortium Database. *Nat. Biotechnol.* 26, 162–164.
22. Pauling, L., Robinson, A. B., Teranishi, R., and Cary, P. (1971) Quantitative analysis of urine vapor and breath by gas-liquid partition chromatography. *Proc. Natl. Acad. Sci. USA* 68, 2374–2376.
23. Styczynski, M. P., Moxley, J. F., Tong, L. V., Walther, J. L., Jensen, K. L., and Stephanopoulos, G. N. (2007) Systematic identification of conserved metabolites in GC/MS data for metabolomics and biomarker discovery. *Anal. Chem.* 79, 966–973.
24. Forster, J., Famili, I., Fu, P., Palsson, B. O., and Nielsen, J. (2003) Genome-scale reconstruction of the *Saccharomyces cerevisiae* metabolic network. *Genome Res.* 13, 244–253.
25. Kanehisa, M., and Goto, S. (2000) KEGG: Kyoto encyclopedia of genes and genomes. *Nucleic Acids Res.* 28, 27–30.
26. Raamsdonk, L. M., Teusink, B., Broadhurst, D., Zhang, N., Hayes, A., Walsh, M. C., Berden, J. A., Brindle, K. M., Kell, D. B., Rowland, J. J., Westerhoff, H. V., van Dam, K., and Oliver, S. G. (2001) A functional genomics strategy that uses metabolome data to reveal the phenotype of silent mutations. *Nat. Biotechnol.* 19, 45–50.
27. Fessenden, M. (2016) Metabolomics: Small molecules, single cells. *Nature* 540, 153–155.
28. Fu, S., Yang, L., Li, P., Hofmann, O., Dicker, L., Hide, W., Lin, X., Watkins, S. M., Ivanov, A. R., and Hotamisligil, G. S. (2011) Aberrant lipid metabolism disrupts calcium homeostasis causing liver endoplasmic reticulum stress in obesity. *Nature* 473, 528–531.
29. Witherspoon, M., Chen, Q., Kopelovich, L., Gross, S. S., and Lipkin, S. M. (2013) Unbiased metabolite profiling indicates that a diminished thymidine pool is the underlying mechanism of colon cancer chemoprevention by alpha-difluoromethylornithine. *Cancer Discov.* 3, 1072–1081.
30. Srere, P. A. (1987) Complexes of sequential metabolic enzymes. *Annu. Rev. Biochem.* 56, 89–124.
31. Cleland, W. W. (1970) Steady state kinetics. In *The Enzymes* (Boyer, P. D. ed.), Academic Press, New York, pp. 1–65.
32. Lim, W., Mayer, B., and Pawson, T. (2016) *Cell Signaling: Principles and Mechanisms.*

ENDNOTE

i. There are many other *omics*; it is a designation that has been overused. The notion of adding *omics* or *omic* to the end of virtually anything has spawned a vast number of new fields of inquiry. For example, the cataloging study of lipid species is *lipidomics*; that of foods is *foodomics*. Not to mention *microbiomics*, *pharmacogenomics*, *epigenomics*, *kinomics*, *organomics*, *transcriptomics*, *antibodyomics*, and *speechomics*.

10 Medical Issues Related to Metabolism

In this final chapter, I consider a few disease states that have distinct metabolic connections, both in the underlying condition and in the experimental approach. Metabolism is not the only consideration in the analysis of medical issues, but these examples illustrate a growing interest in the area and the need to consider the fundamentals of metabolic principles.

10.1 DIABETES AND METFORMIN

Diabetes is a disease which clearly displays its underlying metabolic disorder. The major energy pathways of carbohydrate and lipid are dysregulated, with excessive glucose and fatty acid concentrations in the blood. The pathways of liver metabolism are similar to severe fasting: gluconeogenesis is increased, glycolysis suppressed, and fatty acid oxidation is increased. In muscle, glucose transport is diminished, due to a poor sensitivity of the tissue to insulin. In fact, insulin insensitivity is the hallmark of the most common form of diabetes, Type II. The less-common form, Type I, has an absolute lack of insulin as a result of destruction of the pancreatic beta cells that produce the hormone. Diabetics have one thing in common: most take the drug metformin.

10.1.1 METFORMIN AND AMPK

Metformin is a biguanide (Figure 10.1). The first structure shown, guanidine, is a natural product of the French lily. This and similar compounds have long been known to have medicinal value, including treatment of diabetes [1,2]. The second compound shown, phenformin, was the first drug used as a specific treatment for diabetes. While effective, it also produced severe acidosis. This results from the activation of glycolysis, a characteristic of all biguanides. Metformin, probably because of its greater water solubility (and thus poorer cell penetration) is effective as a drug and produces far less acidosis. Metformin is presently the most widely prescribed drug for diabetes treatment.

While the full mechanism of metformin remains unknown, it has been shown to activate AMPK [3], which has known metabolic consequences. The first of these was demonstrated before the enzyme was named AMPK: phosphorylation and inhibition of HMG CoA reductase, the key regulatory enzyme of cholesterol biosynthesis [4]. Subsequently, it was shown to inhibit acetyl CoA carboxylase, which has the effect of inhibition of fatty acid synthesis and stimulation of fatty acid oxidation. The latter is the result of a lower malonyl CoA (the product of this enzymatic reaction), which is an inhibitor of the esterification step needed to import fatty acids into the mitochondria for oxidation (Figure 10.2). How metformin leads to AMPK activation is less clear.

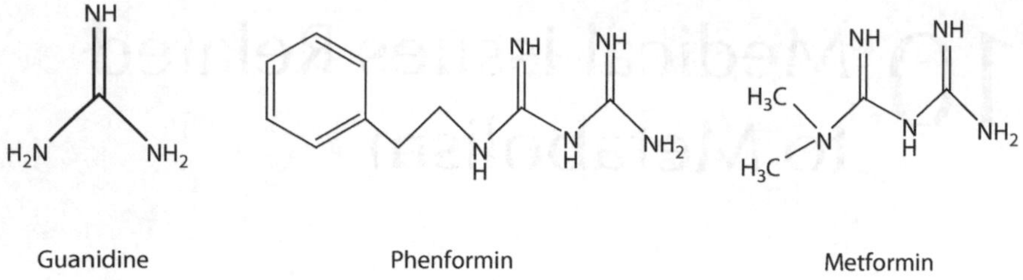

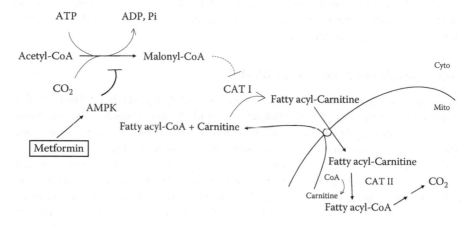

Guanidine   Phenformin   Metformin

FIGURE 10.1   Structures of guanides.

FIGURE 10.2   Acetyl CoA carboxylase and control of fatty acid oxidation. Because fatty acyl-CoA requires transfer to the carnitine ester prior to entry into the mitochondria, activity of the CAT I (carnitine acyl transferase I) is necessary for its oxidation. CAT I is inhibited by malonyl CoA, the product of acetylation of acetyl-CoA as shown, which is catalyzed by acetyl CoA carboxylase, and suppressed by metformin-induced activation of AMPK.

I present two hypotheses here. The first is a widely accepted one: inhibition of mitochondrial Complex I. The second is my own proposal, an inhibition of AMP deaminase.

10.1.2   COMMONLY ACCEPTED MECHANISM FOR METFORMIN: INHIBITION OF MITOCHONDRIAL COMPLEX I

Metformin and other biguanides have long been known to exert an inhibition of Complex I in isolated mitochondria [5,6]. Expanding this observation to intact cells, Owen et al. [7] found that metformin inhibits gluconeogenesis by hepatocytes as well as cellular respiration. They concluded that, like isolated mitochondria, metformin blocks Complex I in intact cells. *In vitro* analysis of AMPK shows that ATP inhibits the enzyme and AMP activates it [8]. Thus, a ratio of AMP/ATP provides an index of activity: the higher this ratio, the higher the enzyme activity. This was (erroneously) referred to as the energy charge,[i] and the investigators postulated that this ratio, set by metformin inhibition of Complex I, controls AMPK.

It has been pointed out, however, that the inhibition observed even in isolated mitochondria is relatively modest [9,10]. A consequence of this mechanism is that mitochondrial respiration—and thus the majority of energy production by the cell – would be diminished by metformin. If that were true *in vivo*, then in diabetics treated with metformin, the oxidation of glucose and lipid substrates would be impaired, leading to even greater accumulation of blood fatty acid levels.

Proponents of the Complex I inhibition hypothesis suggest it could be a link between metformin and AMPK. An inhibition of mitochondrial respiration would cause a drop in ATP formation, a rise in ADP through the near-equilibrium action of adenylate kinase, a rise in AMP, and thus an activation of AMPK. However, this ignores another metabolic activity that is known both for liver and muscle: metformin activates fatty acid oxidation [11,12]. This presents a profound difficulty in accepting Complex I as a mechanism, since that would inhibit the oxidation of NADH. This is inconsistent with metformin stimulation of fatty acid oxidation because this pathway requires turnover of NADH, which is accomplished by Complex I of the respiratory chain.

Recall that, in the pathway view, NADH is not merely a reaction intermediate but a parallel connector of pathways (Figure 10.3). In the case of fatty acid oxidation (discussed in Chapter 6) both NADH and QH_2 are formed. The first requires flow through the entire respiratory chain; the second reacts at Complex III. Focusing on the NADH production by fatty acids, it is clear that if fatty acid oxidation is stimulated overall then all steps of that pathway must also be stimulated. Thus, for this new steady state (metformin-induced increased activity), NADH production by fatty acid oxidation must increase. In order to balance the pathway, NADH utilization must also, and that utilization occurs strictly through the respiratory chain. It is inescapable that the oxidation through the respiratory chain is increased. Since this involves as a first step Complex I, then it is also the case that the rate of Complex I must increase.

10.1.3 Measuring Cellular ATP Production in Response to Metformin

Surprisingly, it is not a simple matter to determine cellular ATP production. Even oxygen consumption is an indirect assay, indicating the rate of electron transport. While often tightly coupled to ATP production, this is not always the case. Many investigators permeabilize the plasma membrane to obtain access to the mitochondria: this functionally produces an isolated mitochondrial preparation, meaning that

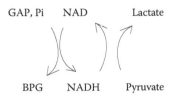

GAP, Pi   NAD   Lactate

BPG   NADH   Pyruvate

FIGURE 10.3 NAD and NADH are parallel connectors. The redox cofactors are illustrated as reactants of separate reactions; here, the conversion of GAP (glyceraldehyde-P) to BPG (bisphosphoglycerate), and lactate to pyruvate.

cell demand no longer controls ATP production. Measurement of isolated enzymes or enzyme complexes similarly does not provide the key information of ATP production of an intact pathway.

There is a distinct method to measure mitochondrial ATP production that was developed for muscle [13]. In this tissue, the creatine phosphokinase reaction achieves near equilibrium, and has been employed to report the formation of mitochondria ATP by first suppressing mitochondrial energy formation, and subsequently measuring the appearance of phosphocreatine when mitochondrial oxidation is resumed. This "creatine phosphate recovery" experiment (introduced in Chapter 8) was first performed in humans *in vivo* as described earlier.

We adopted this method for isolated muscle cell cultures, with the albeit less-sophisticated measurement of creatine phosphate by enzymatic endpoint methods [14]. Using reversible inhibitors of the respiratory chain, we demonstrated that metformin increased phosphocreatine recovery, providing evidence that it stimulates respiration in the cell. While metformin is known to stimulate fatty acid oxidation, the phosphocreatine recovery experiment was performed in the absence of fatty acids, so it likely represents a property of the respiratory chain itself.

A stimulation of respiration by metformin would explain the known metabolic targets, in particular the increase in fatty acid oxidation. It supports the general notion that metformin acts to increase energy formation by cells. It is also consistent with the clinically observed weight-loss in patients taking this drug [15].

10.1.4 POSSIBLE TARGET FOR METFORMIN: AMP DEAMINASE

The question of how metformin acts—apart from its ability to activate AMPK—remains an open question. We have evidence for one possible action: an inhibition of AMP deaminase (AMPD). Such an action would be a simple way to explain the accumulation of AMP and subsequent AMPK activation (Figure 10.4). The hypothesis is supported by experiments in isolated L6 cells, in which metformin stimulation of glucose transport is mimicked by known AMPD inhibitors [16]. Metformin also caused a decrease in ammonia accumulation, and inhibited the activity of isolated AMPD directly. Suppressing AMPD with selective siRNA led to the loss of

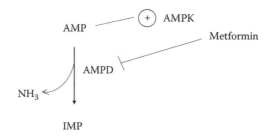

FIGURE 10.4   Proposal for metformin action. Metformin may act in cells as an inhibitor of the demanination of AMP, to form IMP and ammonia. Subsequently, increased AMP concentration can activate AMPK.

metformin action. Moreover, the free concentration of AMP was shown to rise in the presence of metformin or the AMPD inhibitor.

The question of whether AMPD inhibition can be an *in vivo* target has also been addressed. Admyre et al. [17] found that knockout of AMPD1, the muscle-specific isoform, in mice did not lead to observed changes in blood glucose in control or high fat-fed animals. On the other hand, Cheng et al. [18] did find an improvement in glycemia of high fat-fed mice rendered deficient in AMPD1. It is possible that the difference relates to the experimental system of fat feeding [18], so this discrepancy awaits resolution. However, the group finding no difference also found that several AMPD1 inhibitors did not improve glucose levels under their conditions [17]. While this may also be the result of experimental conditions, a separate problem exists in the use of inhibitors of AMPD. As shown in Figure 10.5, the catabolism of AMP in different cells has two pathways: deamination first and then dephosphorylation, or the reverse. If dephosphorylation is first, adenosine can leave the cell, which itself complicates analysis of blood glucose changes [19]. Moreover, inhibitors of AMPD also inhibit adenosine deaminase (ADA); these cannot be discriminated [20].

10.1.5 FREE VERSUS BOUND AMP

Recognition of free versus bound metabolites has already been discussed in prior chapters. In our own studies, we have observed about a three-fold increase in free AMP concentration in the presence of metformin [14]. However, some investigators measure only total AMP contents.[ii] In the study of Cheng et al. [18], total levels of AMP were assessed, and they were unchanged in muscle tissue, despite an increase

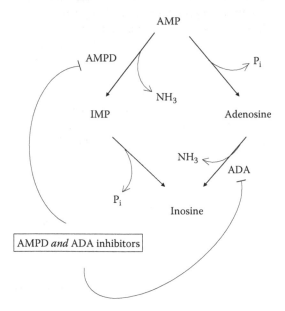

FIGURE 10.5 Catabolism of AMP and inhibitor sites. Inhibitors of both the deamination and dephosphorylation steps are shown. All of the known inhibitors act at both sites.

in AMPK activity. The investigators did find that the level of IMP dropped to unde-tectable levels after AMPD knockout. Thus, half of the crossover was observed as a diminution of IMP; the other half was obscured by the analysis of the total pool of AMP in place of the free concentration. The findings are nonetheless consistent with a diminution of cellular AMPD levels.

10.2   METFORMIN FOR OTHER DISEASES

Metformin, possibly through its ability to activate AMPK, leads to the phosphoryla-tion and inactivation of mTORC1 [21,22]. This is a key protein kinase involved in cell growth and replication. This action may be responsible for the anticancer action of metformin that was uncovered in retrospective studies of the large number of patients taking this drug [23]. While other diseases appear to be positively influ-enced by metformin as well, such as Parkinson's [24], Alzheimer's [25], and polycys-tic ovary syndrome [26], the finding that cancer stem cells appear to be preferentially targeted by metformin has led to a strong interest in this disease state in particular [27]. In recent years, there has been a strong interest in the metabolic connections to cancer, which we consider next.

10.3   CANCER AND METABOLISM

One reason for the renewed interest in cancer is the intersection of molecular biol-ogy with cell metabolism in the disease. There has been considerable interest in an observation made in the early twentieth century known as the Warburg Effect [28]. Warburg's original observation was simply that cancer cells show unusually strong glycolytic activity. The hypothesis of Warburg was that glycolysis was needed to supply cellular ATP because of mitochondrial damage. The hypothesis has long since been set aside. Further progress required looking beyond the energy contribu-tion of glycolysis versus oxidative phosphorylation and instead considering pathway fluxes that might be needed for rapid cell growth [29].

As a direct connection to glycolysis, the laboratory of Cantely found that a dis-tinct isoform of pyruvate kinase (PKM2) is expressed in some cancer cells. This promotes conversion of the substrate PEP into serine and subsequently glycine, important amino acid precursors for nucleotide synthesis [30].

Other metabolic pathways are needed for rapidly growing cells, in particular those with insufficient glucose. Thus, glutamine can be used both as building block for key intermediates as well as directly for energy [31].

Lipid pathways also play a key role in cancer and understanding their regula-tion is essential in developing therapeutic strategies. In particular, hyperactivity of the mevalonate pathway is established as having a role in cancer [32]. Some drugs known for their action in other diseases have been shown to be beneficial in can-cer therapy, opening new metabolic targets. Metformin (see previous section) is the most widely taken treatment for Type II diabetes, but recently has been shown to be anticancer [33]. While its action is likely multifactorial, one target is the HMG-CoA reductase step of the mevalonate pathway; in fact, this was the first target of AMPK, the enzyme activated by metformin [4,34].

Inhibition of the mevalonate pathway impedes not just the pathway product cholesterol, but intermediates such as farnesyl pyrophosphate, which modifies proteins such as *ras*, targeting it to membranes, and driving cell replication. As further evidence, statin drugs used primarily to lower cholesterol levels in the body also show anticancer activity [35]. Another drug with surprising anticancer activity is the class of bisphosphonates, the pyrophosphate analogs used to inhibit bone resorption and thereby stem osteoporosis. These compounds are believed to inhibit certain pyrophosphate producing enzymes of the mevalonate pathway, such as farnesyl diphosphate synthase [32].

Recently, a report by Bernstein and colleagues demonstrated a link between the oncogenic activity in gliomas to a mutation in isocitrate dehydrogenase (IDH) [36]. In a genetic screening of tumors, a correlation was established between mutation of IDH—even of a single amino acid—and an increased formation of an unusual product, 2-hydroxyglutarate (2HG) [37]. The mutation changes the kinetics of the enzymatic reaction [38], favoring a binding of NADPH first, and enabling 2-ketoglutarate to be reduced rather than carboxylated (Figure 10.6). The mutated form is not unique: several IDH mutants can result in an increased 2HG formation [39].

The links to cancer can be considered to flow from the 2HG formation, which inhibits at least three enzymes involved in cell transformation: prolyl hydroxylase, TET methylcytosine hydroxylases, and histone demethylase. These enzymes normally suppress the formation of HIF1α, DNA methylation, and histone-DNA binding, respectively [40].

There are three isoforms of IDH. IDH1 is cytosolic and NADP-specific. IDH2 is mitochondrial and is also NADP-specific. IDH3 is also mitochondrial and NAD-specific. As pointed out in a review of the role of the IDH and its isoforms in gliomas [41] the isozymes mutated are IDH1 and IDH2, the NADP-linked enzyme forms. IDH1 and IDH2 are described as catalyzing "reversible reactions," with no modes of regulation considered. If "reversible reactions" mean near-equilibrium ones, then they are not subject to regulation. To be a regulatory point, enzymes must be mIRR, as presented in prior chapters. There are indications that at least the NADPH-specific IDH forms catalyze mIRR reactions [42]. In order to make sense of the growing

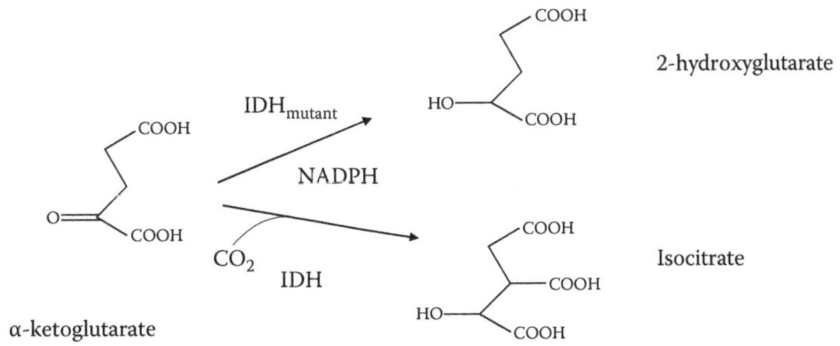

FIGURE 10.6 Mutations in IDH produce a distinct metabolite. Certain mutants of IDH (isocitrate DH) produce 2-hydroxyglutarate instead of isocitrate.

literature concerning rapid cell growth and altered flow through the Krebs cycle (e.g., ref. [40]) it is important to construct a pathway that can explain how flows through distinct IDH forms during rapid growth might proceed.

One possibility consistent with the findings of "reversed flow" through the first portion of the Krebs cycle is outlined in Figure 10.7. Glutamine conversion first to glutamate (catalyzed by the mitochondrial glutaminase) and then to α-ketoglutarate (via glutamate dehydrogenase or aspartate aminotransferase), there are two subsequent routes for this Krebs cycle intermediate. One is the "reductive pathway," in which the flow proceeds to succinyl CoA and beyond. This is the major route usually considered when glutamine metabolism is outlined, for example in studies in which cancer cells are shown to utilize large amounts of this amino acid for both energy and nitrogen [43]. However, the "oxidative pathway" is also shown, in which α-ketoglutarate traverses the Krebs cycle in a reverse direction, converted first to isocitrate and then citrate. As indicated, this is possible by utilizing IDH2, with NADP as the cofactor. The subsequent step is aconitase, which is NEQ and thus citrate can be produced, and subsequently transported to the cytosol to serve as a precursor to fatty acids.

The equilibrium status of IDH isoforms in muscle and liver was examined in studies summarized by Newsholme (1973). In muscle, IDH was considered mIRR based on total enzyme content measurements. Assuming all pools of IDH substrates and products equilibrate across the mitochondrial membrane, ^{14}C-bicarbonate incorporation into isocitrate suggests significant flux in both directions, that is, near-equilibrium status. However, this does not discriminate between NADP-linked IDH (IDH2) and NAD-linked IDH (IDH3), both of which are in mitochondria. While at the time it was not clear what the function of the two isoforms might be, with the newer information

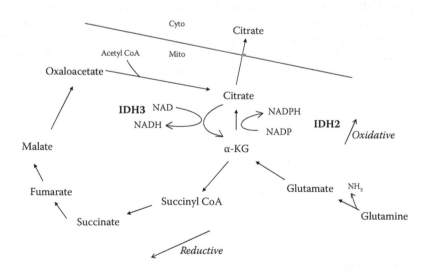

FIGURE 10.7 Oxidative and reductive pathways of glutamine metabolism. After entry of the carbon from glutamine into the Krebs Cycle, the pathway can diverge to citrate (oxidative path) or succinyl CoA (reductive path).

demonstrating the "oxidative Krebs cycle" pathways (e.g., ref. [44]), a consistent scheme would be two separate mIRR reactions. These would use the two separate IDH isoforms of mitochondria, each with its distinctive redox couple in order to produce the observed flux in both directions, as indicated in Figure 10.7.

A separate question has been raised that expresses a different view from the one considered throughout this work: that the redox states are themselves controlled by enzymes that use them as a substrate. Thus, Waitkus et al. [41] state: "IDH1 and IDH2 play important roles ... [including] regulation of cellular redox status." The basis of this assertion appears to be merely the fact that these are in fact dehydrogenases. What exactly is the cellular redox status? One aspect of the cytosolic redox state, the NADP/NADPH couple, was analyzed by Veech et al. [45]. Several distinct enzymes were considered in order to evaluate this redox couple. Some, like glutathione reductase, were eschewed because of the difficulty in accurately assessing free reduced and oxidized glutathione. Two enzymes fit the criteria: malic enzyme and IDH1. These showed a very similar calculated NADP/NADPH ratio, and the conclusion was that they were in near-equilibrium with the same redox potential. Similar studies have been cited in prior chapters concerning the redox states of NAD/NADH, and the same two redox couples in mitochondria.

Thus, there is no entity that can be identified as "the redox state," nor is there any evidence that any dehydrogenase enzyme can be said to regulate it or any one of the redox cofactor ratios. Rather, it is the combination of the available substrates to the cell which dictate the metabolic pathways that are used to handle those substrates, and the need for energy production. NADH is an intermediate in the formation of energy: it is the link between the production in mitochondrial reactions removing electrons and hydrogens from substrates on the one hand and feeding electron transport on the other. The redox state is a balance between the activities of these two in a given steady state. The control of energy production or utilization is not vested in the redox cofactors themselves but rather the multitude of pathways that produce and consume them.

10.4   EPILEPSY AND THE KETOGENIC DIET

An ancient treatment of epilepsy is starvation [46]. In fact, fasting has been proposed as a treatment for a wide number of disorders, including diabetes [47]. By the 1920s a ketogenic diet was used in place of starvation, and was successful in diminishing the incidence of seizures [48]. The diet, consisting of high fat and low carbohydrate content, mimics the altered metabolic fuel supply of starvation, and yet provides nutrition. As these conditions lead to ketogenesis, it is considered a "ketogenic diet." However, the development of drugs that target neural ion channels in the mid-twentieth century, starting with diphenylhydantoin [46] led to diminished interest in the dietary approach. More recently there has been a resurgence in the popularity of the approach due to the recognition of drug resistant epilepsy cases [49].

The mechanism for the ketogenic diet remains unresolved. One possibility is that the ketone bodies themselves are antiepileptic [50]. It has also been suggested that ketone bodies are utilized by astrocytes [51,52], although the quantitative contribution in fasting or ketogenic diet has not been established. In any event, among the various hypothesis cited by McNally and Hartman [50], a lowering of aspartate was proposed; this would presumably relieve inhibition of glutamate decarboxylase, leading to enhanced GABA (γ-aminobutyrate, principal negative neurotransmitter). While possible, it was immediately pointed out that this conflicts with other findings, including a lack of correlation of GABA levels in epilepsy models. Possible effects on the ATP-inhibited potassium (KATP) channel have also been proposed, making this similar to the metabolic control of insulin and glucagon release by pancreatic endocrine cells [53].[iii]

Clearly, a new appreciation for the metabolic underpinnings of disease has emerged in recent times. A further connection to metabolism from a different direction is the epilepsy syndrome of Lafora disease. A metabolic defect is the absence of a phosphatase (laforin) that is active on phosphorylated glycogen. While glycogen metabolism is abnormal [54] a ubiquitin ligase targeting laforin, known as malin, is also commonly defective and its activity may be the principal defect leading to epilepsy [55]. The ketogenic diet approach does not, however, alleviate seizures resulting from Lafora body syndrome [56].

10.5   MICROBIOME AND METABOLIC SYNDROME

The notion that the collective bacterial population of the large intestine—the *microbiome*—controls human metabolism and is responsible for the sudden appearance of metabolic syndrome is presented in the recent monograph *Missing Microbes* [57]. Collecting various arguments including the recent increased use of antibiotics for livestock and humans and the more frequent Cesarean sections, Blaser argues that the emergence of new diseases in a relatively short period of time (the last few decades) cannot be an evolutionary change, and the alteration in the microbial population is responsible. A review of three more books on the human genome [58] views the same phenomenon from epidemiological, research, and sociological viewpoints. The review is less sanguine about the missing microbe hypothesis. Apart from the treatment of specific microbe-related diseases of the gut (*Clostridium difficile*), transplant studies to restore the balance have not shown reversal of diseases in the metabolic syndrome spectrum. The fact that farm animals are routinely fed antibiotics in order to fatten them up does lend support to the notion that the microbiome at least has the power to influence obesity. Against this is the fact that many patients who have had their large intestine removed (in order to counter one of the diseases of the metabolic syndrome itself, Crohn's) do not consequently acquire metabolic syndrome. A more measured conclusion to the connection between microbiome and metabolic syndrome is that there is a link, but it is multifactorial. It is difficult to dismiss the ready availability of high-calorie foods and decreased mobility.

10.6   INBORN ERRORS

Classically, many disease states related to metabolism are those with associated "inborn errors," usually attributed to one but sometimes more missing genes. The classic compendium, originally the "Metabolic Basis of Inherited Diseases" is presently available as a web site [59]. While it is not always obvious how to correlate metabolic defects to clinical manifestations, this compendium is a rich trove of insights into metabolic behavior and a way of helping to resolve the conflict between purely reductive experiments on isolated systems and ultimate disease expression.

REFERENCES

1. Schafer, G. (1983) Biguanides. A review of history, pharmacodynamics and therapy. *Diabetes Metab.* 9, 148–163.
2. Witters, L. A. (2001) The blooming of the French lilac. *J. Clin. Invest.* 108, 1105–1107.
3. Zhou, G. C., Myers, R., Li, Y., Chen, Y. L., Shen, X. L., FenykMelody, J., Wu, M., Ventre, J., Doebber, T., Fujii, N., Musi, N., Hirshman, M. F., Goodyear, L. J., and Moller, D. E. (2001) Role of AMP-activated protein kinase in mechanism of metformin action. *J. Clin. Invest.* 108, 1167–1174.
4. Beg, Z. H., Allmann, D. W., and Gibson, D. M. (1973) Modulation of 3-hydroxy-3-methylglutaryl coenzyme A reductase activity with cAMP and wth protein fractions of rat liver cytosol. *Biochem. Biophys. Res. Commun.* 54, 1362–1369.
5. Pressman, B. C. (1963) The effects of guanidine and alkylguanidines on the energy transfer reactions of mitochondria. *J. Biol. Chem.* 238, 401–409.
6. Schafer, G., and Bojanowski, D. (1972) Interaction of biguanides with mitochondrial and synthetic membranes. *Eur. J. Biochem.* 27, 364–375.
7. Owen, M. R., Doran, E., and Halestrap, A. P. (2000) Evidence that metformin exerts its anti-diabetic effects through inhibition of complex 1 of the mitochondrial respiratory chain. *Biochem. J.* 348 Pt 3, 607–614.
8. Hardie, D. G., Ross, F. A., and Hawley, S. A. (2012) AMPK: A nutrient and energy sensor that maintains energy homeostasis. *Nat. Rev. Mol. Cell. Biol.* 13, 251–262.
9. Hoek, J. B. (2006) Metformin and the fate of fat. *Gastroenterology* 131, 2234–2237.
10. Hoek, J. B. (2006) Reply. *Gastroenterology* 131, 974–975.
11. Collier, C. A., Bruce, C. R., Smith, A. C., Lopaschuk, G., and Dyck, D. J. (2006) Metformin counters the insulin-induced suppression of fatty acid oxidation and stimulation of triacylglycerol storage in rodent skeletal muscle. *Am. J. Physiol. Endocrinol. Metab.* 291, E182–E189.
12. Tokubuchi, I., Tajiri, Y., Iwata, S., Hara, K., Wada, N., Hashinaga, T., Nakayama, H., Mifune, H., and Yamada, K. (2017) Beneficial effects of metformin on energy metabolism and visceral fat volume through a possible mechanism of fatty acid oxidation in human subjects and rats. *PloS one* 12, e0171293.
13. Blei, M. L., Conley, K. E., and Kushmerick, M. J. (1993) Separate measures of ATP utilization and recovery in human skeletal muscle. *J. Physiol. (Lond.)* 465, 203–222.
14. Vytla, V. S., and Ochs, R. S. (2013) Metformin increases mitochondrial energy formation in L6 muscle cell cultures. *J. Biol. Chem.* 288, 20369–20377.
15. Kitabchi, A. E., Temprosa, M., Knowler, W. C., Kahn, S. E., Fowler, S. E., Haffner, S. M., Andres, R., Saudek, C., Edelstein, S. L., Arakaki, R., Murphy, M. B., Shamoon, H., and Diabetes Prevention Program Research, G. (2005) Role of insulin secretion and sensitivity in the evolution of type 2 diabetes in the diabetes prevention program: Effects of lifestyle intervention and metformin. *Diabetes* 54, 2404–2414.
16. Ouyang, J., Parakhia, R. A., and Ochs, R. S. (2011) Metformin activates AMP kinase through inhibition of AMP deaminase. *J. Biol. Chem.* 286, 1–11.
17. Admyre, T., Amrot-Fors, L., Andersson, M., Bauer, M., Bjursell, M., Drmota, T., Hallen, S., Hartleib-Geschwindner, J., Lindmark, B., Liu, J. M., Lofgren, L., Rohman, M., Selmi, N., and Wallenius, K. (2014) Inhibition of AMP deaminase activity does not improve glucose control in rodent models of insulin resistance or diabetes. *Chem. Biol.* 21, 1486–1496

18. Cheng, J., Morisaki, H., Toyama, K., Sugimoto, N., Shintani, T., Tandelilin, A., Hirase, T., Holmes, E. W., and Morisaki, T. (2014) AMPD1: A novel therapeutic target for reversing insulin resistance. *BMC Endocr. Disord.* 14, 7.
19. Rusing, D., Muller, C. E., and Verspohl, E. J. (2006) The impact of adenosine and A(2B) receptors on glucose homoeostasis. *J. Pharm. Pharmacol.* 58, 1639–1645.
20. Bojack, G., Earnshaw, C. G., Klein, R., Lindell, S. D., Lowinski, C., and Preuss, R. (2001) Design and synthesis of inhibitors of adenosine and AMP deaminases. *Org. Lett.* 3, 839–842.
21. Zito, C. I., Qin, H., Blenis, J., and Bennett, A. M. (2007) SHP-2 regulates cell growth by controlling the mTOR/S6 kinase 1 pathway. *J. Biol. Chem.* 282, 6946–6953.
22. Green, A. S., Chapuis, N., Maciel, T. T., Willems, L., Lambert, M., Arnoult, C., Boyer, O., Bardet, V., Park, S., Foretz, M., Viollet, B., Ifrah, N., Dreyfus, F., Hermine, O., Moura, I. C., Lacombe, C., Mayeux, P., Bouscary, D., and Tamburini, J. (2010) The LKB1/AMPK signaling pathway has tumor suppressor activity in acute myeloid leukemia through the repression of mTOR-dependent oncogenic mRNA translation. *Blood.*
23. Pryor, R., and Cabreiro, F. (2015) Repurposing metformin: An old drug with new tricks in its binding pockets. *Biochem. J.* 471, 307–322.
24. Wahlqvist, M. L., Lee, M.-S., Hsu, C.-C., Chuang, S.-Y., Lee, J.-T., and Tsai, H.-N. (2012) Metformin-inclusive sulfonylurea therapy reduces the risk of Parkinson's disease occurring with Type 2 diabetes in a Taiwanese population cohort. *Parkinsonism Relat. Dis.* 18, 753–758.
25. Chen, Y. M., Zhou, K., Wang, R. S., Liu, Y., Kwak, Y. D., Ma, T., Thompson, R. C., Zhao, Y. B., Smith, L., Gasparini, L., Luo, Z. J., Xu, H. X., and Liao, F. F. (2009) Antidiabetic drug metformin (Glucophage(R)) increases biogenesis of Alzheimer's amyloid peptides via up-regulating BACE1 transcription. *Proc. Natl. Acad. Sci.* 106, 3907–3912.
26. Lebinger, T. G. (2007) Metformin and polycystic ovary syndrome. *Curr. Opin. Endocrinol. Diab. Obes.* 14, 132–140.
27. Rattan, R., Ali Fehmi, R., and Munkarah, A. (2012) Metformin: An emerging new therapeutic option for targeting cancer stem cells and metastasis. *J. Oncol.* 2012, 928127. Epub 922012 Jun 928124.
28. Warburg, O. (1956) On respiratory impairment in cancer cells. *Science* 124, 269–270.
29. Vander Heiden, M. G., Lunt, S. Y., Dayton, T. L., Fiske, B. P., Israelsen, W. J., Mattaini, K. R., Vokes, N. I., Stephanopoulos, G., Cantley, L. C., Metallo, C. M., and Locasale, J. W. (2011) Metabolic pathway alterations that support cell proliferation. *Cold Spring Harb. Symp. Quant. Biol.* 76, 325–334.
30. Anastasiou, D., Poulogiannis, G., Asara, J. M., Boxer, M. B., Jiang, J. K., Shen, M., Bellinger, G., Sasaki, A. T., Locasale, J. W., Auld, D. S., Thomas, C. J., Vander Heiden, M. G., and Cantley, L. C. (2011) Inhibition of pyruvate kinase M2 by reactive oxygen species contributes to cellular antioxidant responses. *Science* 334, 1278–1283.
31. Deberardinis, R. J., Sayed, N., Ditsworth, D., and Thompson, C. B. (2008) Brick by brick: Metabolism and tumor cell growth. *Curr. Opin. Genet. Dev.* 18, 54–61.
32. Mullen, P. J., Yu, R., Longo, J., Archer, M. C., and Penn, L. Z. (2016) The interplay between cell signalling and the mevalonate pathway in cancer. *Nat. Rev. Cancer* 16, 718–731.
33. Ben Sahra, I., LeMarchandBrustel, Y., Tanti, J. F., and Bost, F. (2010) Metformin in cancer therapy: A new perspective for an old antidiabetic drug? *Mol. Cancer Ther.* 9, 1092–1099.
34. Beg, Z. H., Stonik, J. A., and Brewer, H. B. (1985) Phosphorylation of hepatic 3-hydroxy-3-methylglutaryl coenzyme A reductase and modulation of its enzymic activity by calcium-activated and phospholipid-dependent protein kinase. *J. Biol. Chem.* 260, 1682–1687.

35. Yeganeh, B., Wiechec, E., Ande, S. R., Sharma, P., Moghadam, A. R., Post, M., Freed, D. H., Hashemi, M., Shojaei, S., Zeki, A. A., and Ghavami, S. (2014) Targeting the mevalonate cascade as a new therapeutic approach in heart disease, cancer and pulmonary disease. *Pharmacol. Ther.* 143, 87–110.
36. Flavahan, W. A., Drier, Y., Liau, B. B., Gillespie, S. M., Venteicher, A. S., Stemmer-Rachamimov, A. O., Suvà, M. L., and Bernstein, B. E. (2015) Insulator dysfunction and oncogene activation in IDH mutant gliomas. *Nature*, advance online publication.
37. Dang, L., White, D. W., Gross, S., Bennett, B. D., Bittinger, M. A., Driggers, E. M., Fantin, V. R., Jang, H. G., Jin, S., Keenan, M. C., Marks, K. M., Prins, R. M., Ward, P. S., Yen, K. E., Liau, L. M., Rabinowitz, J. D., Cantley, L. C., Thompson, C. B., Vander Heiden, M. G., and Su, S. M. (2009) Cancer-associated IDH1 mutations produce 2-hydroxyglutarate. *Nature* 462, 739–744.
38. Pietrak, B., Zhao, H., Qi, H., Quinn, C., Gao, E., Boyer, J. G., Concha, N., Brown, K., Duraiswami, C., Wooster, R., Sweitzer, S., and Schwartz, B. (2011) A Tale of Two Subunits: How the Neomorphic R132H IDH1 mutation enhances production of αHG. *Biochemistry (Mosc.)* 50, 4804–4812.
39. Rendina, A. R., Pietrak, B., Smallwood, A., Zhao, H., Qi, H., Quinn, C., Adams, N. D., Concha, N., Duraiswami, C., Thrall, S. H., Sweitzer, S., and Schwartz, B. (2013) Mutant IDH1 enhances the production of 2-hydroxyglutarate due to its kinetic mechanism. *Biochemistry (Mosc.)* 52, 4563–4577.
40. Sameer Agnihotri, Kenneth D. Aldape, and Gelareh Zadeh. (2014) Isocitrate dehydrogenase status and molecular subclasses of glioma and glioblastoma. *Neurosurg. Focus* 37, E13.
41. Waitkus, M. S., Diplas, B. H., and Yan, H. (2016) Isocitrate dehydrogenase mutations in gliomas. *Neuro-Oncol.* 18, 16–26.
42. Alp, P. R., Newsholme, E. A., and Zammit, V. A. (1976) Activities of citrate synthase and NAD+-linked and NADP+-linked isocitrate dehydrogenase in muscle from vertebrates and invertebrates. *Biochem. J.* 154, 689–700.
43. Wise, D. R., and Thompson, C. B. (2010) Glutamine addiction: A new therapeutic target in cancer. *Trends Biochem. Sci.* 35, 427–433.
44. Griss, T., Vincent, E. E., Egnatchik, R., Chen, J., Ma, E. H., Faubert, B., Viollet, B., DeBerardinis, R. J., and Jones, R. G. (2015) Metformin antagonizes cancer cell proliferation by suppressing mitochondrial-dependent biosynthesis. *PLoS Biol.* 13, e1002309.
45. Veech, R. L., Eggleston, L. V., and Krebs, H. A. (1969) The redox state of free nicotinamide-adenine dinucleotide phosphate in the cytoplasm of rat liver. *Biochem. J.* 115, 609–619.
46. Wheless, J. W. (2008) History of the ketogenic diet. *Epilepsia* 49, 3–5.
47. Bliss, M. (1982) *The Discovery of Insulin*, McClelland & Stewart, Toronto.
48. Peterman, M. G. (1925) The ketogenic diet in epilepsy. *J. Am. Med. Assoc.* 84, 1979–1983.
49. Kwan, P., Schachter, S. C., and Brodie, M. J. (2011) Drug-resistant epilepsy. *N. Engl. J. Med.* 365, 919–926.
50. McNally, M. A., and Hartman, A. L. (2012) Ketone bodies in epilepsy. *J. Neurochem.* 121, 28–35.
51. Auestad, N., Korsak, R. A., Morrow, J. W., and Edmond, J. (1991) Fatty acid oxidation and ketogenesis by astrocytes in primary culture. *J. Neurochem.* 56, 1376–1386.
52. Blazquez, C., Woods, A., de Ceballos, M. L., Carling, D., and Guzman, M. (1999) The AMP-activated protein kinase is involved in the regulation of ketone body production by astrocytes. *J. Neurochem.* 73, 1674–1682.
53. Rorsman, P., Ramracheya, R., Rorsman, N. J., and Zhang, Q. (2014) ATP-regulated potassium channels and voltage-gated calcium channels in pancreatic alpha and beta cells: similar functions but reciprocal effects on secretion. *Diabetologia* 57, 1749–1761.

54. Pederson, B. A., Turnbull, J., Epp, J. R., Weaver, S. A., Zhao, X., Pencea, N., Roach, P. J., Frankland, P., Ackerley, C. A., and Minassian, B. A. (2013) Inhibiting glycogen synthesis prevents Lafora disease in a mouse model. *Ann. Neurol.* 74, 297–300.
55. Knecht, E., Criado-Garcia, O., Aguado, C., Gayarre, J., Duran-Trio, L., Garcia-Cabrero, A. M., Vernia, S., San Millan, B., Heredia, M., Roma-Mateo, C., Mouron, S., Juana-Lopez, L., Dominguez, M., Navarro, C., Serratosa, J. M., Sanchez, M., Sanz, P., Bovolenta, P., and Rodriguez de Cordoba, S. (2012) Malin knockout mice support a primary role of autophagy in the pathogenesis of Lafora disease. *Autophagy* 8, 701–703.
56. Cardinali, S., Canafoglia, L., Bertoli, S., Franceschetti, S., Lanzi, G., Tagliabue, A., and Veggiotti, P. (2006) A pilot study of a ketogenic diet in patients with Lafora body disease. *Epilepsy Res.* 69, 129–134.
57. Blaser, M. J. (2014) *Missing Microbes: How the Overuse of Antibiotics Is Fueling Our Modern Plagues*, 1st edition. Henry Holt and Company, New York.
58. Shah, S. (2016) Inside Job. Three books report on the microscopic creatures that live in our bodies, and their influence on how we function. *New York Times Book Review* 9.
59. Vale, D. (2017) *The Online Metabolic and Molecular Bases of Inherited Disease*, https://ommbid.mhmedical.com/book.aspx?bookID=971.

ENDNOTES

i. Energy charge, proposed as a means of summarizing adenine nucleotide energy provision as a ratio, was described in Chapter 4.
ii. The question of *why* some molecules have very low free concentrations is not known. However, in the case of AMP in particular, it provides an opportunity for AMP to be both an intermediate in reactions that form it, such as pyrophosphate-forming enzymes and phosphodiesterases, and also be a signal. The former conditions contribute only to total AMP and would not be expected to alter free concentrations. Yet, under conditions when large shifts in ADP occur, as in changes in energy delivery, it can be expected that free AMP would rise and serve, at least in part, as a signal to AMPK.
iii. Recall that ATP concentration itself is constant in cells, so the naming of this channel is at best unfortunate, and potentially produces an incorrect view. This channel may be regulated by changes in free ADP, which do vary with metabolic state, but it is not possible to control any metabolic step by a compound that itself is unchanged in metabolic conditions.

Index

Page numbers followed by f and t indicate figures and tables, respectively.